Gall, Spurzheim, and the Phrenological Movement

During the 1790s in Vienna, German physician Franz Joseph Gall (1758–1828) came forth with a new doctrine dealing with mind, brain and behavior—one that could account for individual differences. He maintained that there are many independent faculties of mind, each associated with a separate part of the brain. He fine-tuned his ideas and published two sets of books presenting them after he and his assistant, Johann Gaspar Spurzheim, settled in Paris in 1807.

Gall's ideas had many supporters but were controversial and unsettling to others. In particular, the opposition ridiculed his belief that skull features reflect the growth of specific, underlying cortical organs, and hence correlate with personality traits (i.e., his 'bumpology'). Gall's fundamental ideas about the mind and organization of the brain were debated across the globe, and they also began to be exploited by unscrupulous businessmen, 'professors' who 'read skulls' for a living. But, as some historians have shown, his ideas about mind, brain and behavior led to the modern neurosciences.

The chapters collected in this volume provide new insights into Gall's thinking and what Spurzheim did, and the faddish movement called 'phrenology', which originated as a science of humankind but became a popular source of entertainment. All chapters were originally published in various issues of the *Journal of the History of the Neurosciences*.

Paul Eling is Associate Professor Emeritus at Radboud University, Nijmegen, the Netherlands. He published on a wide variety of topics in the field of cognitive neuropsychology and various textbooks. He is an Editor for the *Journal of the History of the Neurosciences*.

Stanley Finger is Professor Emeritus at Washington University, St. Louis, USA. He has published more than 250 articles and 20 books, including *Franz Joseph Gall: Naturalist of the Mind, Visionary of the Brain* (with Paul Eling in 2019). He was also an Editor for the *Journal of the History of the Neurosciences*.

Gall, Spurzheim, and the Phrenological Movement

Insights and Perspectives

Edited by
Paul Eling and Stanley Finger

Routledge
Taylor & Francis Group

LONDON AND NEW YORK

First published 2021
by Routledge
2 Park Square, Milton Park, Abingdon, Oxon, OX14 4RN

and by Routledge
605 Third Avenue, New York, NY 10158

Routledge is an imprint of the Taylor & Francis Group, an informa business

Chapters 1, 2, 4 and 7–21 © 2021 Taylor & Francis

Chapter 3 © 2020 Paul Eling and Stanley Finger. Originally published as Open Access.

Chapter 5 © 2020 Paul Eling and Stanley Finger. Originally published as Open Access.

Chapter 6 © 2020 Paul Eling and Stanley Finger. Originally published as Open Access.

British Library Cataloguing-in-Publication Data
A catalogue record for this book is available from the British Library

ISBN13: 978-0-367-49781-1 (hbk)
ISBN13: 978-0-367-49785-9 (pbk)
ISBN13: 978-1-003-04736-0 (ebk)

Typeset in Minion Pro
by codeMantra

Publisher's Note
The publisher accepts responsibility for any inconsistencies that may have arisen during the conversion of this book from journal articles to book chapters, namely the inclusion of journal terminology.

Disclaimer
Every effort has been made to contact copyright holders for their permission to reprint material in this book. The publishers would be grateful to hear from any copyright holder who is not here acknowledged and will undertake to rectify any errors or omissions in future editions of this book.

Contents

Citation Information

The following chapters were originally published in different issues of the *Journal of the History of the Neurosciences*. When citing this material, please use the original citations for each article, as follows:

Chapter 2
Phrenology: Scheherazade of etymology
Régis Olry and Duane E. Haines
Journal of the History of the Neurosciences, volume 29, issue 1 (2020) pp. 150–157

Chapter 3
Franz Joseph Gall on hemispheric symmetries
Paul Eling and Stanley Finger
Journal of the History of the Neurosciences, volume 29, issue 3 (2020) pp. 325–338

Chapter 4
Matters of sex and gender in F. J. Gall's organology: A primary approach
Tabea Cornel
Journal of the History of the Neurosciences, volume 23, issue 4 (2014) pp. 377–394

Chapter 5
Franz Joseph Gall on the "deaf and dumb" and the complexities of mind
Paul Eling and Stanley Finger
Journal of the History of the Neurosciences, DOI 10.1080/0964704X.2020.1780545

Chapter 6
Gall's German enemies
Paul Eling and Stanley Finger
Journal of the History of the Neurosciences, volume 29, issue 1 (2020) pp. 70–89

Chapter 7
An early description of Crouzon syndrome in a manuscript written in 1828 by Franz Joseph Gall
Stephan Heinrich Nolte, Werner Hansen, Paul Eling, and Stanley Finger
Journal of the History of the Neurosciences, volume 29, issue 3 (2020) pp. 339–350

For any permission-related enquiries please visit:
http://www.tandfonline.com/page/help/permissions

Contributors and Affiliations when Published

Simon Beierholm Independent Researcher, Copenhagen, Denmark.

Jaco Berveling Independent Researcher, Rotterdam, The Netherlands.

Matthijs Conradi Science and Technology Writer, Groningen, The Netherlands.

Tabea Cornel Department of History and Sociology of Science, University of Pennsylvania, Philadelphia, USA.

Douwe Draaisma Department of Psychology, State University of Groningen, The Netherlands.

Paul Eling Department of Psychology and Donders Institute for Brain, Cognition and Behaviour; Radboud University, Nijmegen, The Netherlands.

Stanley Finger Department of Psychological and Brain Sciences, and Program in History of Medicine, Washington University, St. Louis, USA.

Frank R. Freemon Department of Neurology, Vanderbilt University and Department of Veteran Affairs Medical Center, Nashville, USA.

Michael Hagner Department of Humanities and Social and Political Sciences, ETH Zurich, Switzerland.

Duane E. Haines Department of Neurobiology and Anatomy, Wake Forest School of Medicine, Winston-Salem, USA.

Werner Hansen 2nd Medical Department, Technical University of Munich, Germany.

Gonia Jarema Département de linguistique et de traduction, Université de Montréal, Canada.

Stephan Heinrich Nolte Medical Faculty, Philipps University Marburg, Germany.

Régis Olry Département d'Anatomie, Université du Québec à Trois-Rivières, Canada.

Marc Renneville Centre National de la Recherche Scientifique (CNRS), Paris, France.

Gül A. Russell Humanities in Medicine Department, College of Medicine, Texas A&M University, Bryan, USA.

Eglė Sakalauskaitė-Juodeikienė Department of Neurology and Neurosurgery, Center for Neurology, Vilnius University, Lithuania.

Catherine E. Storey Sydney Medical School, University of Sydney, Australia.

Jacob Lauge Thomassen Independent Researcher, Copenhagen, Denmark.

Harry Whitaker Department of Psychology, Northern Michigan University, Marquette, USA.

John van Wyhe Department of Biological Sciences, National University of Singapore, Singapore.

Introduction

Gall, Spurzheim, and the phrenological movement

Paul Eling and Stanley Finger

[Franz] Joseph Gall, as he was called prior to his trip to Paris (Finger and Eling, 2019), was born in 1758 the small town of Tiefenbronn, near the border of German's *Schwarzwald* (Black Forest). After primary school, although pressured to become a priest, he went to Strasbourg to study medicine. In 1785, he completed his medical education in Vienna and began a private practice catering to the well-to-do in the Austrian capital. But when not seeing patients, he delved into nature, studying flora and fauna, including the plants growing in his garden and the dogs, birds, and other animals he housed.

Gall was stimulated by German minister and philosopher Johann Gottfried Herder's (1744–1803) views. Herder's writings and Gall's own experiences turned him even more into a naturalist, a man with a passion to probe Nature's deepest secrets, especially in new ways. His astute gaze and probing mind stimulated him to publish his first book in 1791, one that promoted a new approach to understanding and treating the sick (Gall, 1791). Neglected by most historians, some of the roots of Gall's later thinking about brain, mind, and behavior can be discerned in this medical treatise. Notably, he was already concluding from clinical cases that the classic philosophical view of the mind as a kind of a general or unitary system for receiving, processing, storing, and utilizing information could not be correct. He did not, however, provide a new list of higher faculties of mind or mention cranioscopy in this volume.

Gall began to articulate his now-famous, more encompassing doctrine in his public lectures in 1796. These talks took place in his stately home, where he housed the growing collection of human and animal skulls and casts he used to make his points. Two years later, he presented his organology, as well as the methodologies that would guide his research program, in a published letter (Gall, 1798). Among his guiding assumptions were that different cortical areas are necessary for different functions, and that the shape of the skull can reveal much about these functionally and anatomically separate parts of the cortex.

Gall was taken by surprise by a decree from the Emperor of the Holy Roman Empire on Christmas Eve in 1801. It effectively put an end to all such lecturing in homes, unless official permissions were granted. Others received permissions, but he did not. Realizing what he was up against in politically charged, conservative Vienna, he decided to leave Austria temporarily to present his ideas to German colleagues and collect more case studies and feedback, while also taking time to reunite with his family. He began his scientific tour in

1805, not realizing that it would take him over much of what is now Germany, as well as into Denmark, The Netherlands, and Switzerland – and that he would never return to Vienna.

After having been received with great enthusiasm by the Germans and Danish (though less so by the Dutch and Swiss), Gall entered Paris in 1807. A year after arriving, he and his most important assistant, Johann Gaspard Spurzheim (1776–1832), presented their anatomical discoveries (not their more controversial *organologie*) to the highly respected *Académie des Sciences*. George Cuvier (1769–1832), the Permanent Secretary, reviewed the manuscript and rejected it, seemingly influenced by his xenophobic emperor, Napoleon Bonaparte (1769–1821), who had little respect for Gall's ideas and was also President of the *Académie*.

Despite this initial setback, Gall decided to settle in Paris, where he began a private practice and gave lectures. More importantly, this is where he completed his four-volume *Anatomie et Physiologie du Système Nerveux en Général, et du Cerveau en Particulier* and accompanying atlas. The first volume and atlas were published in 1810 and the last volume in 1819 (Gall and Spurzheim, 1810–19). With most people finding these volumes unaffordable, however, he felt compelled to move his organology into a less expensive and consequently more accessible set of books (without the atlas and detailed anatomy) in 1825 (Gall, 1825; translated into English in 1835). Three years later, he died in his house at Montrouge and was buried in Paris' sprawling new Père Lachaise Cemetery.

Gall and his doctrine were always highly controversial. On the one hand, his lectures, whether in Vienna, Berlin, Paris, or elsewhere, always attracted the interest of colleagues and the general public, with many people accepting some or all of what he had to say. But on the other, conservatives were afraid of his "enlightened" approach, which drew heavily on nature while circumventing metaphysical notions and entities, including the human soul. The clergy formulated two major objections to his doctrine: first, his theories would lead to materialism, and, second, they supported fatalism or determinism, meaning even heinous criminal acts could be attributed to faulty brain organization, not free will, hence challenging long-held legal precedents. These interpretations of Gall's doctrine made him a dangerous person, not only in conservative Austria but also in Catholic France, where many people agreed with René Descartes (1596–1650) that the soul or mind is an indivisible entity unique to humans, in contrast to what Gall was proposing with his 27 independent faculties of mind, 19 of which we share with other animals, with each dependent on a different parcel of cortex. Gall knew his doctrine was controversial, but he was also convinced he was right about the tenets of his new science. After all, his views were empirical and based on observable nature, the ultimate arbitrator, in contrast to the loose theories of mind and brain that armchair philosophers had been promoting for centuries.

Documents show that Gall was also ridiculed for another reason, particularly by men of medicine. Some found that they were unable to confirm his associations between notable behavioral traits or propensities and specific cranial bumps. Gall's work also had another glaring weakness, one assailed by physician Peter Mark Roget (1779–1869) later of *Thesaurus* fame (Roget, 1818/1824). While willing to accept all sorts of "facts" supportive of his craniological beliefs, Roget recognized that Gall and Spurzheim had a habit of ignoring or finding reasons to reject contradictory cases and other sorts of evidence opposing their beliefs.

Several years ago, when we began working on our biography of Gall as a controversial naturalist of the mind and visionary of the brain (Finger and Eling, 2019), we discovered that practically all of his biographical sketches were based on a short piece written by his friend, Italian physician Giovanni Fossati (1786–1874) (Fossati, 1857). Consulting various primary sources, we noticed that many parts of his Gall storyline were incorrect and that more were simply ignored or glossed over. These errors and omissions stimulated us to delve deeper into the overlooked early history of what would later be called the phrenological movement.

In this endeavor, we became acutely aware of the dubious objections formulated against Gall in Vienna and during his scientific tour (see the chapter in this book on Gall's "enemies"), as well as many other things about him and his doctrine. And we became convinced that, although frequently portrayed today as a fraud, a charlatan, or a quack, Gall was none of these things. This is not to say that his "bumpology" was based on good science, for it was not. But Gall also had many good methods, and he fervently believed in and was devoted to coming forth with a new, empirical science of man – one based on observable differences in behavior anchored in brain anatomy and physiology, and notably devoid of metaphysics. Tirelessly observing humans and animals in a healthy and disordered conditions, Gall had set forth to blaze new paths, hoping to make a name for himself while also serving humanity. No other scientist had even tried to explain different kinds of memory (for music, places, people, etc.) or other individual differences with reference to brain organization, and throughout his life he was determined to leave no stone unturned.

Gall's fundamental assumption, that the mind consists of a set of distinct functions, is, of course, no longer controversial. Nor is his assumption that different parts of the cerebral cortex are crucial for mathematics, color perception, music, and other higher functions. Further, how he approached psychiatric and criminal behaviors, and tried to explain artistic talents, had the effect of altering the courses of more than the basic brain sciences and medicine.

For these reasons, Gall's influence on the histories of the hard and social sciences, medicine, and more, merit a fresh look. In this context, what Gall strove to achieve must be distinguished from how phrenology evolved into a cultural fad with people paying itinerant "professors" for head readings, particularly in America but also in other parts of the world. Further, more attention has to be given to how and why Gall and his immediate followers had more or less influence in different places, and how (at the least) the craniological facets of his doctrine came to be ridiculed in general magazines and other public venues: in poetry, plays, novels, editorial cartoons, and more.

This book began as a special (2020) issue of the *Journal of the History of the Neurosciences* that looked at a variety of issues concerning Gall, how phrenology was received, and the doctrine's cultural impact. Knowing this journal had previously published other articles on these subjects and that some newer ones were being prepared for press, we opted to include them in a more inclusive book.

We trust our readers will learn many new things about how Gall and his doctrine were treated in Germany, France, Denmark, and how his ideas were conveyed to countries he did not visit, notably Lithuania and Australia. In addition, they will find new information about how the word "phrenology" (a term Gall despised) caught on, and how phrenologists and phrenology were portrayed in select cartoons and lampooned by some authors (e.g., Mark Twain) of popular literature.

We hope readers will find these studies interesting and informative, and that they will offer new insights and perspectives concerning Gall's thinking and the wide-spread movement(s) he fathered. We thank our authors for joining us on this venture and the publisher for the production of this special book.

References

Finger, S., and Eling, P. 2019. *Franz Joseph Gall: Naturalist of the Mind, Visionary of the Brain*. New York: Oxford University Press.

Fossati, G. A. L. 1857. Gall. In *Nouvelle Biographie Général*, Vol. XIX, J.-C. Hoefer (Ed.). Paris: Firmin Didot, pp. 271–283.

Gall, F. J. 1791. *Philosophisch-medicinische Untersuchungen über Natur und Kunst im kranken und gesunden Zustande des Menschen*. Wien: Rudolph Graffer [also http://www.deutschestextarchiv. de/book/view/gall_untersuchungen].

Gall, F. J. 1798. Schreiben über seinen bereits geendigten Prodromus über die Verichtungen des Gehirns der Menschen und der Thiere an Herrn Jos. Fr. von Retzer. *Neue Teutsche Merkur*, 3, 311–323.

Gall, F. J. 1835. *On the Functions of the Brain and Each of Its Parts: With Observations on the Possibility of Determining the Instincts, Propensities, and Talents, or the Moral and Intellectual Dispositions of Men and Animals, by the Configuration of the Brain and Head* (6 vols.). N. Capen (Ed.), W. Lewis (trans.). Boston: Marsh, Capen and Lyon.

Gall, F. J., and Spurzheim, J. G. 1810–19. *Anatomie et Physiologie du Système Nerveux en Général, et du Cerveau en Particulier* (4 vols. with an atlas). Paris. [Gall was the sole author on the last two vols.]

Roget, P. M. 1818/1824. Cranioscopy. In *Supplement to the Fourth, Fifth, and Sixth Editions of the Encyclopaedia Britannica*, Vol. 3. Edinburgh: Printed for A. Constable and Co, pp. 419–437.

Phrenology: Scheherazade of etymology

Régis Olry and Duane E. Haines

Phrenology belongs to a never-ending trend—sometimes verging on fanaticism (Noel and Carlson 1970, 694)—that, for many centuries, strives to correlate anatomy with behaviors and ultimately with destiny (Kern 1975). This trend aims, whatever the cost, at ascribing behaviors and abilities to limited and specific macroscopical or, more recently (although not really more successful), biological features (Grüsser 1990). As the centuries passed, these presumably-involved features included, among others, the general characteristics of the face (physiognomony: Aristotle, see Fœrster 1893; lavaterism, see Jaton 1988), lines of the forehead (metoposcopy, see Cardano 1658), bumps of the skull (phrenology, see Gall and Spurzheim 1810-1819), and the number of longitudinal frontal gyri (quaternary theory of the frontal lobe, see Benedikt 1879). Some biological features were of a genetic (e.g., supernumerary chromosome, see Jacobs et al. 1965) or molecular (e.g., neuronal nitric oxide synthase, see Nelson et al. 1995) nature.

Although this trend dates back at least to Aristotle, it was still usual during the second half of the eighteenth century on "to assert the existence of a necessary (although much debated) link between the body and the soul" (Pogliano 1990, 144). Unfortunately, this trend, particularly as developed in criminology, has sometimes been used to justify the darkest sides of racial anthropology, whether allegedly scientific or—perhaps even more dangerous—popular. In a letter to his cousin, Marie de Rabutin-Chantal, marquise de Sévigné (1626–1696), dated August 11, 1675, Roger de Rabutin, comte de Bussy (1618–1693), wrote:

> Ne vous souvenez-vous pas, Madame, de la physionomie funeste de ce grand homme (M. de Turenne)? Du temps que je ne l'aimais pas, je disais que c'était une physionomie patibulaire. (Monmerqué 1820, 377)

> [Don't you remember, Madam, the lugubrious physiognomy of this great man (M. de Turenne)? At the time I did not like him, I said it was a suspicious-looking physionomy.]

The French adjective *patibulaire* refers to the *fourches patibulaires*, that is:

> (des) piliers de pierre, au haut desquels il y a une pièce de bois posée en-travers sur deux de ces piliers, à laquelle pièce de bois on attache les criminels qui sont condamnés à être pendus & étranglés. Soit que l'exécution se fasse au gibet même, ou que l'exécution ayant été faite ailleurs, on apporte le corps du criminel pour l'attacher à ces fourches, & l'y laisser exposé à la vûe des passans. (Diderot and d'Alembert 1757, 224)

> [stone pillars, at the top of which there is a piece of wood, put crocked on two of these pillars: the piece of wood to which are tied people sentenced to be hanged & strangled. Whether the execution takes place at the gibbet itself, or the execution has been conducted elsewhere, one

brings the body of the criminal to be tied to these forks & to be left at the sight of the passersby.]

To have a *physionomie patibulaire* therefore means that your *physionomie* shows your evil tendencies. Despite the fact that, according to some authors, "neurobiology has given up ...looking for any 'centre' or neuronal system which would be the generator of ... aggressiveness" (Karli 1987, 28–29), old habits die hard, and many people keep on mistrusting someone merely looking suspicious.

Let us briefly summarize the origins of the terms "metoposcopy," "physiognomony," and "lavaterism" before tackling the topic of this article: the amazingly complex meanings of the term *phren* at the roots of the term phrenology.

Metoposcopy

Metoposcopy, from the Greek μέτωπον (*metopon*: forehead) and σκοπέω (*scopeo*: to examine), is "l'art de connaître les hommes par les rides du front" [the art of knowing humans by the lines of the forehead] (Collin de Plancy 1863, 458). Although to be found as far back as 1615 in Samuel Fuchs's (1588–1630) *Metoposcopia & Ophthalmoscopia*, and 1626 in Ciro Spontone's (ca. 1552–ca. 1610) *La Metoposcopia ouero Commensuratione delle Linee della Fronte* (see Spontone 1626), the term metoposcopy was quite likely coined by Italian physician and polymath Girolamo Cardano (1501–1576), sometimes referred to as a "forerunner of Lombroso" (Ore 1967, 889). Although the Latin first edition of Cardano's *Metoposcopia libris tredecim* was published in 1658 (a French translation by Claude Martin de Laurendière was made available in the same year), this book had been written during the author's stay in Milan in 1550 (Hirsch 1884, 663) and, hence, prior to both above-mentioned publications. Surprisingly, the French Renaissance writer François Rabelais (between 1483 and 1494–1553) used the term *métaposcopie* (underlining by authors) in his 1546 *Tiers livre* (Rabelais 1546, 181, chapter 25, line 29). Yet we could not find any etymology or satisfying explanation for this term.

Physiognomony and lavaterism

The term "physiognomony," from the Greek φύσις (*phusis*: nature) and γνώμων (*gnomon*: one who knows, who detects), is sometimes dated back to Aristotle (384–322 BCE; Zucker 2006, 2), who taught, "It will be possible to infer the character from the features of the face" (Aristotle, cited in Froment 2013, 126). We could find five variant spellings of the term physiognomony: *physonomia* in Lucas Brandis's 1473 edition of the *Lapidarius* (Hain 1826, 220, no. 1777; Osler 1923, 52, no. 30), *physionomia* in Arnold ther Hoernen's 1474 edition of the *Tractatus de pomo* (Hain 1826, 221, no. 1786; Osler 1923, 60, no. 48; Klebs 1938, 51, no. 94.1), *physiognomia* in Joan Paquet's 1611 edition of the *Physiognomonica* (Krivatsy 1989, 42, no. 394), *physiogno-monica* in Gottlob Emanuel Richter's 1780 edition of the *Scriptores Physiognomoniae* (Osler 1969, 22, no. 243), and *physiognomica* in the Immanuel Bekker's 1831–1870 edition of *Aristotelis Opera* (Schwab 1967, 6, no. 15).

The term was taken up by Italian scholar Giovanni Battista Della Porta (1535–1615) in 1586 and later by Swiss theologian Johan Kaspar Lavater (1741–1801) in 1772 (see Della Porta 1586; Lavater 1772), leading to a new (quasi-)synonym for physiognomony: "lavaterism." We could not find who coined the term lavaterism, but it was already being widely used before Lavater's death on January 2, 1801. For example, we found this term in a postscript of a letter dated December 20, 1777, from Dutch philosopher and writer François Hemsterhuis (1721–1790) to Princess Adelheid Amalie Gallitzin (1748–1806; see van Sluis 2011, 235). Shortly afterward, in a letter written on March 25, 1785, to her friend—and soon-to-be first editor of her *Memoirs* (Cornut-Gentille 2004, 31)—Louis-Augustin Bosc d'Antic (1759–1828), the famous revolutionary Manon Roland (1754–1793) wrote,

> Mais, à propos, dites-moi donc quelque chose du lavatérisme [...] Vous espérez donc exercer votre science sur ma figure? (Dauban 1864, lxxvii)

> [But, incidentally, tell me something about lavaterism [...] So, you intend to practice your science on my face?]

We could also find the term "lavaterism" in a letter dated August 21, 1791, from a Mr. Brand to English writer Mary Berry (1763–1852; see Lewis 1865, 351).

Phrenology

The term "phrenology" was coined by physician, medical educator, and philanthropist Benjamin Rush (1746–1813) in his November 21, 1805, *Lectures Upon the Mind*. He wrote, "Very different is the state of phrenology, if I may be allowed to coin a word, to designate the science of the mind" (Rush 1811, 271). These *Lectures* are among the 39 documents and archives Rush bequeathed to the Library of The College of Physicians of Philadelphia (Hirsch 1983, no. 917, 189–190). Importantly, Rush's definition of phrenology encompassed much more than:

> Die Lehre, welche sich damit beschäftigt, den Charakter aus der äusseren Form des Schädels (in Verbindung mit den Temperamenten) zu beurteilen. (Ullrich 1898)

> [The teaching which aims at judging the character from the outer morphology of the skull (in connection with the temperament).]

At first sight, the etymology of the term phrenology might look simple: It derives from the Greek φρήν (*phren*: mind) and λόγος (*logos*: knowledge). But appearances can be deceptive: The many meanings of *phren* (Sullivan 1988, 1997, 13–94), as well as the links between the diaphragm and the mind, are both highly complex issues.

The polysemy of *phren*

The meaning, and therefore the translation of the Greek terms "phren" and its derivation *phresin "pose de nombreux problèmes, souvent discutés"* [raises many problems, often debated] (Chantraine 1977, 1227). They may refer to intelligence (Stappers 1885, no. 2693, 352), *chœur* [chorus] (Pradier 1997, 45), physical localization of the *thūmos* (to be understood as the concept of spiritedness) for "in some Homeric contexts, *thūmos* is used as a synonym of *phrenes* " (Nagy 2013, 282). They were also used to designate some parts—anatomical or spiritual—of the human body: the diaphragm (see below); the lungs—or, more broadly, the viscera contained in the upper part of the body (Onians 1951, 23–42); and the âme végétative [the vegetative soul] (Magnien 1927, 122).

The situation becomes even more complicated with the two occurrences of the term "phresin" in the New Testament (First Epistle to the Corinthians 14: 20):

Ἀδελφοί, μὴ παιδία γίνεσθε ταῖς φρεσίν· ἀλλὰ τῇ κακίᾳ νηπιάζετε, ταῖς δὲ φρεσὶν τέλειοι γίνεσθε. (Nestle 1923, 450)

[Brothers and sisters, stop thinking like children. In regard to evil be infants, but in your thinking be adults.]

In its successive translations, the biblical meaning of *phresin* has been "sage, sans malice" [reasonable, without malice] (D'Allioli 1884, 510), "mûrs" [mature] (École biblique de Jérusalem 1961, 1420), and "*en adulte*" [like an adult] (Grosjean 1971, 559), to name a few.

From phren to *metaphren*

Etymologically, the term "metaphren" (or *metaphrenon*) refers to something located near to the phren—that is, near to the putative seat of the mind. Its first mention probably goes back to Homer's *Iliad* (Pierron 1884, 177) in which metaphrenon refers to "le dos et les épaules" [the back and the shoulders] (Flacelière 1955, XVI–791, 386). Since its spreading in scientific vocabulary during the sixteenth century (Adrados 2005, 276), its meaning has varied considerably. It referred to the back and/or the 12 thoracic vertebrae (Vassé 1555, 90; Paré 1633, 164), the region located "entre les deux épaules" [between both shoulders] (James 1746, col. 17), the part of the back opposite the breast (Duverney 1761, 8; Taylor 1809, 438; Rabelais 1823, 494), the "dorsum" [whole back] (Dunglison 1868, 616), the region behind the diaphragm (Tecusan 2004, 333), and even the region about the kidneys (Friel 1974, 945).

From phren to frenzy

Frenzy (or its archaic variant "phrenzy") shares the same etymology as phrenic. Both spellings can even be found in the first edition of the early *Dictionnaire de l'Académie françois*, which lists *frénésie* in the index for Volume 1, but *phrénésie* in the index for Volume 2 (*Dictionnaire de l'Académie françoise*, 1694).

The term "frenzy" has been defined as "Unsinnigkeit, Taubsucht" [foolishness] (Frisius 1723, 519), "forte et violente alteration d'esprit" [strong and brutal deterioration of the mind] (Joubert 1738, 863), and "égarement d'esprit, alienation d'esprit, fureur violente" [distraction of the mind, insanity of the mind, brutal fury] (*Dictionnaire de l'Académie Françoise 1765*, 547). *The etymological connection between the diaphragm and some kinds of mental derangement therefore rests on the hypothesis of a so-called "psychical idea" of the diaphragm muscle (Petit 1922).*

The diaphragm and the mind

There are many diaphragms in anatomical terminology: *diaphragma bulbi, oris, pelvis, secundarium* Luschka, *sellae*, and *urogenitale*, among others (Terra 1913, 107), all of them referring to a structure of septal nature and/or function. The diaphragm in question here is, of course, the "thoraco-abdominal" diaphragm (Dumas 1797, 128–129).

Although sometimes attributed to Galen (Lauth 1815, 211), the term "diaphragm" might have been coined by Plato from the verb διαφράσσω (*diaphrasso*: to separate; Furetière 1727;

Morin 1809, 278; Stappers 1885, no. 2692, 352). Indeed, Plato considered the diaphragm as "a septum which should isolate in the abdomen the lowest soul and prevent it from too severely disturbing the superior soul" (trans. fom Joly 1961, 448).

Rufus of Ephesus (ca. 70–ca. 110), "probably the first author to have submitted an anatomical nomenclature" (Olry 1989, 92), confirmed this etymology, although with an emphasis on thoracic rather than abdominal, viscera: "On le nomme diaphragme (cloison) parce qu'il sépare les viscères contenus dans le thorax de ceux qui sont au dehors" [It is called diaphragm (septum) because it separates the viscera contained in the thorax from those which are outside] (d'Éphèse 1879, 178). Chinese medicine also refers to *Huang* (one of the terms for diaphragm) as "a thin membrane located above the diaphragm … which prevents the ascent of troubled breath towards the upper heater" (Huchet 2006, 214).

Since the 1895 *Basle Nomina Anatomica*, the adjective phrenic has replaced diaphragmatic in anatomical terminology (His 1895, 72). Thus, all structures—10 in number—related to the diaphragm are now referred to as phrenic (Federative Committee on Anatomical Terminology 1998, 270). The notion of septum has therefore been cast aside in favor of the notion of mind. But why?

Denis Diderot (1713–1784) helps us answer this question. In his 1769 *Le Rêve de d'Alembert*, French physician and anatomist Théophile de Bordeu (1722–1776) explained to his friend Julie de Lespinasse (1732–1776): "Qu'est-ce qu'un être sensible? Un être abandonné à la discrétion de son diaphragme" [What is a sensitive being? A being abandoned at the discretion of his diaphragm] (Assézat 1875, 171; Siess 1990, 190).

A putative link between the diaphragm and troubles of the mind was raised by many anatomists. Jean Riolan (1577–1657) wrote, "il est souvent cause de la phrénésie" [It is often cause of frenzy] (Riolan 1672, 323), whereas Bernardo Santucci thought that the inflammation of the diaphragm produces a "*continos delirios*" [relentless delirium] (Santucci 1739, 380–381). Isbrand van Diemerbroeck (1609–1674) provided more details:

Parce que quand il (le diaphragme) est offensé, l'esprit & les sens sont troublés par communication, & que c'est dans son inflammation qu'arrive cet [sic] espèce de délire qu'on nomme Paraphrénesie. (Diemerbroeck 1695, vol. 2, 43)

[because when it (the diaphragm) is offended, the mind and the senses become troubled by communication, and that it is in its inflammation that the kind of delirium one refers to as parafrenzy happens.]

It is therefore because it was believed to be involved in the soul/mind/mood disorders that the diaphragm shared the Greek root phren with phrenology.

So, why did we call phrenology the "Scheherazade of etymology"? Because one could easily spend a thousand and one nights studying all the meanings and interpretations of the term phren!

Disclosure statement

No potential conflict of interest was reported by the authors.

References

Adrados F. R. 2005. *A history of the Greek Language. From its origin to the present.* Leiden, Boston: Brill.

Assézat J. 1875. Le Rêve de d'Alembert. In *Œuvres complètes de Diderot revues sur les éditions originales*, vol. 2, 122–81. Paris: Garnier frères.

Benedikt M. 1879. *Anatomische Studien an Verbrecher-Gehirnen für Anthropologen, Mediciner, Juristen und Psychologen bearbeitet.* Wien: Wilhelm Braumüller.

Cardano G. 1658. *Metoposcopia libris tredecim, et octigentis faciei humanae eiconibus complexa. Cui accessit Melampodis De naevis corporis tractatus, Graece & Latine nunc primum editus. Interprete Claudio Martino Laurenderio …*. Lutetiae Parisiorum, Apud Thomam Jolly.

Chantraine P. 1977. *Dictionnaire étymologique de la langue grecque. Histoire des mots*, vol. IV–1. Paris: Klincksieck.

Collin de Plancy J. 1863. *Dictionnaire infernal.* 6th ed. Paris: Henri Plon.

Cornut-Gentille P. 2004. *Madame Roland. Une femme en politique sous la Révolution.* Paris: Perrin.

D'Allioli J. F. 1884. *Nouveau commentaire littéral, critique et théologique avec rapport aux textes primitifs sur tous les livres des divines écritures*, vol. 7, 8th ed. Paris: Louis Vivès.

d'Éphèse R. 1879. *Œuvres. Texte collationné sur les manuscrits (…) publication commencée par le Dr Ch. Daremberg, continuée et terminée par Ch. Émile Ruelle.* Paris: à l'Imprimerie nationale.

Dauban C. A. 1864. *Étude sur madame Roland et son temps.* Paris: Plon.

Della Porta G. B. 1586. *De humana physiognomonia libri IIII.* Vici Aequensis, apud I. Cacchium.

Dictionnaire de l'Académie françoise. 1694. Paris: chez la veuve de Jean Baptiste Coignard et chez Jean Baptiste Coignard.

Dictionnaire de l'Académie françoise, vol. 1. 1765. Paris: Chez les libraires associés, nouvelle édition.

Diderot D., and J. L. R. d'Alembert, eds. 1757. *Encyclopédie, ou Dictionnaire raisonné des sciences, des arts et des métiers*, vol. 7. Paris: Briasson, David, Le Breton, Durand.

Diemerbroeck I. D. 1695. *L'anatomie du corps humain.* Lyon: Anisson & Posuel.

Dumas C. L. 1797. *Système méthodique de nomenclature et de classification des muscles du corps humain.* Montpellier: Bonnariq, Avignon et Migueyron.

Dunglison R. 1868. *A dictionary of medical science.* Philadelphia, PA: Henry C. Lea.

Duverney J. G. 1761. *Œuvres anatomiques*, vol. 1. Paris: Charles-Antoine Jombert.

École biblique de Jérusalem, ed. 1961. *La Sainte Bible.* Montréal: Les Éditions Leland Ltée.

Federative Committee on Anatomical Terminology. 1998. *Terminologia Anatomica. International anatomical terminology.* Stuttgart: Thieme.

Flacelière R., ed. 1955. *Iliade. Odyssée*, vol. 115. Paris: Gallimard, Bibliothèque de la Pléiade.

Fœrster R., ed. 1893. *Scriptores Physiognomonici et Latini*, vol. 2. Lipsiae: in aedibus B.G. Teubneri.

Friel J. P., ed. 1974. *Dorland's illustrated medical dictionary.* 25th ed. Philadelphia, PA: W.B. Saunders.

Frisius J. 1723. *Dictionarium Latino-Germanicum nec non Germanico-Latinum.* Coloniae Agrippinae: Sumptibus Wilhelmi Metternich.

Froment A. 2013. *Anatomie impertinente. Le corps humain et l'évolution.* Paris: Odile Jacob.

Fuchs S. 1615. *Metoposcopia & Ophthalmoscopia.* Argentinae, excudebat Theodosius Glaserus, sumptibus Pauli Ledertz.

Furetière A. 1727. *Dictionnaire universel, contenant généralement tous les mots François, tant vieux que modernes … Nouvelle edition par … Basnage de Beauval … Brutel de la Rivière*, vol. 1. La Haye, Pierre Husson, Thomas Johnson, Jean Swart, Jean van Duren, Charles Le Vier, la Veuve Van Dole.

Gall F. J., and G. Spurzheim. 1810-1819. *Anatomie et physiologie du système nerveux en general, et du cerveau en particulier, avec des observations sur la possibilité de reconnoitre plusieurs dispositions intellectuelles et morales de l'homme et des animaux, par la configuration de leurs têtes*, vol. 4 umes and atlas. Paris: F. Schoell.

Grosjean J. 1971. *La Bible. Nouveau Testament.* Paris: Gallimard. Bibliothèque de la Pléiade 226.

Grüsser O. J. 1990. Vom Ort der Seele. Cerebrale Lokalisationstheorien in der Zeit zwischen Albertus Magnus und Paul Broca. *Aus Forschung Und Medizin* 5 (1):75–96.

Hain L. 1826. *Repertorium bibliographicum in quo libri omnes ab arte typographica inventa usque ad annum MD*. Stuttgartiae: sumtibus J.G. Cottae.

Hirsch A. 1884. *Biographisches Lexikon hervorragenden Aerzte aller Zeiten und Völker*, vol. 1. Wien und Leipzig: Urban & Schwarzenberg.

Hirsch R., ed. 1983. *A catalogue of the manuscripts and archives of the library of the college of physicians of Philadelphia*. Philadelphia: University of Philadelphia Press, Francis Clark Wood Institute, College of Physicians of Philadelphia.

His W. 1895. Die anatomische Nomenclatur, Nomia anatomica, Verzeichniss der von der Commission der anatomischen Gesellschaft festgestellten Namen, eingeleitet und im Einverständniss mit dem Redactionsausschuss erläutert. *Archiv Für Anatomie Und Physiologie* (Supplement-band).

Huchet A. 2006. Geshu (17V), étymologie et indications. *Acupunture & Moxibustion* 5 (3):213–20.

Jacobs P. A., M. Brunton, M. M. Melville, R. P. Brittain, and W. F. Clemont. 1965. Aggressive behaviour, mental subnormality and the XYY male. *Nature* 208:1351–52. doi:10.1038/2081351a0.

James R. 1746. *Dictionnaire universel de médecine*, vol. 2. Paris: Briasson, David l'aîné, Durand.

Jaton A. M. 1988. *Jean Gaspard Lavater*. Lucerne, Lausanne: Éditions René Coeckelberghs.

Joly R. 1961. Platon et la médecine. *Bulletin De l'Association Guillaume Budé: Lettres D'humanité* 20:435–51. doi:10.3406/bude.1961.4200.

Joubert J. 1738. *Dictionnaire françois et latin, tiré des auteurs originaux et classiques de l'une et l'autre langue*. Lyon: Louis et Henry Declaustre.

Karli P. 1987. *L'homme agressif*. Paris: Odile Jacob.

Kern S. 1975. *Anatomy & destiny. A cultural history of the human body*. Indianapolis: The Bobbs-Merrill Company, Inc.

Klebs A. C. 1938. *Incunabula Scientifica et Medica*. Bruges: The Saint Catherine Press Ltd.

Krivatsy P. 1989. *A catalogue of seventeenth century printed books in the national library of medicine*. Bethesda, MD: National Library of Medicine.

Lauth T. 1815. *Histoire de l'anatomie*, Vol. 1. Strasbourg: F.G. Levrault.

Lavater J. C. 1772. *Von der Physiognomik*. Leipzig: Weidmanns Erben.

Lewis T. 1865. *Extracts of the journals and correspondence of Miss Berry from the year 1783 to 1852*, vol. 1. London: Longmans, Green and Co.

Magnien V. 1927. Quelques mots du vocabulaire grec exprimant des opérations ou des états de l'âme. *Revue Des Études Grecques* 40 (184–188):117–41.

Monmerqué L., ed. 1820. *Lettres de madame de Sévigné, de sa famille et de ses amis*, vol. 3. Paris: J.J. Blaise.

Morin J. B. 1809. *Dictionnaire étymologique des mots françois dérivés du grec*, vol. 1, 2nd ed. Paris: de l'Imprimerie Impériale.

Nagy G. 2013. *The ancient Greek hero in 24 hours*. Cambridge, MA: Harvard University Press.

Nelson R. J., G. E. Demas, P. L. Huang, F. C. Fishman, V. L. Dawson, T. M. Dawson, and S. H. Snyder. 1995. Behavioural abnormalities in male mice lacking neuronal nitric oxide synthase. *Nature* 378:383–86.

Nestle E. 1923. *Novum Testamentum Graece et Latine. Editio septima*. Stuttgart: Privilegierte Württembergische Bibelanstalt.

Noel P. S., and E. T. Carlson. 1970. Origins of the word "Phrenology". *The American Journal of Psychiatry* 127 (5):694–97.

Olry R. 1989. Histoire des nomenclatures anatomiques. *Documents pour l'Histoire du Vocabulaire Scientifique du CNRS* 9:91–98.

Onians R. B. 1951. *The origins of European thought about the body, the mind, the soul, the world, time, and fate*. Cambridge, UK: Cambridge University Press.

Ore O. 1967. Cardano, Geronimo. In *Encyclopedia Britannica*, vol. 4, 889–90. Chicago, IL: William Benton.

Osler W. 1923. *Incunabula medica. A study of the earliest printed medical book 1467-1480*. Oxford, UK: Oxford University Press.

Osler W. 1969. *Bibliotheca Osleriana. A Catalogue of books illustrating the history of medicine and science.* Kingston and Montreal: McGill-Queen's University Press.

Paré A. 1633. *Les Œuvres.* 9th ed. Lyon: chez la veuve de Claude Rigaud et Claude Obert.

Petit G. 1922. Sur la conception ancienne – Anatomique, physiologique et psychique – Du muscle diaphragme. *Bulletins et Mémoires de la Société d'Anthropologie de Paris* 3:48–54.

Pierron A. 1884. *L'Iliade d'Homère. Texte grec. Chants XIII-XXIV.* 2nd ed. Paris: Hachette et Cie.

Pogliano C. 1990. Entre forme et fonction: Une nouvelle science de l'homme. In *La fabrique de la pensée*, ed. P. Corsi, 144–57. Milano: Electa.

Pradier J. M. 1997. *La Scène et la fabrique des corps. Ethnoscénologie du spectacle vivant en Occident (Vᵉ siècle av. J.-C. – XVIIIᵉ siècle).* Bordeaux: Presses Universitaires de Bordeaux.

Rabelais F. 1546. *Tiers liure des faictz et dictz Heroïques du noble Pantagruel.* Paris: Cheftien wechel.

Rabelais F. 1823. *Œuvres complètes*, Vol. 8. Paris: Dalibon.

Riolan J. 1672. *Manuel anatomique et pathologique.* new ed. Lyon: Antoine Laurens.

Rush B. 1811. *Sixteen introductory lectures, to courses of lectures upon the institutes and practice of medicine, with a syllabus of the latter.* Philadelphia: Bradford and Innskeep.

Santucci B. 1739. *Anatomia do corpo humano.* Lisboa Occidental: Antonio Pedrozo Galram.

Schwab M. 1967. *Bibliographie d'Aristote.* New York: Burt Franklin (reprint of the 1896 edition).

Siess J. 1990. Lespinasse, ou l'ancienne Julie, petite étude à plusieurs voix, à propos d'une edition récente des Lettres à Condorcet. *Recherches Sur Diderot Et Sur l'Encyclopédie* 9:190–91.

Spontone C. 1626. *La Metoposcopia ouero Commensuratione delle Linee della Fronte.* Venetia: Evangelista Deuchino.

Stappers H. 1885. *Dictionnaire synoptique d'étymologie française.* Bruxelles: Merzbach & Falk.

Sullivan S. D. 1988. *Psychological activity in homer: A study of Phren.* Ottawa: Carleton University Press.

Sullivan S. D. 1997. *Aeschylus' use of psychological terminology. Traditional and new.* Montreal & Kingston: McGill-Queen's University Press.

Taylor T. 1809. *The history of animals. Aristote and his treatise on physiognomony*, vol. 5. London: Printed for the Translator.

Tecusan M. 2004. *The fragments of the methodists. Volume one: Methodists outside Soranus.* Leiden: Brill.

Terra P. D. 1913. *Vademecum anatomicum. Kritisch-etymologisches Wörterbuch der systematischen Anatomie.* Jena: Verlag von Gustav Fischer.

Ullrich M. W. 1898. Phrenologie (Schädellehre). In *Nachdruck anlässig der 18. Göttinger Neurobiologentagung "Gehirn – Wahrnehmung – Kognition"*, ed. N. Elsner. Göttingen: Juni 1990, unpaginated.

van Sluis J. ed. 2011. *François Hemsterhuis. Ma toute chère Diotime. Lettres à la princesse de Gallitzin, 1775-1778*, 234–35. Berltsum: van Sluis, letter 1.196.

Vassé L. 1555. *Tables anatomicques du corps humain universel: Soit de l'homme, ou de la femme.* Paris: Jean Foucher.

Zucker A. 2006. La physiognomonie antique et le langage animal du corps. *Rursus* 1:1–24.

Overlooked Gall and
Features of his Doctrine

Franz Joseph Gall on hemispheric symmetries

Paul Eling and Stanley Finger

ABSTRACT

Franz Joseph Gall believed that the two cerebral hemispheres are anatomically and functionally similar, so much so that one could substitute for the other following unilateral injuries. He presented this belief during the 1790s in his early public lectures in Vienna, when traveling through Europe between 1805 and 1807, and in the two sets of books he published after settling in France. Gall seemed to derive his ideas about laterality independently of French anatomist Marie François Xavier Bichat (1771–1802), who formulated his "law of symmetry" at about the same time. He would, however, later cite Bichat, whose ideas about mental derangement were different from his own and who also attempted to explain handedness, a subject on which Gall remained silent. The concept of cerebral symmetry would be displaced by mounting clinical evidence for the hemispheres being functionally different, but neither Gall nor Bichat would live to witness the advent of the concept of cerebral dominance.

Prior to the nineteenth century, the prevailing view was that, in one way or another, the soul or mind controlled the brain through underlying structures, such as the ventricles or the pineal body and, at later times, brain stem areas or subcortical white matter. German-born anatomist Franz Joseph Gall (1758–1828) began to challenge these views late in the eighteenth century by studying such things as the brains and behaviors of humans and animals on the great chain of being, developmental landmarks, and individuals with brain damage (Finger and Eling 2019).

Gall's focus on the cortex resonated with many other physicians and anatomists, as did how he circumvented long-held metaphysical constructs and emphasized empiricism in his science (Ackerknecht and Vallois 1956; Finger and Eling 2019; Rawlings and Rossitch 1994). More controversial to his enlightened audiences was his claim of many independent cortical organs, each associated with a highly specific, "concrete" function (e.g., language, music)—that is, his organology. The latter began to take form early in the 1790s, when Gall listened to a five-year-old girl named Bianchi who, without training, could sing and memorize music exceptionally well, while appearing ordinary in every other trait (Eling, Finger, and Whitaker 2017). When he remembered how well some of his classmates with bulging eyes (presumably pushed outward by the growth of the underlying part of the brain) did when memorizing verbal material, his revolutionary theory linking faculties of mind, cortical areas, and cranial features was born.

In 1796, Gall started lecturing on these ideas at his stately home in Vienna, where he used some of his funds as a physician to support his growing collection of skulls and casts, which focused on exceptional people (e.g., criminals, geniuses, and the insane) and animals. He continued to modify his list of faculties and their associated cortical organs, before settling on 27 faculties for humans, 19 shared with animals, these being what he presented in Paris, where he arrived in 1807 and settled.

Gall believed the faculties are duplicated, one on each side of the cerebral and cerebellar cortices (which housed a single faculty, reproductive drive). After all, the two sides of the brain look very much alike, even when one ascends the ladder of life, give or take small differences that seem inconsistent from one brain to the next when there is no evidence of disease. Even René Descartes (1596–1650) had written how he observed "the brain to be double" (Descartes 1649/1958, 275). Indeed, no one studying the gross anatomy of the brain during the eighteenth century had drawn sustained attention to hemispheric differences in healthy brains, although French physician and anatomist Vicq d'Azyr (1748–1794) had noticed some asymmetries. This remained the case when, in 1800, French anatomist Marie François Xavier Bichat (1771–1802) published the first edition of a book discussing, among other things, what would be called his "law of symmetry."

The attention given to Bichat in his own day and by later historians raises the question of whether Gall derived his ideas about brain symmetry from Bichat or was influenced by him when constructing his own doctrine. In an attempt to answer this question, we have examined what Gall said about both symmetry and Bichat in his early lectures, when presenting his anatomy and his revolutionary doctrine in Berlin in 1805, in his four-volume *Anatomie et Physiologie* of 1810–1819, and in his less-expensive *Sur les Fonctions* of 1825, which deviated in only minor ways from what he wrote in his "great work" (Gall and Spurzheim 1810-1819, 1825). In addition, we looked for what the two men wrote about handedness, as it too could be related to the concept of brain symmetry. We conclude this survey with some discoveries leading to the death of symmetry, a development neither Gall nor Bichat lived to witness.

Early accounts

Gall attracted enthusiastic audiences when he began lecturing at his home during the mid-1790s. Physicians, philosophers, government officials, clerics, students, writers, and others came to hear him. Some took extensive notes and reworked them into publications presenting Gall's doctrine with examples, evidence, and logic (Finger and Eling 2019). There is considerable similarity in the longer accounts, which follow how he covered the material in his courses.

Ludwig Heinrich Bojanus (1776–1827), who studied medicine at the University of Jena, provided one of these accounts. Bojanus had traveled to Vienna immediately after graduating, and he practiced in the General Hospital from 1797 to 1798, where he might have met Gall. He also attended Gall's lectures, taking copious notes that he published in 1801 (Bojanus 1801; see also Sakalauskaitė-Juodeikienė, Eling, and Finger 2017, 2020). His article shows that Gall mentioned hemispheric symmetry when discussing unilateral brain lesions. As stated in the 1802 English translation:

It might be here objected, that in several cases individuals have lost a considerable portion of the substance of the brain without the faculties being sensibly diminished; but it is to be observed that the greater part of the cerebral organs exist double, and that the observations mentioned are not exact. (Bojanus 1802, 78)

German physician Ludwig Friedrich von Froriep (1779–1847; Figure 1) was another physician who attended Gall's lectures in Austria. After getting his medical degree in Jena, he went to Vienna for six months during 1799. He published his "*Kurze Darstellung*" [Short Presentation] in *Voigts Magazine der neuensten Zustand der Naturkunde* a year later, and two years afterward he came forth with a more detailed account in his *Darstellung der ganzen auf Untersuchungen der Verrichtungen des Gehirns gegründeten Theorie der Physiognomik des Dr. Gall in Wien* [Presentation of the New Theory of the Physiognomy of Dr. Gall in Vienna Based on Investigations of the Brain], which we consulted (Froriep 1800, 1802).

Froriep began by comparing Gall's views to those of Swiss physiognomist Johann Kaspar Lavater (1741–1801), whose ideas about character were well known at the time but were not tied to brain areas. He then discussed the two fundamental assumptions underlying Gall's organology: multiple inborn faculties of mind and distinct specialized organs in the cerebral cortex. Next, he turned to Gall's methods, bringing up diseases and lesions of the brain, and making the point that "a lesion to specific areas may enhance the

Figure 1. Ludwig Friedrich von Froriep (1779–1847).

activity of a given organ or erase it completely" (pp. 44–45). He then presented counter-arguments, one being instances of brain damage that do not seem to affect mental powers.

Froriep (1802) presented Gall's response to this objection in the same way as Bojanus had done:

> The entire brain is split up in half and most parts of the brain are duplicated. If now some tissue of one brain half is lost and the substance on the other side remains intact, it can replace the function of the organ on the other side without it being noticeable. One can object to this double representation that one is superfluous or that our representation should be double, one can point to the external senses where the same is happening. (p. 46)

There is also a footnote at the start of Gall's counterargument about how the organs are duplicated on the right and left cortices. On the basis of this assumption, readers are informed, one can also explain a phenomenon that has been observed several times—namely, that some patients think in a delirious manner with one side of the brain while thinking correctly with the other, such that they can recognize the incorrect thinking from the faulty side (p. 46).

Based on the Bojanus and Froriep accounts, it seems clear that Gall was already convinced about anatomical and functional cortical symmetry in 1799. The two hemispheres are comparable to our two eyes or ears. If only one is damaged, we can still see, hear, or (as now shown) engage higher functions of mind, the surviving organ providing the redundancy that allows this to happen. Froriep faithfully followed what Gall was saying, and there is no mention here of Bichat's name or ideas, or of why most people favor one hand, usually the right over the left.

Philipp Franz von Walther (1782–1849; Figure 2) provided a third account of what Gall was covering in his Vienna lectures. Born in the small German village of Burrweiler near Karlsruhe, Walther went to Vienna in 1800 to study ophthalmology. He published his *Critische Darstellung der Gall'schen anatomisch- physiologischen Untersuchungen des Gehirnund Schädel-baues* [Critical Account of Gall's Anatomical-Physiological Investigations of the Brain and the Form of the Skull] in 1802, a year before obtaining his doctorate from the University of Landshut.

Although covering the same material, Walther's essay is somewhat more philosophical than Froriep's and it uses more of his own wording. After introducing the notion of multiple organs, he turned to organ "dualism," writing:

> For the construction of the brain, nature followed the laws of dualism. All individual brain parts are represented double so that the entire brain mass can be split in two identical halves. Even there, where these connect to each other one can demonstrate the dichotomy of the apparent whole, for instance in the formation of the corpus callosum, the fornix, etc. Consequently, the function of a specific brain site will not be always lost after its disorganization: the specific organ in the opposite hemisphere replaces the defective function. (Walther 1802, 45)

He continued by stating that, as is often the case with double viscera (*Gingeweide*; organs), one is stronger than the other. For instance, the left eye often shows weaker acuity, and the left lung and kidney tend to become infected more often than their counterparts on the right. Similarly, the left side of the head is more likely to be affected by headaches, and the left side of the body by hemiplegia. Walther recognized that the two hemispheres might differ, but he assumed this only relates to the extent to which each might be activated. Indeed, even the two halves of the skull might show some asymmetries in the absence of disease.

Figure 2. Philipp Franz von Walther (1782–1849).

Walther now presented the same argument about the independence of the two hemi-spheres to which Froriep alluded in his footnote, writing:

> How much the function of a given brain organ on one side is independent from that on the other side, can be observed particularly in the remarkable phenomenon of the delirious speech of some patients with nervous fever, who are only delirious with one hemisphere, and think efficiently with the other, and they are therefore perfectly conscious of the perverted condition of the association of their ideas. (Walther 1802, 46)

Hence, we find Gall continuing to state that the two hemispheres are functionally identical and, with this being the case, a function need not be affected by unilateral damage. Furthermore, the two organs are able to act independently from one another even in brains free of disease. As before, there is no mention of Bichat or of handedness.

In 1805, Christian Heinrich Ernst Bischoff (1781–1861; Figure 3) came forth with his account of how Gall presented his doctrine. Bischoff had been born in Hannover and obtained his medical degree at Jena in 1801. Three years later, he was appointed Extraordinary Professor of Physiology in Berlin, where he worked with his mentor and friend, Christoph Wilhelm Hufeland (1762–1836). When Gall arrived in Berlin (the first stop on his European tour) in 1805 and gave his lectures and demonstrations, Bischoff took notes for a book, his *Darstellung der Gall'schen Gehirn- und Schädel- Lehre*

Figure 3. Christian Heinrich Ernst Bischoff (1781–1861).

[Presentation of Gall's Brain and Skull Theory], which he published that year (an English edition followed in 1807).

After covering the basic assumptions underlying Gall's doctrine, Bischoff addressed the subject of the hemispheres containing duplicate organs. He did this in the same way his predecessors had done, showing Gall was following a fixed schema for presenting his theory and supportive arguments, although the finer details about some of his faculties and organs were still being developed.

He specifically mentioned how considerable parts of the brain could be destroyed, either from an external wound or from some disease, with little effect on the faculties. This phenomenon was explained, as before, by the duplicity of the organs of mind, with a healthy organ maintaining its function following injury to its counterpart. Anyone, he related, can see that the organs of sense and other organs of animal life are double (e.g., the eyes, ears, and muscles), unlike those maintaining vegetative life (e.g., the stomach and liver), which are single. "It is true, the lungs, kidneys, &c. may seem to be an exception, yet they are not, from their inequality, to be considered as completely double, and these organs form a transition from the lower and organic, to the higher and animal life" (Bischoff 1807, 30).

Bischoff continued:

Against this notion of a duplicity of organs in the brain, the unity of perception and consciousness has been brought forwards, But the analogy of the external senses is a sufficient reply to this objection; the organ is not in the one case, any more than in the other, considered as the principle of sensation or perception, it is but the material condition of their exercise.

    G. digressed here concerning the use of the double organs; it is enough briefly to observe that he is of the opinion only one eye, one ear, &c, is employed at a time; and that these succeed each other in their operation. Probably, he said, the right side of the brain is the more active, as the right side of the body throughout, head, breast, eye, hand, arm, foot, &c. are generally the stronger. Eight tenths of those, he says, who have a hump, have it on the right shoulder, as the muscles on this side are the most active and strong. He carried these remarks (without laying any stress on them) so far as to observe that, when a boy, he used to ask himself how it came that men seldom walk quite straight; and that he imputed it to the successive use of each eye, by means of which the point of vision is changed. (pp. 31–32)

Thus, Gall did not envision significant anatomical or physiological differences between the cerebral hemispheres, or what we would now call lateralization of function. But he did seem to believe that, for some things at least, one side might be a little stronger than the other. It should also be recognized that what he was saying about duplicate organs of mind made this part of his doctrine unchallengeable. When there are no signs or symptoms accompanying unilateral damage, it is because of organ redundancy, whereas when there are abnormal behaviors, they could always be attributed to real or imagined brain abnormalities or to the unilateral pathology supposedly affecting the natural balance between the two sides. With this position, Gall had a fallback, allowing him to brush off the opposition and explain everything, as some critics would later recognize (Finger and Eling 2019).

Gall and Bichat

Marie François Xavier Bichat was and still is recognized far more than Gall for bringing the concept of symmetry to the fore at the beginning of the nineteenth century. This was partly due to where he worked, his standing in the scientific and medical communities, his widely read books, how he framed his ideas, and the fact that he was interested in the gross anatomy and histology of all organs, unlike Gall, who was focused only on the brain.

    Born in the French village of Thoirette, Bichat studied mathematics and physical sciences at the University in Lyon before turning to anatomy and surgery at Lyon's Hôtel-Dieu. In 1793, he moved to Paris, where he was appointed chief physician at its oldest and most famous hospital, Hôtel-Dieu. Bichat's life was, however, short: He died at age 30, just four years after Gall entered Paris, and was buried in the Père Lachaise Cemetery, where Gall's body (although not his head) would be placed 17 years later.

    Bichat first presented his treatise on symmetrical and asymmetrical organs in 1800. His venue was a book, his *Recherches physiologiques sur la vie et la mort*, which had a second edition in 1805, and which was translated as *Physiological Researches on Life and Death* (Bichat 1799, 1805/1809). In these editions, he made a sharp distinction between two types of life. Organic life, he opined, is the life of the heart, intestines, and other singular organs regulated via the ganglionic nervous system. In contrast, animal life involves the symmetrical organs of sensation, the passions (emotion), and cognition (understanding, intellect, etc.).

Two perfectly similar globes [eyes] receive the impression of the light. Sound and odours have each also their double analogous organ. A single membrane is the seat of savours, but in it the

median line is manifest; and each division marked by it resembles that of the opposite side. …
The nerves which transmit the impression made by sounds, such as the optic, the acoustic,
the lingual and olfactory are evidently assembled in symmetrical pairs. (Bichat 1805/1809, 8)

Bichat wrote that the brain is symmetrical when he included it with the other organs of
animal life. "Those parts [of the brain] not in pairs," he maintained, "are all symmetrically
divided by the median line, of which several afford visible traces, as the *corpus callosus*, the
fornix, tuber annulare, &c." (p. 8).

Bichat linked his anatomy to physiology when he brought up the long-held principle
that symmetry in structure indicated symmetry in function.

Harmony is to the *functions* of the organs what symmetry is to their *conformation*; it supposes
a perfect equality of force and action, as symmetry indicates an exact analogy in the external
forms and internal structure. It is a consequence of symmetry; for two parts essentially alike
in their structure, cannot be different in their mode of acting. (Bichat 1805/1809, 14)

Lauren Harris (1991), who also provided some of these quotations, wrote that Gall
followed the same line of reasoning as Bichat, and we agree. Harris did not explicitly
attribute Gall's thinking about duplicate organs of mind to Bichat in his 1991 article, but
eight years later he seemed to do so, contending: "In this [Bichat's] law of symmetry, Gall
(1835) saw an important implication: If one hemisphere is injured, all normal functions
could go on as before, supported by the other hemisphere" (Harris 1999, 13). But did Gall
draw on Bichat's *Recherches physiologiques sur la vie et la mort* when formulating his own
theory?

Gall did not mention Bichat's name in his Vienna lectures, which some of his
attendants published as books. This was true for Bojanus, who was in Vienna in
1797–1798, and for Froriep, whose book was based on a lecture series Gall gave in
1799. And from all indications, Gall had been lecturing on the same material since 1796.

As for Walther, he arrived in Vienna in 1800 and could have listened to Gall at that
time or during the next year, when Bichat's ideas about duplicate organs were starting to
circulate. The book Walther published in 1802 was like Froriep's, in that it did not
mention Bichat's name or bring up two types of life. But what he wrote about "the
formation of the corpus callosum, the fornix, etc." is reminiscent of Bichat, who had
written that these structures "are all symmetrically divided by the median line, of which
several afford visible traces, as the *corpus callosus*, the *fornix, tuber annulare*, &c." Hence,
there is reason to think that Gall (or perhaps Walther?) might have become familiar with
Bichat's treatise soon after it was published.

Bichat's name remained missing from Bischoff's 1805 publication, which was based on
Gall's Berlin lectures (see Bischoff 1805). Nonetheless, they now show more evidence for
Gall having read Bichat's book. How Gall mentioned that some "organs form a transition
from the lower and organic, to the higher and animal life," would strongly suggest that he
had become acutely aware of Bichat's dichotomy and had determined it was in accord
with his own thinking about hemispheric duplication, although each of his hemispheres
(and not Bichat's) contained multiple organs of mind.

From the chronology of the renditions consulted here, we can conclude that Gall was
not initially inspired or guided by Bichat. He had reached his own conclusions about
symmetry in the cerebral and cerebellar cortices during the 1790s, prior to reading
Bichat's (1800) book. Bichat's two-fold classification of asymmetrical and symmetrical

organs linked to organic and animal life was, in fact, largely irrelevant or tangential to the fundamentals of Gall's doctrine. Still, when he learned what the Frenchman wrote about symmetry early in the new century, he found it supportive of his own observations and thinking about the two sides of the brain. Hence, Gall began to integrate some of what Bichat wrote into his own lectures, although his name did not appear in the published reports of his early lectures.

But did Gall, who was egotistical and protective of his own ideas, mention symmetry and present Bichat's name in this context in his later books? His first set of books, published between 1810 and 1819, and his second set, completed in 1825 and presenting what was essentially the same material without the atlas and detailed neuroanatomy, reveal that he continued to make the case that the cerebral organs of one hemisphere are perfectly duplicated on the other (Gall and Spurzheim 1810-1819; 1825, 1835 trans.). In his words:

I have proved, in the first volume of my large work, that the nervous systems of the spinal marrow, of the organs of sense, and of the brain, are double, or in pairs. But, as, when one of the optic nerves or one of the eyes is destroyed, we continue to see with the other eye, so when one of the hemispheres of the brain, or one of the brains, has become incapable of executing its functions, the other hemisphere or the other brain may continue to perform those belonging to itself; in other words, the functions may be disturbed or suspended on one side, and remain perfect on the other. (Gall 1835, Vol. 2, 164)

Gall also presented cases of his own and others showing that just one side of the brain could be affected in cases of "mental alienation" before concluding:

Since, therefore, the state of one hemisphere of the brain may be wholly different from that of the other, this difference must extend to the functions of these hemispheres also; and since all the organs of the primitive faculties of the mind are double, it is possible that, in the severest diseases and injuries of the brain, all those faculties may exist, whose organs have not been paralyzed or destroyed, at the same time, on both sides. (1835, Vol. 2, 166–167)

With regard to providing Bichat's name, Gall referred to him 22 times in his "great work," although some of these citations involved Bichat's (1801, 1801-1803) other writings on histology and pathological anatomy, not what he had to say about symmetrical parts of the brain. Furthermore, some were no more than references or in lists that included the names of other physicians. Even his two direct quotations from Bichat seemed unrelated to hemispheric differences. Bichat was, however, mentioned in connection to symmetry in one of the nine places where we found the word "symmetry" used.

These searches brought us to a section in Gall's works titled, "On the Means of Finding, By the Aid of the Cerebral State, a Measure for the Intellectual Faculties, and the Moral Qualities" (Gall 1835, Vol. 2, 182–222). Here Gall discussed the relations between brain volume and functions, correlations between brain and body volumes, proportions between brain and nerves, and the like. He then asked, rhetorically: What can be inferred from the different forms of the head? More specifically: "Does there exist a form of the head from which the existence of mania can be inferred?" (p. 202).

He began his answer to this question by mentioning what Philippe Pinel (1745–1826) wrote about mania:

"The opinion is pretty general," says Pinel, "that mental alienation is to be attributed to defects in the brain, and especially to defects and disproportions in the cranium. ... But observation is far from confirming these specious conjectures; for we sometimes find the most beautiful forms of the head, accompanying the most limited degree of intelligence, and even perfect mania; and, on the other hand, strange varieties of conformation co-exist with all the attributes of talent and genius." (Gall 1835, Vol. 2, 202-203)

Gall contended that Pinel's conclusion about there being no relation between the form of the head and mental alteration was incorrect. Interestingly, he now argued that what is important is not the form of the head but the brain itself. Here he presented some of his own observations, while mentioning Bichat's own asymmetric head and praising him as a genius.

> Want of symmetry in the head is frequently a consequence of rickets, sometimes also of particular cerebral maladies, such as effusion of the cavities of the brain, &c. Hence in an equal number of heads not symmetrical and symmetrical, a larger proportion of the former will be found to have belonged to deranged persons. Haller and Bichat thought, that a want of symmetry in the two halves of the head, was one of the principal causes of mania. But it must not be forgotten, that frequently the most healthy heads, I mean those whose form has not been in the least influenced by disease, have the two halves unequal. ... There was considerable inequality between the two halves of Bichat's head, as is shown by the cast taken after his death. Probably he himself was not aware of this deformity: but, who will maintain that Bichat was not a man of genius? (Gall, 1835, Vol. 2, 206-207)

It is important to note that Gall was discussing the relationship between mental disorders and head symmetry here, and that his main objection to Pinel was that he was not examining brains. Pointing to Bichat's own head, Gall warned that, although asymmetry could result from specific diseases, it should not immediately be interpreted as a physical sign of mental derangement.

Accounting for handedness

During this era, human handedness did not necessarily present a challenge to the principle that the two hemispheres, being similar in structure, must function similarly. A popular explanation did not even involve the hemispheres. Instead, it was based on a peripheral difference, one pertaining to different sizes of the subclavian arteries to the two limbs.

Harris (1991, 1999) mentioned this explanation when discussing Bichat's theories, writing that Bichat recognized that the right subclavian artery's "slight excess of diameter" could affect limb use. Nonetheless, Bichat considered this anatomical difference to be inconsequential compared to the symmetry of the limbs themselves, which showed "perfect equality of volume, number of fibers, and nerves." Instead, Bichat argued, "this discordance [between the left and right sides] is seldom or never in Nature, but is the manifest consequence of our social habits." He pointed specifically to writing, maintaining that the way that we write from left to right makes the use of the right hand "better adapted than the left to the formation of letters in this direction" (Bichat 1805/1809, 22; see Harris 1991, 14; 1999, 6–7; quotes from Bichat 1805/1809, 22, 24).

We now know that habits cannot explain handedness, although education could play a limited role in supporting or counteracting natural preferences. Among various sorts of evidence favoring nature over nurture, it has been found that right-hand preferences occur even in societies in which the writing system deviates from the left-to-right direction.

Gall did not concern himself with handedness, although he had to have been aware of this asymmetry common to humans that was not yet thought to be a feature of other animals. His neglect of the subject in his lectures and later sets of volumes would indicate that he did not think handedness was a higher brain function. Whether he might have associated hand differences with blood vessel sizes, as did some others at the time, and what he might have believed about hand preferences being learned or innate were separate issues that did not bear on the edifice he was intent on constructing.

The death of symmetry

The idea that the two sides of the brain are anatomically and functionally similar received few challenges during the first half of the nineteenth century. François Magendie (1783–1855), however, raised one rather muted objection in a footnote to the fourth edition of Bichat's book. He wrote that he found anatomical differences between the hemispheres, but he stopped short of correlating these asymmetries with functional divergences and elsewhere continued to attribute handedness to vascular differences (Magendie 1822/1827, 21, 33; also see Magendie 1838). Another challenge came several decades later from French surgeon Joseph-François Malgaigne (1806–1865). Malgaigne (1859) also noted differences between the hemispheres, which he maintained would make the "organs" of the two hemispheres different; but what he wrote had little or no impact.

Jean Baptiste Bouillaud (1796–1881), who revered Gall (unlike Magendie, who criticized his doctrine and dubbed it a "pseudo-science") is more interesting in this context. The mostly cerebrovascular patients in the Paris hospitals he began to write about in 1825 tended to have a much higher percentage of left-hemispheric than right-hemispheric brain lesions (Bouillaud 1825a, 1825b, 1830, 1839, 1848). Yet, perhaps due to Bichat's deification in the city, Bouillaud remained focused on anterior vs. posterior damage, choosing not to attend to right- vs. left-hemispheric differences that would later be found to be statistically significant, even in his 1825 publication (Benton 1976, 1984).

Marc Dax (1770–1837), a physician in the south of France, was more perceptive of hemispheric differences (Joynt and Benton 1964). He became intrigued with hemispheric damage and speech disorders in 1800, and he continued to amass cases revealing that language impairments are much more likely to involve damage to the left than the right hemisphere. He reported his findings at the Congrès Meridional de Montpellier in 1836, contending that, although not every illness of the left hemisphere will alter verbal memory, when it is affected the cause must be sought in the left hemisphere (Dax 1865b; Joynt and Benton 1964). For reasons unknown, Marc Dax failed to publish this paper; as a result, his compelling evidence for cerebral dominance remained unknown for several more decades.

The concept of symmetry met its demise in the mid-1860s. The single most influential event was Paul Broca's (1824–1880) publications between 1861 and 1865 on what would soon be called *aphasie* or aphasia. As his sample size grew and to his surprise, he recognized that the brain damage associated with severe speech defects in his sample involved the anterior part of the left hemisphere (Broca 1861, 1863, 1865).

The second event was the publication of Marc Dax's *mémoire* by his son Gustave Dax (1815–1893) in 1865. The younger Dax began to present additional support (although, like his father, again without autopsies) for language being a left-hemispheric function at the same time (Dax 1865a, 1865b; Finger and Roe 1996, 1999; Joynt and Benton 1964; Leblanc

2017; Roe and Finger 1996). Neither of the Daxes, however, attempted to account for this phenomenon. In contrast, Broca thought that the left hemisphere receives more oxygen-rich blood than its counterpart, and therefore will develop sooner and assume the leading role in language functions.

Broca did not associate the left hemisphere's supremacy for language with hand preference. Like everyone else, he viewed language as a cognitive function, but not so handedness. Moreover, he could think of no reason why two such different functions should be connected to each other (Eling 1984). Nonetheless, the "entire world" seemed to connect speech in the left hemisphere with right-handedness, at least until relatively recently (Harris 1991).

The most important point for us is that the idea of two equal hemispheres fell by the wayside during the 1860s (Finger 1994; Harrington 1985, 1987; Leblanc 2017; Young 1970). Both Gall and Bichat had been wrong about the hemispheres being functionally (and anatomically) symmetrical.

The earlier literature on symmetry is small relative to that on cerebral dominance, but it is nonetheless intriguing. It is important because it provides information needed for a more thorough and detailed picture of how the neurosciences evolved from the end of the eighteenth century through the long and eventful nineteenth century.

Disclosure statement

No potential conflict of interest was reported by the authors.

References

Ackerknecht E. H., and H. V. Vallois. 1956. *Franz Joseph Gall, inventor of phrenology and his collection.* Madison: University of Wisconsin Press.

Benton A. 1976. Historical development of the concept of hemispheric cerebral dominance. In *Philosophical dimensions of the neuro-medical sciences*, ed. S. F. Spicker and H. T. Engelhardt, 35–57. Dordrecht: D. Reidel.

Benton A. 1984. Hemispheric dominance before Broca. *Neuropsychologia* 22:807–11. doi:10.1016/0028-3932(84)90105-2.

Bichat M. F. X. 1800. *Recherches physiologiques sur la vie et la mort.* Paris: Brosson, Gabon et Cie.

Bichat M. F. X. 1801. *Anatomie générale appliquée à la physiologie et à la médecine.* Paris: Brosson, Gabon et Cie.

Bichat M. F. X. 1801-1803. *Traité d'anatomie descriptive.* Paris: Brosson, Gabon et Cie.

Bichat X. 1805/1809. *Physiological researches upon life and death.* T. Watkins, Trans. of the 2nd Paris edition. Mort. Philadelphia, PA: Smith & Maxwell, 1809.

Bischoff C. H. E. 1805. *Darstellung der Gall'schen Gehirn- und Schädel-Lehre; nebst Bemerkungen über diese Lehre von Christoph Wilh. Hufeland.* Berlin: Wittich.

Bischoff C. H. E. 1807. *Some account of Dr. Gall's new theory of physiognomy founded upon the anatomy and physiology of the brain and the form of the skull with the critical strictures of C. W. Hufeland.* Trans. and Preface H. C. Robinson. London: Longman, Hurst, Rees, and Orme.

Bojanus L. 1801. Encephalo-cranioscopie. *Magazin Encyclopédique, Année* VIII (1):445–72.

Bojanus L. 1802. A short view of the craniognomic system of Dr. Gall, of Vienna. *Philosophical Magazine* 14:77–84, 131–138. doi:10.1080/14786440208676165.

Bouillaud J.-B. 1825a. Recherches cliniques propres à démontrer que la perte de la parole corre-spond à la lésion des lobules antérieurs du cerveau et à confirmer l'opinion de M. Gall sur le siège de l'organe du langage articulé. *Archives Générale de Médecine* 8:25–45.

Bouillaud J.-B. 1825b. *Traité Clinique et Physiologique de l'Encéphalite ou Inflammation du Cerveau.* Paris: J. B. Ballière.

Bouillaud J.-B. 1830. Recherches expérimentales sur les fonctions du cerveau (lobes cérébraux) en général, et sur celles de sa portion antérieure en particulier. *Journal Hebdomadaire de Médecine* 6:527–70.

Bouillaud J.-B. 1839. Exposition de nouveaux faits à l'appui de l'opinion qui localise dans les lobules antérieurs du cerveau le principe législateur de la parole; examen préliminaire des objections dont cette opinion à été sujet. *Bulletin de l'Académie Royale de Médecine* 4:282–328.

Bouillaud J.-B. 1848. *Recherches Cliniques Propres à Démontrer que le Sens du Langage Articulé et le Principe Coordinateur des Mouvements de la Parole Résident dans les Lobules Antérieurs du Cerveau.* Paris: J. B. Ballière.

Broca P. 1861. Remarques sur le siège de la faculté du langage articulé, suivies d'une observation d'aphémie (perte de la parole). *Bulletins de la Société Anatomique de Paris* 6:330–57.

Broca P. 1863. Localisation des fonctions cerebrales. Siege de la faculté du langage articulé. *Bulletins de la Société d'Anthropologie de Paris* 4:200–04.

Broca P. 1865. Sur le siège de la faculté du langage articulé. *Bulletins de la Société d'Anthropologie de Paris* 6:377–93. doi:10.3406/bmsap.1865.9495.

Dax G. 1865a. Sur le même sujet. *Gazette Hebdomadaire de Médecine et de Chirurgie* 2:259–60.

Dax M. 1865b. Lésions de la moitié gauche de l'encéphale coïncidant avec l'oubli des signes de la pensée (lu au Congrés Méridional tenu à Montpellier en 1836). *Gazette Hebdomadaire de Médecine et de Chirurgie* 2:259–60.

Descartes R. 1649/1958. *Treatise on the passions of the Soul. In descartes: Philosophical writings.* Trans. N. K. Smith. New York: Modern Library.

Eling P. 1984. Broca on the relation between handedness and cerebral speech dominance. *Brain and Language* 22:158–59. doi:10.1016/0093-934X(84)90085-3.

Eling P., S. Finger, and H. Whitaker. 2017. On the origins of organology: Franz Joseph Gall and a girl named Bianchi. *Cortex* 86:123–31. doi:10.1016/j.cortex.2016.11.010.

Finger S. 1994. *Origins of neuroscience: A History of explorations into brain function.* New York: Oxford University Press.

Finger S., and D. Roe. 1996. Gustave Dax and the early history of cerebral dominance. *Archives of Neurology* 53:806–13. doi:10.1001/archneur.1996.00550080132021.

Finger S., and D. Roe. 1999. Does Gustave Dax deserve to be forgotten? The temporal lobe theory and other contributions of an overlooked figure in the history of language and cerebral dominance. *Brain and Language* 69:16–30. doi:10.1006/brln.1999.2040.

Finger S., and P. Eling. 2019. *Franz Joseph Gall. Naturalist of the mind, visionary of the brain.* New York: Oxford University Press.

Froriep L. F. 1800. Kurze Darstellung der vom Herrn D. Gall in Wien auf Untersuchungen über die Verrichtungen des Gehirns gegründeten Theorie der Physiognomik. *Magazin für den neuesten Zustand der Naturkunde mit Rücksicht auf die dazu gehörigen Hülfswissenschaften* 2:411–68.

Froriep L. F. 1802. *Darstellung der ganzen auf Untersuchungen der Verrichtungen des Gehirnes gegründeten Theorie der Physiognomik des Dr. Gall in Wien.* 3rd ed. Weimar: Industrie Comtoir.

Gall F. J. 1825. *Sur les Fonctions du Cerveau et sur Celles de Chacune de ses Parties*, Vol. 6. Paris: J.-B. Baillière.

Gall F. J. 1835. *On the functions of the brain and each of its parts: With observations on the possibility of determining the instincts, propensities, and talents, or the moral and intellectual dispositions of men and animals, by the configuration of the brain and head.* Ed. N. Capen and Trans. W. Lewis, Vol. 6. Boston: Marsh, Capen and Lyon.

Gall F. J., and J. G. Spurzheim. 1810-1819. *Anatomie et Physiologie du Système Nerveux en Général, et du Cerveau en Particulier*, Vol. 4 and an atlas. Paris: Schoell. [Gall was sole author of the last two volumes in this series.].

Harrington A. 1985. Nineteenth-century ideas on hemisphere differences and "duality of mind.". *Behavioral and Brain Sciences* 8:617–60. doi:10.1017/S0140525X00045337.

Harrington A. 1987. *Medicine, mind and the double brain.* Princeton: Princeton University Press.

Harris L. J. 1991. Cerebral control for speech in right-handers and left-handers: An analysis of the views of Paul Broca, his contemporaries, and his successors. *Brain and Language* 40:1–50. doi:10.1016/0093-934X(91)90115-H.

Harris L. J. 1999. Early theory and research on hemispheric specialization. *Schizophrenia Bulletin* 25:11–39.

Joynt R. A., and A. L. Benton. 1964. The memoir of Marc Dax on aphasia. *Neurology* 14:851–54. doi:10.1212/WNL.14.9.851.

Leblanc R. 2017. *Fearful asymmetry: Bouillaud, Dax, Broca, and the localization of language, Paris, 1825–1879.* Montreal: McGill University.

Magendie F. 1822/1827. *Bichat's physiological researches upon life and death.* Trans. F. Gold. Boston: Richardson and Lord. [From the 4th ed. of Bichat's *Recherches physiologique sur la vie et la mort.* Paris: Gabon.].

Magendie F. 1838. *Précis Elementaire de Physiologie.* 5th ed. Bruxelles: Société Typographique Belge.

Malgaigne J.-F. 1859. *Traité d'anatomie chirurgicale et de chirurgie experimentale.* 2nd ed. Paris: Ballière et Fils.

Rawlings C., and E. Rossitch Jr. 1994. Franz Joseph Gall and his contribution to neuroanatomy with emphasis on the brain stem. *Surgical Neurology* 42:272–75. doi:10.1016/0090-3019(94)90276-3.

Roe D., and S. Finger. 1996. Gustave Dax and his fight for recognition: An overlooked chapter in the history of cerebral dominance. *Journal of the History of the Neurosciences* 5:228–40. doi:10.1080/09647049609525672.

Sakalauskaitė-Juodeikienė E., P. Eling, and S. Finger. 2017. The reception of Gall's organology in early-nineteenth-century Vilnius. *Journal of the History of the Neurosciences* 26 (4):385–405. doi:10.1080/0964704X.2017.1332561.

Sakalauskaitė-Juodeikienė E., P. Eling, and S. Finger. 2020. Ludwig Heinrich Bojanus (1776–1827) on Gall's craniognomic system, zoology, and comparative anatomy. *Journal of the History of the Neurosciences* 29:29–47. doi:10.1080/0964704X.2019.1684752.

Walther P. F. [using W_R]. 1802. *Critische Darstellung der Gallschon anatomisch-physiologischen Untersuchungen des Gehirn- und Schädel-baues.* Zürich: Ziegler.

Young R. M. 1970. *Mind, brain and adaptation in the nineteenth century: Cerebral localization and its biological context from Gall to Ferrier.* Oxford: Clarendon.

Matters of sex and gender in F. J. Gall's organology: A primary approach

Tabea Cornel

The originator of phrenology, F. J. Gall (1758–1828), saw himself as a natural scientist and physiologist. His approach consisted of brain anatomy but also of palpating skulls and inferring mental faculties. Unlike some of the philosophical principles underlying Gall's work, his conception of sex/gender has not yet been examined in detail. In this article, I will focus on Gall's treatment of men and women, his idea of sex differences, and how far an assumed existence of dichotomous sexes influenced his work. In examining his primary writings, I will argue that Gall held some contradictory views concerning the origin and manifestation of sex/gender characteristics, which were caused by the collision of his naturalistic ideas and internalized gender stereotypes. I will conclude that Gall did not aim at deducing or legitimizing sex/gender relations scientifically, but that he tried to express metaphysical reasons for a given social order in terms of functional brain mechanisms.

Introduction

> What would become of the propensity to propagation, if there were not two sexes?—[B]ut two sexes exist. (Gall, 1835f, p. 243)

In his 2005 article on the "neuronal nature of femininity," Frank W. Stahnisch pointed out a remarkable desideratum in the historiography of eighteenth- and nineteenth-century medicine concerning the mutual influences of sex or gender concepts and the neurosciences on one another (Stahnisch, 2005). This mutual influence is particularly notable in the case of Franz Joseph Gall (1758–1828), who is very well known as the originator of a "cult" (Clarke, 1970, p. 22) or "fleeting fashion" (Regal & Nanut, 2008, p. 314) erroneously named "phrenology" (Noel & Carlson, 1970; van Wyhe, 2002, p. 22). With his so-called *organology*, Gall tried to localize distinct functions in several organs of the brain and to develop a scientific theory of the mind and propensities in humans and nonhuman animals. His biography, the philosophical foundations of and influences on his work as well as his contributions to contemporary neurobiology have already been treated with historiographical attention, but, apart from Stahnisch's work, there is a significant absence of research on Gall's concept regarding sex/gender.[1] I do not know of a single study focused on the question of how Gall posed innovative views regarding the neuronal nature of sex/gender dichotomies with his doctrine about the brain.

---

[1] In his French writings, Gall used the term *sexe* relating to male/female. I have chosen "sex/gender" to follow the German expression *Geschlecht*, which includes not only sex and gender but can also relate to genitals and entire lineages. The double term is also in line with current debates on the neuroscientific usage of "sex" and "gender" (cf., e.g., Kaiser, 2012).

A source-based reflection tracking conceptions of sex/gender in the early days of brain localization is feasible because there is a considerable amount of primary literature that can be analyzed on this issue. Moreover, an analysis of Gall's work is particularly desirable in this connection if we believe that Gall's "immediate neuroanatomical contributions . . . have been insensibly absorbed into the corpus of our knowledge" (Critchley, 1965, p. 778). Hence, I will spell out Gall's notion of this category on several levels: femininity and masculinity in society (androcentrism, stereotypes), sexed matters in bodies (brains, skulls), as well as sexuality and propagation. It is neither my intention to give a thorough account of Gall's personality and research nor to discuss in how far the latter can be considered as revolutionary or scientifically valuable. Instead, I will focus on his work with and on women and men in order to argue that his new psychological system that was meant to link mental properties and anatomy included the quest for the material foundations of human sex/gender identity. To this end, I will elaborate on his opinion of the hard-wired nature and absoluteness of a male/female dichotomy on various anatomical and behavioral levels.

This enterprise is very promising, as Gall repeatedly made surprising remarks regarding femininity that seem contrary to the role of women in his time: he refuted the contemporary findings of sex-dependent nerve differences in brains (Gall, 1835b, p. 190); he fought the ban against women attending his lectures (Vereinte Hofstelle, 1802, pp. 19–20; Walther, 1804, pp. 6–7); he bemoaned how fashion bound women's bodies (Gall, 1791, pp. 307–310); he demanded sex education for women and institutions for anonymous births (Gall, 1806a, pp. 360–361); he did not introduce an organ in the brain for sex/gender identity; he stored dozens of skulls without noting the sex of the donator (Ackerknecht & Vallois, 1956, pp. 42–60); finally, he believed in a periodical distraction for men that was similar to the female menstrual cycle (Gall, 1835a, p. 300, 1835d, pp. 217–225).

In a defense of his scientific method in response to criticisms by contemporaries, Gall made his awareness of the importance of observations relating to sex/gender differences very clear. He claimed reliability by listing his methods, which were, among others, the following: comparing the brains and skulls of species and individuals with a strikingly strong or weak instinct for sexual propagation—including "erotic mania"—and a good or bad sense for distinguishing the sexes within their species; examining cerebella of humans with deviant sexual habits; evaluating mutual stimulations of animals during coitus; investigating effects of differently severe lesions of genitals on the cerebellum and vice versa; and analyzing apoplexies during sexual intercourse in humans (Gall, 1835f, pp. 104–105). In the following, I will exploit such excerpts dealing purposefully or implicitly with sex/gender. I will consider parts of Gall's own published writings[2] as well as two

---

[2] His first monograph (Gall, 1791); a published letter containing the first printed account of his doctrine (Gall, 1798); an anonymously published work (Gall, 1806a) of which he confessed his authorship in a letter (B. Heintel & Heintel, 1985, pp. 14–18); a diary on his lecture tour through Germany (Gall, 1806b); an anthology containing a memoir on brain anatomy, its review by some members of the *Institut de France*, as well as Gall's comments on their text (Gall & Spurzheim, 1809); two encyclopedia entries on the brain and the skull respectively (Gall & Spurzheim, 1813a, 1813b); another anthology containing English translations of a piece on the functions of the cerebellum and their relation to sexual reproduction, remarks by Gall on some of his critics doubting this very connection, and a reply to a decree of 1801 (Vereinte Hofstelle, 1802) issued by Franz II that prohibited his lecturing in Vienna and the rest of the Holy Roman Empire (Gall, Vimont, & Broussais, 1838); another publication of the latter in German (Walther, 1804); the English translation of a second edition of Gall's main work of 1822 (Gall, 1835a, 1835b, 1835c, 1835d, 1835e, 1835f).

summaries that were printed earlier (Arnold, 1805; Blöde, 1806).[3] I will group the excerpts regarding subject matter after summarizing Gall's biography and organological system in the next section.

F. J. Gall and Organology

Franz Joseph Gall[4] was born as a merchant's son in 1758 in Tiefenbronn, a small Swabian town, which he left in 1777 to commence his medical studies in Strasbourg. Four years later, he moved to Vienna, where he obtained his MD in 1785 and became a successful physician very quickly. Gall married Maria Katharina Leis(s)ler (1760–1825), whom he had met in Strasbourg.

By 1796 at the latest, Gall began giving lectures open to the public on his "doctrine of the skulls" (*Schedellehre*, Gall, 1806a). It was a system meant to reveal the true nature of life and to lay out the principles of the innate aptitudes and the manifested character of humans and animals through determining the shape of their skull. Gall claimed that the brain of every living creature consisted of several distinct organs, each of which was responsible for a specific faculty central to bodily functions or to moral and intellectual capacities. He named 27 distinct faculties; all of them present in the brain of humans and a smaller number in animals, depending on the species in question. He considered none of these aptitudes to be good or bad per se (Gall, 1835a, p. 216, footnote) as he maintained that every faculty could turn into a devastating drive if its organ was over- or underdeveloped. For example, the ability to raise and love children could turn into both the hatred and murder of children, or into spoiling them tremendously (Gall, Spurzheim, & Fossati, ca. 1830). Gall assumed that the strength of the faculties corresponded with the relative size of the organs, and that their distribution caused the form of the brain in each individual (Gall, 1798). Therefore, Gall also referred to his doctrine as "organology" (*Organologie*, van Wyhe, 2002, p. 22), a fact that stresses his belief that the material parts of heads were able to tell about the natural order of life as well as the predispositions of species and individuals.

To unravel these "truths of nature" (Walther, 1804, pp. 25–26), Gall studied the heads of living humans and animals as well as of the dead. He received brains belonging to deceased inmates from mental asylums and prisons, but he was not equipped with a method to preserve them and their number was not sufficient for his purposes in comparative anatomy. Because he believed that the shape of the cranium followed the profile of the brain (Gall, 1798), however, he palpated skulls of living and dead bodies instead, searching for dents and protuberances even millimeters in size. He was convinced that this procedure allowed him to infer the form of the brain and the size of its organs and therefore the innate aptitudes of each investigated subject. Even so, he often emphasized that these might differ from the actual character of the individual in question. He stated that education and training could and should be of help to suppress some natural drives and to strengthen others in order to

---

[3] Judging by the expressions and examples used by Gall in his own work, these summaries sound like a faithful report of Gall's lectures; and, what is more, Gall recommended them himself until he finished his own treatise (Gall, 1806a, p. 288, 1806b, pp. 270–273, 347–348).

[4] For his biography as displayed in what follows, see Schramm-Macdonald (1878), Hollander (1928), Ackerknecht and Vallois (1956), Ackerknecht (1964), Lesky (1979), Cooter (1984), B. Heintel and Heintel (1985), H. Heintel (1986), and van Wyhe (2002).

perfect oneself and to commit to following the law and to maintaining social order (e.g., Arnold, 1805, p. 173; Gall, 1806b, pp. 73–74; Gall & Spurzheim, 1809, pp. 461–462).

On the one hand, these lectures established his fame, while also causing severe problems for his work on the other hand: in 1802, Gall received a letter in which he was accused by Franz II, Emperor of the Holy Roman Empire, of spreading materialistic ideas that posed a threat to religion and morality (Vereinte Hofstelle, 1802). Gall fought the decree without success. Together with his assistant Johann Gaspar Spurzheim (1776–1832), a servant, a wax molder, two monkeys, and dozens of skulls and brain models he needed for demonstrations, Gall went on tour from March 1805 to October 1807. He taught his doctrine in more than 50 continental European cities, delivered public and private lectures, met with society circles and examined the heads of prison and mental asylum inmates until he settled down in Paris in 1807. His wife remained in Vienna.

Shortly after his arrival in Paris, Gall launched a new medical practice, and he was allowed to start one of many widely attended series of public lectures. Moreover, for the first time in his career, he took the time to work on a treatise on his system of the brain and its functions. In the process of writing the treatise, Spurzheim and Gall split in 1813: Gall wished to set up an anatomically grounded physiological theory, while Spurzheim wanted to establish an applied practice of reading heads and advising people. From their separation on, Spurzheim laid the foundations for what is still known as "phrenology"—an expression never used by Gall for his organology (Giustino, 1975, p. 15).

Shortly after his wife died, Gall remarried in 1825 to his companion Marie Anne "Virginie" Barbe (1795–?). He undertook only one short lecture tour to England in 1823 before his death near Paris in 1828, and he had not appointed any direct successor to continue his teachings and practice. However, in accordance with his instructions, his head was removed from his body by a student of his and added to his collection of heads. Gall's corpse was buried in unhallowed ground at the public cemetery of Père Lachaise because he had rejected a Catholic burial after his works had been censored.

Sex, Gender, and Sexuality

Androcentrism and Stereotypes

In the late eighteenth century, the idea of men and women being dichotomous not only regarding their genitals but also concerning their minds had not yet been questioned in France and the German-speaking European countries (e.g., Jordanova, 1989; Schiebinger, 1990; Honegger, 1992; Stahnisch, 2005). Until several decades after Gall, the concept of women as imperfect men was prevalent in social, religious, and scientific discourses: women's bodies were seen as underdeveloped, weak, and susceptible to illness, their rational capacities and moral judgment were questioned, and female aspiration to live independently from male guidance was denied. Men were regarded essential protectors of women's naturally feeble bodies, timid souls, and defective moral (Lange, 1992; Knibiehler, 1993; Tuana, 1993).

Material differences in male and female bodies have been studied since antiquity. In turning away from humoral medicine around 1800, however, these assertions gained unprecedented biological authority due to studies carried out by prominent anatomists and physiologists (Laqueur, 1992). Gall's contemporary Samuel Thomas von Soemmerring (1755–1830), for instance, assessed sex/gender differences on several levels—from human bones to the nervous system (Schiebinger, 1987). His student Jacob Fidelis Ackermann

(1765–1815) used a similar approach and tried to replicate Soemmerring's findings of hard-wired differences in male and female brains.[5] Although Ackermann did not succeed in doing so, his and other neuro-anatomical endeavors aimed at explaining the contemporary social order scientifically (Laqueur, 1992; Stahnisch, 2007b). These male researchers interpreted their findings as proof of the biologically grounded inferiority of women (Schiebinger, 1989; Tuana, 1993).

The common view of sexuality was derived from this dichotomic sex/gender conception: women were regarded sexually passive and were mostly defined according to their ability to give birth (Lange, 1992). By the time Gall was practicing, however, this view started to change; more pleasure and self-determination were attributed to women with regard to their sexual activity (McLaren, 1974; Matthews-Grieco, 1993). Influenced by his social and intellectual environment, it seems unsurprising that Gall divided humans, animals, and plants into male and female (Gall, 1791, p. 143) and held misogynist views. Most of the detailed characteristics that Gall assigned to men and women can be subsumed under the following stereotypical roles.

Men are archetypical humans. Gall referred to the human species as "man," and his andro-centric use of pronouns sometimes even conceals that several case studies of his involved women. Unless he gave a detailed biographical account, Gall always used "he" for persons of any age who were mentioned regarding their humanness—most of the time in discrimination of animals or when he wrote about sexual stimulation (e.g., Arnold, 1805, pp. 203–204; Blöde, 1806, p. 139; Gall & Spurzheim, 1813a, p. 467; Gall, 1835a, p. 78; Gall, Vimont, & Broussais, 1838, p. 94).

As was common use in scientific writing at the time (Crampe-Casnabet, 1993, pp. 319–324), sections in which Gall wrote explicitly about women are often separated from investigations of humans in general, that is, males. This point is illustrated very well in Gall's consideration of the influence of brain diseases on genitals and propagation (Gall, Vimont, & Broussais, 1838, pp. 16–94): in describing the general relation between sexuality and the brain, he referred only to males; then he proceeded with elaborations on "idiots" and "cretins" (Gall, Vimont, & Broussais, 1838, pp. 66–68), before he concluded with remarks on women (Gall, Vimont, & Broussais, 1838, pp. 69–71). By doing so, Gall positioned females as a curious instance of humanity, even without drawing an explicit comparison between women on the one hand and people with mental handicaps on the other hand. Nevertheless, Gall's efforts to retain the right of lecturing in front of women (Vereinte Hofstelle, 1802, pp. 19–20; Walther, 1804, pp. 6–7) demonstrate that he attached considerable intellectual qualities to them—unlike political and social authorities.

The above-mentioned impression that Gall regarded women as some sort of deviant "men," however, is further fortified by the fact that he displayed only meager interest in studying females to gain more knowledge about human sexuality in general: examining cases of postmortem erections, Gall explicitly compared women and men in their bodily reaction to strokes, but he named only one study of his involving a female stroke patient, and it does not appear that he continued research on this topic (Gall, Vimont, & Broussais, 1838, p. 89).

Women are desired sexual objects. In his work, Gall referred to women as the "fair sex" (*das schöne Geschlecht, le beau sexe*, e.g., Gall, 1791, p. 136; Gall & Spurzheim, 1811, p. 223;

---

[5] For more information on the controversy between Soemmerring, Ackermann, and Gall, see Gall (1806a), van Wyhe (2002), and Stahnisch (2005, 2007a).

Gall, 1835e, pp. 139, 246) whose task in the world was to "create(s) in the male the instinct of generation" (Gall, 1835a, p. 132) and to please men by caring for them and cheering them up (Gall, Vimont & Broussais, 1838, p. 29). Consequently, he drew parallels between females and fruits in the sense that both were able to gain the highest attention of male mammals, just by being presented to them (Gall, 1835a, p. 158).

It seems that Gall regarded especially attractive girls as prone to awakening premarital instincts of generation in men; the more beautiful a young woman was, the more likely Gall considered her to become pregnant unwillingly. According to Arnold's account of his lectures, Gall saw advantages for illegitimate children: because premarital coitus was not permitted, the rendezvous had—from the man's point of view—to be worth going behind the back of the woman's family. This, Gall argued, assured a certain degree of beauty and cleverness in the unmarried woman, which then might be passed on to potential offspring conceived through this very relationship (Arnold, 1805, p. 49).

Women are unreasonable and overwhelmed by emotions. Gall saw a threat to family harmony in women's tendency towards "insane jealousy" (Gall & Spurzheim, 1811, p. 223). In his view, their temper was either so soft that it suppressed their memory and made them lethargic in relationships, or it was very sharp, led them to make many mistakes and caused an ample imagination as well as the ability to concentrate all of their energy on a specific concern. One of his catalogues of male and female stereotypes ends with the picture of a woman as "suffering sinner," that is, as one to trip, to fall, and to regret for life that she succumbed to a fleeting hankering (Gall, 1791, p. 116). This supposed irrationality may be one reason why Gall admitted problems in extending his organology to women: "We know very well that the heads of the women are difficult to unravel," he wrote in a published letter (Gall, 1798; English translation in Gall, 1835a, p. 17).

Men are fighters and protectors. In Gall's worldview, men had to care for women and children—an instinct that included fighting. He considered this predisposition to be so powerful in young men that he thought a horn sounding the charge could cure them from various illnesses spontaneously (Gall, 1791, pp. 561–562). In later years, according to the picture Gall drew of them, men sit at home "absorbed in grief" (Gall, 1791, p. 116)—probably pondering the unreliable personalities of women in their families mentioned above.

Women are weak and malleable. His observation that women usually lived longer than men led Gall to assume this was due to their inherent feebleness (Gall, 1791, p. 136). Gall explained this seemingly contradictory assertion with the following line of reasoning: according to him, the fact that women were weaker than men made them less resistant to exterior influences of any kind. On the one hand, he admitted, this made women susceptible to sickness and deteriorating depressive emotions (Gall, 1791, p. 356). On the other hand, he argued, female liability to influence made them also more susceptible to medical treatment, whereas stronger organisms used to hard work and meagerness—that is, men—were more resistant to exterior forces and, therefore, often unresponsive to medical interventions (Gall, 1791, p. 548).

One exception to the prevalence of female weakness known to Gall, however, was the unimagined strength women could develop through motherhood. He even reported cases when mothers were cured precipitously in response to their children needing them (Gall, 1791, pp. 561–562). Yet, in general, Gall considered female fragility to be so fundamental that it even manifested itself in the process of dying: for women predominantly after fainting, for men usually in death throes (Gall, 1791, p. 314).

Although Gall's case studies and the artifacts of his specimen collection (Ackerknecht Vallois, 1956, pp. 42–60) leave the impression that he examined more men than women, he obviously engaged with the latter in his organological practice. Moreover, Gall is said to have had intimate acquaintances with many women—one of which led to an illegitimate son (Ackerknecht & Vallois, 1956, pp. 7–8). Consequently, his prejudices do not result from isolated reflection; most likely, Gall absorbed contemporary female and male role models and integrated them into his doctrine. Most of the stereotypes presented above, however, lack any organological proof or explanation in Gall's texts. Thus, his ideas of the interconnection between sex/gender and the material body will be presented in the following.

Brain and Skull

Gall declared at least once: "Let anyone [*sic*] present me, in water, the fresh brains of any two adult animals whatever, the one male and the other female, and I will distinguish the two sexes without ever being deceived" (Gall, 1835c, p. 288).[6] But, in contrast to how this quote is at times presented (e.g., Critchley, 1965, p. 780; Hyde, 1990, p. 56; Jordan-Young, 2010, p. 49), this was not an argument for hard-wired differences between the sexes in every instance of their existence. First, there are more passages in which Gall stressed that, in some cases, the variability between *human* individuals of one sex might overcome dichotomies between the two groups (e.g., Gall, 1806a, pp. 358–359; Gall & Spurzheim, 1811, pp. 34–36, 1813a, p. 469; Gall 1835a, p. 182; Gall, Vimont, & Broussais, 1838, pp. 28–32)[7]; second, by demanding *adult* animals, he demonstrated knowledge of an idea resembling what is known today as brain plasticity, that is, that it is possible to strengthen one's organs by making use of them, or to weaken them by not realizing or stimulating the corresponding drives (Gall, 1806b, pp. 244–247).

However, Gall argued that "social life" did not "*produce(s)* certain faculties" (Gall, 1835a, p. 162, italics added) but merely *fortified* them. More specifically, concerning the sex/gender issue, he stated that "[t]he whole education of women tends to confirm this natural modesty [of being timid and bashful]" (Gall, 1835a, p. 299). Thus, Gall was clearly not a constructivist, even though he admitted the possibility of shaping one's character traits by pursuing certain ways of life. For example, he admitted that education might strengthen sex/gender-specific behavior, although only within the boundaries of their assigned female or male nature (Lesky, 1979, pp. 101–103)—apart from exceptions (Gall, Vimont, & Broussais, 1838, pp. 28–32).

It is true that one of Gall's repeatedly stated principles claimed that "often the same fundamental forces are found to exist in different degrees in the two sexes, and that, in this case, the organ of the quality or faculty has a degree of development differing in the two sexes" (Gall, 1835e, p. 250). Nonetheless, there are only a few out of altogether 27 cranial organs of which Gall specified a difference in size and function depending on the sex. For instance, he thought the organ for "distinguishing the relation of colors, talent for painting" (no. 16, *Farbensinn*, Gall, 1835e, pp. 46–50) was larger in women. Hence, he believed to have found an explanation "why female artists, who in every other respect rarely equal men of genius, raise(d) themselves sometimes to the level of the most distinguished painters in the art of

---

[6] A similar passage can be found in Gall (1806a, pp. 318–319).
[7] However, this does not hold true for animals (Gall, 1806a, pp. 318–319).

coloring," why they preferred paintings to monochrome busts, and why they were so fond of colored dresses (Gall, 1835e, p. 54). Similarly, Gall thought the organ of "comparative sagacity" (no. 20, *vergleichender Scharfsinn*, Gall, 1835e, pp. 121–128) was smaller in women and caused their tendency to believe in superstitions as well as their lack of a sense of logic so "they hardly realize that there can be no effect, no event, without a cause" (Gall, 1835e, p. 136).

Furthermore, there were some organs whose size Gall did not declare to depend on sex/gender but that still differed in their outcome, for example, the organ for "vanity, ambition, love of glory" (no. 9, *Eitelkeit, Ruhmsucht, Ehrgeiz*, Gall, 1835d, pp. 184–195): in a man, a strong development of the organ would only lead to unpleasant vanity, while in women the thirst for glory would threaten the domestic peace because wives held clothes and money in higher estimation than caring for children and having a pleasant relationship with their spouse. It even appears that this remark concerned the husbands' clothes (*Anzug*) and not the women's own dresses (which would be *Kleider*, Arnold, 1805, pp. 284–286). This suggests an even larger dissimilarity in the manifestation of several organs in males and females: they would not only differ in their strength or target object but also with regard to its possessing subject—an instance that lines up with Gall's view of the timid woman stated above.

Gall's numerous examinations of 3- to 16-year-olds assured him that these differences could even be found in children. Furthermore, he had skulls belonging to children younger than 3 years at his disposal, some of newborns and animal embryos. He aimed to prove that differences concerning sex could already be recognized prenatally but also, more importantly, that brain development was not yet complete at the age of 2. He grounded this view on the absence of sexual drives in children that were accompanied by a significantly smaller size of their cerebellum. In this part of the brain, Gall located the organs for sexual reproduction and the love for offspring (e.g., Gall & Spurzheim, 1813a, pp. 467–468; Gall, Vimont, & Broussais, 1838, pp. 16–28). In examining different skulls, Gall suggested comparing girls' and boys' crania separately since a protuberance corresponding to their organ for propagation should already be "far more remarkable" in boys—even if they were younger than the girls (Gall, 1806a, pp. 314–315). Otherwise, he feared, one might obtain no substantial results from the comparisons. By delivering this recommendation, Gall acknowledged severe variances between the skulls of different individuals within groups of either boys or girls (Gall, 1822, pp. 303–304), and he did not want these to be mixed up with dissimilarities caused by sex.

So far, I have shown that Gall's perspective of living beings was imbued with contemporary stereotypes and that he integrated these biased illustrations in his scientific publications. Moreover, I have provided evidence that Gall's organology was not only about descriptions and "feeling 'bumps'" (Ackerknecht & Vallois, 1956, p. 8) but about assigning meaning to his observations and using them to distinguish between male and female heads—no matter the species or age.

Concerning the quote on male/female brain differences at the beginning of this section, it is very likely that Gall was referring to the assumption that the protuberances on the surface of male brains belonged to the organ of propagation and, even more significantly, that the female cerebella corresponded to the organ for love of offspring (Gall, 1806a, pp. 358–359, 1835c, pp. 281–284). Thus, I will examine these two important faculties linked to sexuality and reproduction in more detail below.

Propagation and Heteronormativity

One aspect of Gall's doctrine that bears close resemblance to Romantic thought in general and to the idea of "the Great Chain of Being" specifically (Lovejoy, 1936; Hall, 1977) was his distinction between less and "more perfect animals" (e.g., Gall 1835a, p. 79); only the latter used a sexual act for propagation of their species, and in all of them Gall had found a structure that resembled the human cerebellum (*das kleine Gehirn*). He declared this part of the brain to be the only "connexion [*sic*] between the sexes" (Gall, 1835f, p. 102), meaning this was the single reason for copulation. Furthermore, Gall inferred that the cerebellum was the location of the "most noble" organ (Gall, 1806a, pp. 181–182, 312–313)—that is, the one for propagation—and that the ability to differentiate between the sexes was one of its outcomes (Arnold, 1805, p. 178; Gall, Vimont, & Broussais, 1838, pp. 18–19, 33). This organ's existence and position occurred to Gall for the first time when he met a "hysterical" widow suffering from epileptic fits because she could no longer fulfill her sexual desires (Gall, 1806a, p. 344). In accordance with subsequent animal studies, he claimed to have found a direct connection between her hot, painful neck and her sexual longing (Gall, Vimont, & Broussais, 1838, pp. 12–15).[8]

In further experiments and observations, Gall tried to deepen his understanding of the interconnection between the cerebellum, genitals, and sexual activity (e.g., Gall, 1806a, pp. 346–352; Gall, Vimont, & Broussais, 1838, pp. 16–71). Even though he admitted that there were individually different life cycles that awakened the sexual drive in some individuals earlier or made it last longer, Gall had no room for investigations regarding nonheterosexual desire—quite unsurprisingly, considering the heteronormative society at the time (Matthews-Grieco, 1993; Herrn, 1995). In his view, the desire for intercourse overwhelmed children from the outside at a time predetermined by the maturity of their brain:

> The sexual instinct develops itself in a corresponding order of progression. It glides imperceptibly into the mind of the young of either sex; the eyes become more brilliant; the look more expressive; the gait acquires an air of increased pretension; they are liable to be seized with an inexplicable infantine melancholy; they feel a want for which they cannot account; they experience confused desires; until at last the presence of a beloved object solves the enigma, and spreads a vivid pleasure over the whole soul. (Gall, Vimont, & Broussais, 1838, pp. 18–19)

In the context of the quoted passage, Gall did not question that this beloved object belonged to the "opposite sex" (Gall, Vimont, & Broussais, 1838, p. 21) and that exploring the own infantile sexuality via "onanism" prior to engaging in sexual intercourse was a "pernicious habit(s)" that ought to be prevented by parents (Gall, Vimont, & Broussais, 1838, p. 22). Nonetheless, Gall also laid out that mental illnesses and obsessive masturbation could be caused by the same brain disease—instead of naming masturbation as a reason for losing one's mind (Gall, 1806a, pp. 339–340). Be that as it may, Gall did not break with the predominance of reproductive sexuality in his time.

In his system, Gall distinguished between two "natures": the specific assembly of the organs of an individual versus an underlying order of the world. Consequently, unnatural or perverted behavior, in the sense that it deviated from social norms, could still be caused by

---

[8] The idea of female hysteria as result of infrequent sexual intercourse or other emotional causes became prevalent towards the middle of the nineteenth century (see Micale, 1991; Tuana, 1993, pp. 93–107).

the natural drive of an individual. Thus, Gall demanded children who were masturbating be stopped from doing so even though he did not blame them for their desire to engage in this activity. Likewise, Gall committed himself to the improvement of prisons and mental asylums because he found many of the inmates irresponsible for what they had done: for him, the "perverted drive" of the rapists, pedophiles, homosexuals, and zoophiles arrested in prison was due to a "perverted development" of their organ of propagation, and not the personal fault of the individuals (Gall, 1806a, pp. 345–346).

If the organ was but "too little developed," Gall wrote, it might cause "impotence, indifference, or even aversion to the other sex" (Gall, 1835a, p. 216). To him, these instances were natural consequences of the perverted constitution of the organs of the brain. Simultaneously, he feared that these developments might endanger "[t]he propensity to propagation . . . , the most necessary institution of the Creator" (Gall, 1835a, p. 216). Thus, he made clear that concepts of sexuality apart from heterosexuality leading to reproductive intercourse with another person were a threat to the *natural* order. Yet, Gall did not consider all of these to affect the *social* order in a way that allowed imprisoning the deviating individuals. This leniency also affected individuals who would be known as *transgender* in today's society:

> There are cases, where, by an alteration of the organs, the me is transformed into another me; for instance, when a man believes himself transformed into a woman, a wolf, [et] c[etera]; . . . not an uncommon accident after severe disease, especially in cerebral affections. (Gall, 1835c, p. 76)

With these observations, Gall wanted to illustrate the interdependency of mind and brain that made people irresponsible for their disposition. Even so, he implicitly stated his acceptance of a nineteenth-century aversion to abnormalities of unchanging sexes in heterosexual relationships; to him, deviations from this order were mental illnesses akin to any other delusion.

According to Gall, the disinterest of immature children in "the other sex" had been developed by natural forces to make breastfeeding and further features of parental care possible on a nonsexual basis (Gall & Spurzheim, 1811, p. 95). The same holds true regarding the love of offspring, which he supposed to be significantly more developed in females than in males and "ennobled in man; . . . [in a way that it] becomes, in women, the amiable virtue which inspires their tenderness for their children" (Gall, 1835a, p. 103). Gall noted the pre-disposition in females to bear and raise children even in small girls whom he observed to experience a unique pleasure in playing with their dolls (Gall, 1806a, p. 358).[9] On the one hand, he stated that this female affection for children—not necessarily their own flesh and blood—resembled the natural behavior of monkeys (Gall, 1835c, pp. 264–266); and, on the other hand, that it outweighed even the passion for "moral love" unique in humans (Gall, 1835a, p. 103). Gall illustrated the latter with examples of unhappily married women who remained with their husbands for the benefit of having children (Gall, 1835c, pp. 279, 284) and willingly unmarried women[10] who adopted children (Gall, 1806a, p. 358).

---

[9] The most advantageous physical characteristics to carry out this task can be found in Gall (1791, pp. 307–310).
[10] Gall formulated possible reasons for staying unmarried: fear of financial responsibility, self-centeredness, and the inability to sustain emotional relationships. Unlike Spurzheim, he believed that there were no material foundations for monogamy in humans and animals. He admitted, however, that the two of them had been engaged in a search for an organ for marriage (see Gall, 1835c, pp. 305–306).

Gall reported to have observed the continuance of characteristic sex differences even during the considerable transformation of male and female bodies during puberty. In his view, this could only be explained by a resistance in brain physiology to the many changes in the external organs (Gall, Vimont, & Broussais, 1838, p. 3). When children were finally ready for propagation, he maintained that the boys played the "attacking part" in the game of love, not unlike male animals (Gall, 1791, p. 29, 1806a, p. 318). Gall pointed out that men were generally more enthusiastic about coitus than women (Gall, 1806a, p. 320). However, he found exceptions in women with a stronger sexual drive than most men (Gall, Vimont, & Broussais, 1838, p. 24) and in men who were not interested in coitus at all—most of them true genii (Gall, 1806a, pp. 318–319). Still, the ideal of humans and animals without "incomplete"—that is, ambiguous—genitals (Gall, 1806a, p. 320) and with joyful engagement in heterosexual intercourse (Gall, 1791, pp. 29, 47, 576, 636, 650, 1806a, pp. 312–313, 1835c, pp. 141–142) persisted. It is important to note that this depiction belonged to the role model for both sexes; a notion I have found in the way Gall shared his observations of the mutual stimulation of males and females during animal coitus (Gall, 1835f, p. 104). Additionally, in several passages in texts on the human sexual drive, he expressed sympathy for women not being sexually aroused or satisfied by their husbands (Gall, Vimont, & Broussais, 1838, p. 17), which suggests that he regarded the opposite of their situation as desirable.

As mentioned above, playing the active part in sexual intercourse, however, was a sex/gender-specific behavior to Gall, as was caring for children: "There are a few men who show female love towards children" (Gall, 1806a, pp. 358–359), he named one of the exceptions he had found in humans but not in animals. In a similar way, Gall described a woman with a feebly developed organ for the love of offspring: "[S]he has hardly the character of her sex. Her principal destination is wanting" (Gall, 1835c, p. 282). From the quoted passages, it becomes clear that Gall declared behavior to be sex/gender specific; a very close look might even suggest that he imagined female traits of character to be additive in men, while resembling a man caused the loss of femininity in women. Gall did not make it explicit, but this interpretation is in line with the analysis of Gall's view of men as archetypical humans provided above: men might show atypical features, but if women did so, this meant a loss of integral features of their already deviant humanness.

Nevertheless, in the same passage quoted above, Gall also admitted that a woman should not be reduced to her ability to bear children and that even a sterile wife could be a "companion who is very estimable in all other relations" (Gall, 1835c, p. 282). In addition to this unexpected acknowledgement of women who seemed not to fulfill their allegedly principal role, Gall even wrote about the advantages of androgynous humans: he explained that a "happy union" of male and female characteristics produced an "enviable disposition" in the sense that the resulting body was neither too weak to stand the trials of life, nor too strong to resist medical treatment—therefore, it was predestined for a long and healthy life (Gall, 1791, p. 136).

These puzzling instances of convergence of men and women on the one hand and pointing out differences on the other hand do not unravel Gall's idea of sex/gender dichotomies thoroughly. Gall obviously presented stereotypes to which he had imparted a scientific language and explained their realization using the idea of sex/gender-specific parts of the brain and its organization. I have also shown, however, that he held some anti-androcentric views. The extent to which Gall thought that dichotomous sex differences were innate is not entirely clear. Gall's conception of a divine creation and natural forces, as outlined in the next section, will lend a hand in solving this enigma.

Inherent Reality versus Exterior Force

In some case studies, Gall paid special attention to observations in which no sex/gender differences were found in the performance or bodily structure of the examined individuals (e.g., Gall, 1822, p. 341, 1835b, pp. 218–219). These instances point to his notion of differences between two sexes as an underlying truth of nature. It was this view that led Gall to the assumption that individuals were unambiguously sexed, even though they might be too young or too limited to notice their own or others' affiliation to the sexes (Gall, 1835b, p. 213, 1835f, p. 257). Judging by the examples he used, his differentiation between two sexes among individuals of the same species seemed the most obvious to him. In addition, although I found many places where Gall listed sex as one of several influences on the development of living beings, it was nearly always the first or most elaborated one (e.g., Gall & Spurzheim, 1811, pp. 44–48; Gall, 1835a, pp. 150–151, 1835b, 271). Gall consistently placed the organs for propagation and the organ for the love of the offspring in the cerebellum, as close as possible to the spinal cord, which he assumed to be the seat of the most natural and important faculties (Gall, 1806a, p. 355).

Congruently, Gall attached so much importance to the category "sex" that he presumed it could inhibit several inherited traits if they did not fit the sex of the growing person. For example, if daughter and son resembled the mother, this similarity had to stop at some degree in the boy—because he was male, and too much resemblance with a female being was not "permitted" in Gall's system (Gall & Spurzheim, 1813a, p. 469). Still, even though he obviously believed in sex-dependent constraints for skills, Gall admitted that he had met the twin sister of a boy of "most humble mediocrity" who was, despite her heritage, able to "raise(s) herself, in many respects, above her sex" (Gall, 1835a, p. 184). Gall was thus able to see and publish instances of behaviors and skills that seemed contrary to a respective sex/gender role. More noteworthy than the obvious possibility of exceeding these constraints in exceptional cases, however, is the organological notion of predetermined limitations depending on the sex, which, again, was well in line with contemporary scientific and religious sex/gender concepts as laid out previously.

With regard to other case studies, however, it becomes clear that Gall did not consistently proclaim the determinative materialism he was criticized for by contemporaries. For example, he named the periodical change of the willingness for reproduction in female animals tethered to mating seasons as the reason for their smaller cerebellum—not the other way round (Blöde, 1806, p. 65). Gall denied the natural or social evolution of faculties in new generations (Gall, 1806b, pp. 244–247) and was convinced that a species could not improve itself fundamentally. On the contrary, he argued, any combination of brain organs had been assigned to a specific group on purpose in order to secure an effective natural system (Gall, 1835a, pp. 162–163). Even though he had fallen into disfavor with the Catholic Church, it persisted to be a common motif in Gall's arguments that

> God has traced for [man] the circle in which he must act, and has directed his steps. It is for this reason that at all times, and among all nations, man presents the same essential qualities of which he could not have conceived the idea, without the predetermination of the Creator. (Gall, 1835a, p. 104)[11]

---

[11] See also Gall (1791, pp. 46, 163–165, 1835a, p. 164, 1835f, p. 310), Gall, Vimont, & Broussais, (1838, p. 29), Temkin (1947, pp. 300–306), Ackerknecht and Vallois, 1956, p. 20, and Tomlinson (2005, p. 63).

If we trust his words, this means that Gall believed in a divine exterior force of creation that had constructed the blueprint of the brains of all beings. As a consequence, the species and role models in the world were supposed to live on as manifestations of this predesigned and unchanging master plan. This concept of intentionally created differences between and within species disabled and disallowed female emancipation as well as male/female convergence. Gall sometimes referred to the cause of predetermined sexed constitution as a divine creator and sometimes as a natural force (Gall, 1806b, p. 84; Lesky, 1970, pp. 307–311; van Wyhe, 2002, pp. 40–41).[12] However confusing this is, at least at an early stage in his work, Gall referred clearly to a metaphysical power named God that directed minds and bodies naturally via the composition of the organs in their brains (Gall, 1791, pp. 163–165).

There is one additional puzzling instance regarding Gall's view of the origins of sex/gender differences: his rather constructivist depiction of sex as an exterior force from outside of the body. This notion can be found in several similar passages in which he stated that just as to the stars, the climate, and the seasons of the year, their environment, education, religion, nourishment, or age, humans were *exposed* to sex (e.g., Walther, 1804, pp. 37–38; Gall, 1808; Gall & Spurzheim, 1811, pp. 161–162). He states it "had an effect on them" (Gall, 1791, p. 2), meaning that dynamics in these matters modified the brain and its functions, the "desires, propensities, passions, and ultimately, the motives and determinations" of humans (Gall, 1835f, p. 272). It is perplexing that the lists contain sex as an external influence on humans since Gall did not believe in the possibility of variations of the sex of people—as would be possible with a diet, for instance—or in a continuously developing sex similar to age. Rather, I have shown that he considered transgender people mentally ill and dichotomic sex differences to be innate even prenatally.

Moreover, Gall used his view of the impact of an unchanging sex on human constitution as an argument for the truth of his organology: if there existed any *metaphysical* laws for the nature of human beings, he argued, they would be the same for all of them and would not differ in beings of different sex, age, or nourishment. Yet, because he had shown that human minds and characters depended on these matters to a strong degree, he concluded, there had to be a *material* predisposition to differences in individuals—including masculinity and femininity—which could only be unraveled physiologically (Gall, 1835a, pp. 61–62). Contrary to his intention, this leads back to where Gall started his argument: to a creative *metaphysical* authority predetermining sex/gender roles and characteristics by making use of specific organizations of brain tissue and functions. This leaves no room for a constructivist view of sex/gender as a performative power on existing beings as implied in the quote above. The contradiction points to an instance of conflict between Gall's naturalistic ideas about human nature and his internalized acceptance of a dichotomic sex/gender concept.

Conclusion

Let me condense Gall's metaphysical-naturalistic approach to sex/gender[13]:

1. Male and female humans and animals *differ obviously* according to physical and inward characteristics.

---

[12] On the unclear demarcation between the *natural* and the *divine* in Romantic thought and its influence on scientific reasoning, see Lovejoy (1936).

[13] These ideas were concisely developed in Gall and Spurzheim (1811, pp. 34–36). English translations of many of the ideas can be found in Gall (1835a, pp. 182–183, 1835d, pp. 229–232).

2. These differences can only be *explained by the composition of material organs* in their brains that are attached to specific functions. Some of the organs are more perfect in one sex than in the other even though the underlying structure is identical.

3. The *ultimate reason* for the dissimilarities, however, lies on a metaphysical level and is specified as a *force of nature* designed by a divine creation to secure contentment and propagation of living beings on earth.

4. To accomplish this order, there exist *specific profiles for females and males* concerning mind and body. These are necessary to secure a successful cohabitation. Education and social order follow these rules, not the other way round.

5. Because humans are the most complex part of creation, many *modifications and exceptions* from the general rules are possible. Yet, they occur too rarely to put an end to the natural order.

6. Since all of the faculties are innate, they cannot be changed fundamentally, and brains are reset to their inherited basic organization with every new generation. In particular, if a deviating individual has acquired abilities belonging to the other sex by accident or training, these will not be passed on. Modifications of the outcome of the allocated organs are possible in individuals, but the *natural order will not change*.

Gall did not try to deduce or legitimize dichotomous sex/gender relations scientifically. Instead, he used his idea of science to make apparently divine principles comprehensible in a physiological language. This way, he naturalized the *implementation* of the differentiation between female and male, but not the *installment* of these categories. Gall's acceptance of manifold exceptions to this perceived rule would have made it, on the one hand, very hard to furnish him with counterevidence against a dichotomous creation. On the other hand, because he held the view of a fundamentally similar constitution of males and females, his system might have collapsed easily if he had realized that, besides his experiences, there were other possible orders of society that guaranteed contentedness of the participating individuals as well as propagation of the species.

As I have pointed out, Gall did not succeed in carrying out unbiased empirical science, but he cannot be called a noteworthy misogynist judged by the standards of his time. To the contrary, he was a researcher who adapted contemporary prejudices in a scientific system without preventing—or perhaps even noticing—that they led to circular arguments that weakened his whole system and made it anything but naturalistic. Thus, Gall's epistemology was more defective than his methodology: he used the latter to get from a starting point of his investigations (an *observed* and, in his eyes, epistemologically certain male/female dichotomy) to the target (an *explained* and localized male/female dichotomy).

Gall's theoretical accounts of dichotomous sex/gender relations and inferior females exceeded his efforts for women's rights and sex/gender equality. Nevertheless, the evidence presented here shows that any presentation of him as an outstanding misogynist with regard to his sex/gender constructs would be inaccurate. Further research could concentrate on archival sources to sharpen the representation of Gall's actual scientific practice in palpating skulls and choosing objects for his investigations. These sources would offer additional insight in his sex/gender concept. Disentangling Gall's attitude towards sex/gender and contrasting it with the conceptions of his direct and indirect successors in phrenology and the neurosciences could help to reveal which parts of Gall's theories and practices involving sex/gender were specific to brain research, and which were mere scientific reflections of the contemporary social order. Finally, various source-based investigations could help trace more recent developments in the scientific treatment of sex/gender to their origins and scent biases in knowledge production that may have been handed down for centuries.

Acknowledgements

I am grateful to Frank Stahnisch for his support and indebted to two anonymous reviewers for helpful suggestions. Teresa Beuscher was indispensable for proper French-German translations; Anelis Kaiser, Liesel Tarquini, and Friedrich Steinle commented on earlier drafts of this article, and so did my History of Medicine class at the University of Pennsylvania under direction of David Barnes—thank you all.

References

Ackerknecht EH (1964): Gall, Franz Joseph. In: Stolberg-Wernigerode O, ed., *Neue Deutsche Biographie*, vol. 6. Berlin, Duncker & Humblot, p. 42.

Ackerknecht EH, Vallois HV (1956): *Franz Joseph Gall, Inventor of Phrenology and His Collection*. Madison, University of Wisconsin Medical School (Wisconsin Studies in Medical History 1).

Arnold ITFK (1805): *Dr. Joseph Gall's System des Gehirn- und Schädelbaues nach den, bis jetzt über seine Theorie erschienenen Schriften*. Erfurt, Henning'sche Buchhandlung.

Blöde KA (1806): *D. F. J. Galls Lehre über die Verrichtungen des Gehirns: Nach dessen zu Dresden gehaltenen Vorlesungen in einer faßlichen Ordnung mit gewissenhafter Treue dargestellt*. Dresden, Arnoldische Buchhandlung, 2nd revised edition.

Clarke E (1970): The history of the neurological sciences. *Proceedings of the Royal Society of Medicine* 63(1): 21–23.

Cooter RJ (1984): *The Cultural Meaning of Popular Science: Phrenology and the Organization of Consent in Nineteenth-Century Britain*. Cambridge/New York/Melbourne, Cambridge University Press (Cambridge History of Medicine).

Crampe-Casnabet M (1993): A sampling of eighteenth-century philosophy. In: Davis NZ, Farge A., ed., *A History of Women in the West III: Renaissance and Enlightenment Paradoxes*. Cambridge/London, Harvard University Press, pp. 315–347.

Critchley M (1965): Neurology's debt to F. J. Gall (1758–1828). *British Medical Journal* 2: 775–781.

Gall FJ (1791): *Philosophisch-Medicinische Untersuchungen über Natur und Kunst im kranken und gesunden Zustande des Menschen*. Vienna, Rudolf Gräffer und Comp.

Gall FJ (1798): Des Herrn Dr. F. J. Gall Schreiben über seinen bereits geendigten Prodromus über die Verrichtungen des Gehirns der Menschen und Thiere an Herrn Jos. Fr. von Retzer. *Neuer Teutscher Merkur*, 1798/12/03: 311–332. Retrieved from http://www.historyofphrenology.org. uk/texts/retzer1.htm, 2014/01/14.

Gall FJ (1806a): *Beantwortung der Ackermannschen Beurtheilung der Gall'schen Hirn- Schedel-und Organen-Lehre vom Gesichtspuncte der Erfahrung: Herausgegeben von einigen Schülern des Hrn. Dr. Gall, und von ihm selbst berichtigt*. Halle, Neue Societäts- und Buch- und Kunsthandlung.

Gall FJ (1806b): *Meine Reise durch Deutschland, nebst pathognomischen Bemerkungen über meine gemachten Bekanntschaften und einzig wahre Darstellung meiner Lehre: Für Freunde und Feinde*. N.p.

Gall FJ (1808): *Discours d'ouverture à la première séance de son cours public sur la physiologie du cerveau*. Paris, F. Didot. Retrieved from http://www.historyofphrenology.org.uk/texts/ gall_discours_d%27ouverture.rtf,2014/01/14.

Gall FJ (1822): *Sur l'organe des qualités morales et des facultés intellectuelles et sur la pluralité des organes cérébraux*. Paris, Anth. Boucher (Sur les fonctions du cerveau et sur celles de chacune de ses parties, avec des observations sur la possibilité de reconnaitre les instincts, les penchans, les tal-ens, ou les dispositions morales et intellectuelles des hommes et des animaux, par la configuration de leur cerveau et de leur tête 2).

Gall FJ (1835a): *On the Origin of the Moral Qualities and Intellectual Faculties of Man, and the Conditions of their Manifestation.* Lewis W, trans. Boston, Marsh, Capen & Lyon (On the Functions of the Brain and of Each of Its Parts: With Observations on the Possibility of Determining the Instincts, Propensities, and Talents, or the Moral and Intellectual Dispositions of Men and Animals, by the Configuration of the Brain and Head 1).

Gall FJ (1835b): *On the Organ of the Moral Qualities and Intellectual Faculties, and the Plurality of Cerebral Organs.* Lewis W, trans. Boston, Marsh, Capen & Lyon (On the Functions of the Brain and of Each of Its Parts: With Observations on the Possibility of Determining the Instincts, Propensities, and Talents, or the Moral and Intellectual Dispositions of Men and Animals, by the Configuration of the Brain and Head 2).

Gall FJ (1835c): *The Influence of the Brain on the Form of the Head: The Difficulties and Means of Determining the Fundamental Qualities and Faculties and of Discovering the Seat of their Organs: Exposition of the Fundamental Qualities and Faculties, and Their Seat, or Organology.* Lewis W, trans. Boston, Marsh, Capen & Lyon (On the Functions of the Brain and of Each of Its Parts: With Observations on the Possibility of Determining the Instincts, Propensities, and Talents, or the Moral and Intellectual Dispositions of Men and Animals, by the Configuration of the Brain and Head 3).

Gall FJ (1835d): *Organology; or, an Exposition of the Instincts, Propensities, Sentiments, and Talents, or of the Moral Qualities, and the Fundamental Intellectual Faculties in Man and Animals, and the Seat of Their Organs.* Lewis W, trans. Boston, Marsh, Capen & Lyon (On the Functions of the Brain and of Each of Its Parts: With Observations on the Possibility of Determining the Instincts, Propensities, and Talents, or the Moral and Intellectual Dispositions of Men and Animals, by the Configuration of the Brain and Head 4).

Gall FJ (1835e): *Organology; or, an Exposition of the Instincts, Propensities, Sentiments, and Talents, or of the Moral Qualities, and the Fundamental Intellectual Faculties in Man and Animals, and the Seat of Their Organs.* Lewis W, trans. Boston, Marsh, Capen & Lyon (On the Functions of the Brain and of Each of Its Parts: With Observations on the Possibility of Determining the Instincts, Propensities, and Talents, or the Moral and Intellectual Dispositions of Men and Animals, by the Configuration of the Brain and Head 5).

Gall FJ (1835f): *Critical Review of some Anatomico-Physiological Works, with an Explanation of a New Philosophy of the Moral Qualities and Intellectual Faculties.* Lewis W, trans. Boston, Marsh, Capen & Lyon (On the Functions of the Brain and of Each of Its Parts: With Observations on the Possibility of Determining the Instincts, Propensities, and Talents, or the Moral and Intellectual Dispositions of Men and Animals, by the Configuration of the Brain and Head 6).

Gall FJ, Spurzheim JG (1809): *Untersuchungen ueber die Anatomie des Nervensystems ueberhaupt, und des Gehirns insbesondere: Ein dem franzoesischen Institute ueberreichtes Mémoire von Gall und Spurzheim, nebst dem Berichte der H. H. Commissaire des Institutes und den Bemerkungen der Verfasser über diesen Bericht.* Paris/Strasbourg, Treuttel und Würtz.

Gall FJ, Spurzheim JG (1811): *Des dispositions innées de l'âme et de l'esprit du matérialisme, du fatalisme et de la liberté morale, avec des réflexions sur l'éducation et sur la législation criminelle.* Paris, Frédéric Schoell.

Gall FJ, Spurzheim JG (1813a): Cerveau. In: Pinel P, ed., *Dictionnaire des sciences médicales par une société de médecins et de chirurgiens,* vol. 4. Paris, Chapart Libraire/C.L.F. Panckoucke, 447–479. Retrieved from http://www.historyofphrenology.org.uk/texts/gall_cerveau.html, 2014/01/14.

Gall FJ, Spurzheim JG (1813b): Crâne. In: Pinel P, ed., *Dictionnaire des sciences médicales par une société de médecins et de chirurgiens,* vol. 7. Paris, Chapart Libraire/C.L.F. Panckoucke,pp. 260–266. Retrieved from http://www.historyofphrenology.org.uk/texts/gall_crane.html, 2014/01/14.

Gall FJ, Spurzheim JG, Fossati GA (ca. 1830): *Das Gallsche System der Schaedellehre (Cranioskopie): Ueber die Fähigkeiten und Kräfte des Menschen und die Verrichtungen des Gehirns, nach den letzten von Dr. Gall kurz vor seinem Tode gemachten Beobachtungen und nach der zweiten von Dr. Fossati mit der grössten Sorgfalt vermehrten und verbesserten Auflage.* Leipzig, Baumgärtner.

Gall FJ, Vimont J, Broussais F (1838): *On the Functions of the Cerebellum.* Combe A, trans. Edinburgh, Maclachlan & Stewart.

Giustino D (1975): *Conquest of Mind: Phrenology and Victorian Social Thought.* London, Croom Helm.

Hall JY (1977): Gall's phrenology: A romantic psychology. *Studies in Romanticism* 16(3): 305–317.

Heintel B, Heintel H (1985): *Franz Joseph Gall: Bibliographie*. Stuttgart, Offizin Christian Scheufele.

Heintel H (1986): *Leben und Werk von Franz Joseph Gall: Eine Chronik*. Würzburg, Richard Mayr.

Herrn R (1995): On the history of biological theories of homosexuality. *Journal of Homosexuality* 28 (1–2): 31–56.

Hollander B (1928): *In Commemoration of Francis Joseph Gall (1758–1828): A Brief Account of His Life and Achievements on the Hundredth Anniversary of His Death*. London, Ethiological Society.

Honegger C (1992): *Die Ordnung der Geschlechter: Die Wissenschaften vom Menschen und das Weib, 1750–1850*. Frankfurt / New York, Campus Verlag, 2nd edition.

Hyde JS (1990): Meta-analysis and the psychology of gender differences. *Signs* 16(1): 55–73.

Jordan-Young RM (2010): *Brain Storm: The Flaws in the Science of Sex Differences*. Cambridge, Harvard University Press.

Jordanova L (1989): *Sexual Visions: Images of Gender in Science and Medicine between the Eighteenth and Twentieth Centuries*. Hertfordshire, Harvester Wheatsheaf.

Kaiser A (2012): Re-conceptualizing "sex" and "gender" in the human brain. *Zeitschrift für Psychologie* 220(2): 130–136.

Knibiehler Y (1993): Bodies and hearts. In: Fraisse G, Perrot M, ed., *A History of Women in the West IV: Emerging Feminism from Revolution to World War*. Cambridge/London, Harvard University Press, pp. 325–368.

Lange S, ed. (1992): *Ob die Weiber Menschen sind: Geschlechterdebatten um 1800*. Leipzig, Reclam.

Laqueur TW (1992): *Auf den Leib geschrieben: Die Inszenierung der Geschlechter von der Antike bis Freud*. Bußmann J, trans. Frankfurt/New York, Campus Verlag.

Lesky E (1970): Structure and function in Gall. *Bulletin of the History of Medicine* 44(4): 297–314.

Lesky E (1979): *Franz Joseph Gall, 1758–1828: Naturforscher und Anthropologe*. Bern/Stuttgart/Vienna, Hans Huber (Hubers Klassiker der Medizin und der Naturwissenschaften 15).

Lovejoy AO (1936): *The Great Chain of Being: A Study of the History of an Idea*. Cambridge, Harvard University Press.

Matthews-Grieco SF (1993): The body, appearance, and sexuality. In: Davis NZ, Farge A, eds., *A History of Women in the West, III: Renaissance and Enlightenment Paradoxes*. Cambridge/London, Harvard University Press, pp. 46–84.

McLaren A (1974): Some secular attitudes toward sexual behavior in France: 1760–1860. *French Historical Studies* 8(4): 604–625.

Micale MS (1991): Hysteria male/mysteria female: Reflections on comparative gender construction in nineteenth-century France and Britain. In: Benjamin M, ed., *Science and Sensibility: Gender and Scientific Enquiry, 1780–1945*. Cambridge, Basil Blackwell, pp. 200–239.

Noel PS, Carlson ET (1970): Origins of the word "phrenology." *American Journal of Psychiatry* 127(5): 694–697.

Regal W, Nanut M (2008): Franz Joseph Gall und seine "sprechenden Schedel" schufen die Grundlagen der modernen Neurowissenschaften. *Wiener Medizinische Wochenschrift* 158(11–12): 314–319.

Schiebinger LL (1987): Skeletons in the closet: The first illustrations of the female skeleton in eighteenth-century anatomy. In: Gallagher C, Laqueur TW, ed., *The Making of the Modern Body: Sexuality and Society in the Nineteenth Century*. Berkeley/Los Angeles, University of California Press, pp. 42–82.

Schiebinger LL (1989): *The Mind Has No Sex? Women in the Origins of Modern Science*. Cambridge, Harvard University Press.

Schiebinger LL (1990): The anatomy of difference: Race and sex in eighteenth-century science. *Eighteenth-Century Studies* 23(4): 387–405.

Schramm-Macdonald H (1878): Gall, Franz Joseph. In: Bayerische Akademie der Wissenschaften, ed., *Allgemeine Deutsche Biographie*, vol. 8. Leipzig, Duncker & Humblot, pp. 315–316.

Stahnisch FW (2005): Über die neuronale Natur des Weiblichen: Szientismus und Geschlechterdifferenz in der anatomischen Hirnforschung (1760–1850). In: Stahnisch FW, Steger F, eds., *Medizin, Geschichte und Geschlecht: Körperhistorische Rekonstruktionen von Identitäten und Differenzen*. Stuttgart, Franz Steiner (Geschichte und Philosophie der Medizin 1), pp. 197–224.

Stahnisch FW (2007a): Neuromorphologie versus Phrenologie? Hirnforschung in Mainz und Wien um 1800. In: Bock WJ, Holdorff B, eds., *Schriftenreihe der Deutschen Gesellschaft für Geschichte der Nervenheilkunde*. Vol. 13. Würzburg, Königshausen und Neumann, pp. 313–342.

Stahnisch FW (2007b): Über die Natur des weiblichen Gehirns: Geschlechterpolitik im Werk des Mainzer Anatomen Jacob Fidelis Ackermann (1765–1815). In: Schultka R, Neumann JN, eds., *Anatomie und Anatomische Sammlungen im 18. Jahrhundert anlässlich der 250. Wiederkehr des Geburtstages von Philipp Friedrich Theodor Meckel (1755–1803)*. Berlin, LIT Verlag Dr. W. Hopf (Wissenschaftsgeschichte 1), pp. 421–435.

Temkin O (1947): Gall and the phrenological movement. *Bulletin of the History of Medicine* 21(3): 275–321.

Tomlinson S (2005): *Head Masters: Phrenology, Secular Education, and Nineteenth-Century Social Thought*. Tuscaloosa, University of Alabama Press.

Tuana N (1993): *The Less Noble Sex: Scientific, Religious, and Philosophical Conceptions of Woman's Nature*. Bloomington, Indiana University Press (Race, Gender and Science).

Van Wyhe J (2002): The authority of human nature: The Schädellehre of Franz Joseph Gall. *British Journal for the History of Science* 35: 17–42. Retrieved from http://www.historyofphrenology.org.uk/texts/2002van_wyhe.htm, 2014/01/14.

Vereinte Hofstelle (1802): *Bericht und Empfehlungen bzgl. des Lehrverbots von Gall*. University of Vienna Archives, UAW 147.37, 1802/02/06.

Walther PF (1804): *Neue Darstellungen aus der Gallschen Gehirn- und Schedellehre: Als Erläuterungen zu der vorgedruckten Vertheidigungsschrift des Doktor Gall eingegeben bey der niederösterreichischen Regierung*. Munich, Scherrersche Kunst- und Buchhandlung.

Franz Joseph Gall on the "deaf and dumb" and the complexities of mind

Paul Eling and Stanley Finger

ABSTRACT

Franz Joseph Gall used a broad variety of phenomena in support of his organology. Well known are his observations on anatomical features of the brain, species-specific behavioral patterns, the observation that some individuals may excel in one faculty while being mediocre in others, changes in the organs with development and aging, and how the organs associated with the faculties might be affected by diseases and acute brain lesions. We here present a widely overlooked source: his observations on individuals then classified as "deaf and dumb." We discuss how these observations were presented by Gall in support of his organology and in his disputes with empiricists and sensationalists about the nature of mind.

Franz Joseph Gall on the "deaf and dumb"

Franz Joseph Gall (1758–1828) is best known for his organology, his theory about a constellation of faculties together constituting the human mind, each associated with a circumscribed organ in the cerebral cortex (for a biography, see Finger and Eling 2019). In the early 1790s, he began to realize that the centuries-old view that the mind consists of the faculties of perception, reason, and memory could not be valid. It cannot explain, for instance, why a prodigy can have a seemingly inborn talent to learn and remember pieces of music but not verbal texts. Gall concluded there cannot be a single, general memory. Rather, there must be several memories, each associated with a specific faculty of mind.

What, then, are the basic faculties, those that are independent of one another? This was the main topic Gall worked on from the 1790s until the publication of his "great works," his term for his four-volume *Anatomie et Physiologie du Système Nerveux en Général, et du Cerveau en Particulier* (Anatomy and Physiology of the Nervous System in General and the Brain in Particular; Gall and Spurzheim 1810-1819).

Gall considered observations from various domains to determine which faculties under-lie behavior. His most important sources were the behavioral patterns he witnessed in humans and animals, both as species and among individuals, at various levels on the ladder of life forms. He also attempted to study anatomical features of the brain, most importantly, the cortical organs associated with the different faculties, again in the contexts of different species, but also within each group.

When examining species-specific and individual differences, Gall took into consideration differences among races, nationalities, and localities. He also paid attention to gender differences and the ages of his subjects. He further studied the effects of brain injuries and diseases in humans, and even did some experiments on animals, although he acknowledged that it could be very difficult to determine which cerebral organs might be directly or indirectly affected by a sword wound, a cerebral hemorrhage, or even the brain damage sustained by rabbits or dogs subjected to such horrors.

Scholars analyzing Gall's theoretical views and empirical observations have long cited the many types of evidence he drew from when presenting his revolutionary doctrine. Yet one of his subgroups has gone almost ignored. Here we will present how Gall also used observations from "deaf and dumb" individuals to help him formulate and substantiate his views. We realize this is an archaic term that is now considered demeaning and derogatory. We have chosen to use it here for historical accuracy, this being the terminology used at the start of the nineteenth century.

For Gall's statements on the deaf and dumb, we turned to the 1825 revised (cheaper) edition of his 1810–1819 *Anatomie et Physiologie*—namely, his *Sur les Fonctions du Cerveau et sur Celles de Chacune de ses Parties* (On the Functions of the Brain and on Those of Each of Its Parts), which in turn was translated into English in 1835 (Gall 1822–1825, 1835). Importantly, he left almost all of the *organologie* (one of the terms he used, avoiding *phrenology*) that had been in his original set of four volumes intact, while adding some new observations, citations, and commentary to the 1825 edition, which did not include his detailed neuroanatomy (too difficult for most nonmedical readers) and the earlier 100-plate atlas.[1]

First experiences

Early in his scientific quest, having earned a medical degree from Vienna, Gall was granted entrance to the city's new "psychiatric institute," the *Narrenturm*, a round, five-story fortress that was the first building in Europe for housing and treating only the insane (earlier asylums had also housed the poor, criminals, etc.; see Stohl 2000). He also went to a Viennese institution specifically taking care of deaf and dumb children. He not only visited the facility out of curiosity and to learn about behavioral peculiarities but acted as physician to that institute. As he stated in his books:

> At this time I was physician to the Deaf and Dumb Institution, where pupils were received, from six to fourteen years of age, without any preliminary education. M. May, a distinguished physiologist, then director of the establishment; M. Venus, the teacher; and myself, had it in our power to make the most exact observations on the primitive moral condition of these children. Some of them were remarkable for a decided propensity for stealing; while others did not show the least inclination to it. The most of those who had stolen at first, were corrected of the vice in six weeks; while there were others, with whom we had more trouble, and some were quite incorrigible. On one of them, were several times inflicted the severest chastisements, and he was put into the house of correction; but it was all in vain. As he felt incapable of resisting temptation, he wished to learn the trade of a tailor; because, as he said, he might then indulge his inclination with impunity. (Vol. 4, p. 129)

The Deaf and Dumb Institute in Vienna, founded in 1779, was one of the earliest designed for this specific population (Fischbach 1832). The institute was founded after Emperor

Joseph II (1741–1790) met Abbé Charles Michel de l'Epée (1712–1789) in Paris. The priest had founded a school for deaf and dumb children in 1765, and he had even developed a sign language for these children. The enlightened Emperor was stimulated by what he learned during his time with the Abbé. When he left Paris he was already thinking about enacting reforms.

Joseph May (1755–1820), an Austrian pedagogue, had been working in France at this time and was looking for an opportunity to return to his motherland. When he asked Joseph II for his help, the Emperor asked him whether he would be willing to learn De l'Epée's teaching methods, hoping to set up a similar institute in Vienna. May was happy to accept the opportunity. May and priest Friedrich Storck (1742–1823) obtained the needed instructions from De l'Epée and set to work to establish such an institute for the deaf and dumb in Vienna.

The institute was first housed in the "Burger Hospital" (or public hospital). It could handle six boys and six girls. Emperor Joseph II decided that a larger building for 30 children was necessary and, in 1784, the General Seminarium at Dominicans Place was employed for this purpose (Figure 1). When Storck retired as director in 1792, May took over, holding this position until 1819, when Michael Venus (1774–1850), a teacher also mentioned by Gall, replaced him.

Fischbach (1832) devoted a section of his history of the institute to how the physicians in Vienna provided their services free of charge. Here he mentioned that "the former

Figure 1. The Institute for Deaf and Dumb, next to the Theresianum in what was later called the Taubstummengasse 13–17, in the center of Vienna.

physicians of the institute were the medical doctors," and included "Joh. Jos. Gall" on his list (p. 371). Unfortunately, that is all he wrote about Gall, and calling him "Joh." was a mistake. He went by Franz Joseph or just Joseph at this time (see Finger and Eling 2019, 5). As Gall wrote that May was the director when he saw the children at the institute, this must have been after 1792.

Skulls of the deaf and dumb

In order to determine the location of the cerebral organs of the faculties, propensities, and instincts, Gall collected skulls and plaster casts from individuals gifted or lacking in an everyday talent or special in any other way. His collections included several items from deaf and dumb individuals. He explained:

> In a few years I thus formed a collection of four hundred casts, of men of all conditions and classes, from the beggar to the prince; the deaf and the dumb; idiots, children of all ages, boys, girls, women, &c. I laid schools, houses of correction, hospitals for the insane, all of them under contributions for this object. I possessed those casts of individuals whose qualities and faculties I had already observed; in this number, there were found persons of the poorest education, as well as those educated with the greatest care. (Vol. 3, p. 115)

But although he collected a number of pieces from the deaf and dumb, he did not seem to discover special skull features that would appear time and again in this population. That is, he could not find reproducible depressions in some regions or compensatory bumps in others. Nonetheless, he did learn more about mind and brain organization and development from these individuals.

The deaf and dumb and inborn faculties

Gall was active during a time of new-found empiricism. Along with this empirical approach, many philosophers were convinced that "everything comes from the senses." In France, this widely held philosophy was referred to as *sensualism*. But Gall did not construe this to mean that our fundamental capacities are located in the senses. The eye is critical for vision, but conscious visual perception is a higher function that involves the cerebral cortex. The same is true of the ears, and so on.

Gall was especially interested in how sensation relates to cognition or understanding, and this is where he focused on deaf and dumb individuals, a group that seemed deficient when it came to understanding or intellect. In his books, he introduced the topic by stating, "Let us first examine what the influence of the senses can be on our moral and intellectual powers, whether Aristotle was correct in saying, '*Nihil est in mente quod non olim fuerit in sensu*'" (There is nothing in the mind which was not first in the senses). His paragraph heading was long but notable: "The senses and the sensations received by external impressions, cannot give truth to any ingenious aptitude, any instinct, propensity, sentiment, or talent, any moral or intellectual faculty" (Vol 1, p 106). He then proceeded to make statements on this issue for each of the senses and, after having dealt with smell and taste, he turned to hearing:

> As for hearing, I have demonstrated, that we have hitherto been mistaken in attributing to this the talent for music, and to the glottis the talent for singing; that it is not the hearing, which

gives the capacity for language; that the languages, however imperfect or perfect they may be, are not the creation of the hearing but of the cerebral organization; that the irresistible and lawless acts of certain deaf and dumb persons should not be attributed to their want of hearing, but to the imbecility of their minds, &c. (Vol. 1, pp. 106–107)

In effect, Gall argued that a lack of understanding among the deaf and dumb is not caused by an impaired sense of hearing but, rather, by deficiencies or disorders of the mind. This condition was called "idiocy" or "imbecility" at this time. During the 1790s, Phillipe Pinel (1745–1826), one of the most recognized nosologists of the period and a physician whose observations Gall respected, used imbecility interchangeably with "idiotism" to indicate both an intellectual and an emotional disorder (Pinel 1800).

After Pinel, nosologies often differentiated imbecility from idiocy in terms of degree and age of onset. Idiocy was usually defined as congenital, whereas imbecility was sometimes congenital and sometimes acquired (De Sanctis 1905). In an essay on the deaf and dumb that was later incorporated into the 1824, fourth edition of the *Encyclopedia Britannica*, British physician Peter Mark Roget (1779–1869) maintained: "All who are deaf from birth must necessarily be dumb; that is, they must be incapable of using language, of the sound of which they have never had the perception, and which they consequently could never attempt to imitate" (Roget 1824, 467). Roget's view was clearly dependent on the empiricist philosophy that Gall forcefully rejected.

The discovery of the faculty of imitation

At this time, it was common for a physician to discuss individual cases. Gall did this with deaf and dumb individuals, reporting several such cases in his books. The following probably took place while Gall was still in Vienna, acting as physician in the Institute for the Deaf and Dumb:

When I was talking with one of my friends, respecting the forms of the head, he assured me, that his own had a very peculiar one. He then directed my hand to the anterior superior part of his head; I found this region considerably bulging; and behind the protuberance, a depression, a cavity, which descended on each side, towards the ear. At this period, I had not observed this conformation. This man had a peculiar talent for imitation. He imitated in so striking a manner the gait, the gestures, the sound of the voice, &c., that the person was immediately recognised. I hastened to the institution for the deaf and dumb, to examine the head of the pupil Casteigner, who had been received into the establishment six weeks previous, and who, from the first, had fixed our attention by his prodigious talent for imitation. (Vol. 5, p. 201)

This citation forms the beginning of Gall's section on the history of the discovery of *Faculty XXV: Faculty of Imitation, Mimicry*. Gall now had an opportunity to determine the localization of another faculty of mind, one for imitation. But did Casteigner's skull show the same characteristics as Gall's friend's? Was there a bulging of the anterior superior part of the forehead? "To my great astonishment," Gall wrote, "I found in him the superior anterior part of the head, as prominent as in my friend Annibal" (Vol. 5, p. 201). Thus, the deaf and dumb pupil Casteigner played a critical role for the determination of one of Gall's faculties, although not one unique to deaf and dumb individuals. In this context, it should be emphasized that Casteigner's capacity for imitation was not interpreted as a symptom of some brain disease.

Loss of control by higher faculties

Some of Gall's cases suggested that deviant, incorrigible behavior might be due to the inability of higher faculties to exert themselves over lower propensities and instincts. Gall assumed that faculties can interact—in fact, that higher-order thinking and behaving are rarely the result of a single faculty. Importantly, he did not conceive of a controlling agent, as many current neuroscientists do when they mention some sort of a central executive or working memory.

In the following citation, Gall mentioned two individuals. The first was a 12-year-old boy he had met in Berne. This must have been in August 1807, when he went to Switzerland at the end of his tour through Germany and neighboring countries, before heading to Paris (Finger and Eling 2019). His second case concerned "the obstinate robber Fesselmayer," whom he had observed in Haina (Kloster Haina), approximately 40 kilometers northeast of the German city of Marburg. Only his second case was deaf and dumb, and he associated his condition with imbecility.

> We saw in the prison of Berne, a boy of twelve years, ill organized and rickety, who could never prevent himself from stealing; with his own pockets full of bread, he still took that of others. At Haina, the overseers gave us a long account of an obstinate robber, named Fesselmayer, whom no corporal punishment could correct. In the prison he stole every thing he saw, and they had put on his arm a card which served as a mark of disgrace, warning others not to trust him. Before seeing him, we anticipated what his organization must be, and our expectation was confirmed at the very first glance. He appeared about sixteen years of age, though in fact he was twenty-six. His head was round, and about the size of a child of one year. This individual was also deaf and dumb, which often happens in cases of mental imbecility. (Vol. 1, p. 318)

Gall referred to this case a second time in the section titled "Natural history of the propensity to theft in diseases, with remarkable weakness of understanding" (Vol. 4, p. 139). Here Gall also presented a case he had seen in a prison in Berlin (Stadt-Vogtey). Among the boys who were presented to him, he noticed one deaf and dumb boy, who he suggested should never be let free, "because he would not be restrained from a continuance of his robberies" (Vol. 4, p. 137).

A "young man, fifteen years old, half imbecile and incorrigible, who died in Vienna's house of correction" (Vol. 4, p. 139) also showed Gall that some propensities may be pervasive to the point that they are incorrigible. He expressed this thought as follows:

> I have already made evident, that we ought to consider man in two points of view; first, as having qualities common with animals; that is to say, those of an inferior order; then, as being endowed with the character of humanity, or with qualities of a superior order. I have also shown that man, in virtue of his superior qualities, is capable of subduing and directing his propensities of an inferior order. But, if the qualities of a superior order are controlled in an extraordinary manner, to such a degree that their free action is prevented, while those of the inferior order, on the contrary, are active, then the animal part of the man predominates exclusively, and the flesh, or the brutal desires, hold in subjection the spirit, or the dispositions of the superior qualities, which are hardly developed. With such an organization for the functions of the soul, which belong to a superior order, the same happens which takes place in regard to each organ whose development is defective; that is, there results a relative imbecility, and, in consequence, the incapability of acting morally; while the propensities of an inferior order act with uncontrolled energy. (Vol. 4, pp. 137–138)

To summarize, if the moral and cognitive capacities are limited, an individual will be dominated by lower tendencies and, should this be the case, no improvement in behavior can be expected. For those deaf and dumb individuals displaying "imbecility," Gall felt forced to conclude that there was little chance of improvement.

The wild boy (man) from Aveyron

A case mentioned several times in Gall's works is the well-known "*Sauvage d'Aveyron.*" He brought "the savage" up in Volume 2 (Section III), when dealing with the plurality of intellectual and moral faculties. Having discussed individuals who behave normally in many ways, he was now ready to discuss in greater detail individuals with limited cognitive capacities, writing, "Even in congenital idiocy, all the moral qualities and intellectual faculties are not paralyzed to the same extent," and, "In the majority of cases, as I have many times remarked, some of the faculties still enjoy a considerable degree of activity" (Vol 2., p. 289). For instance, he had seen two idiotic girls in Paris who sang well, understood what they were singing, and remembered songs for long periods of time.

And here he introduced the so-called wild boy or man of Aveyron:

> The wild man of Aveyron, so called, placed in the institution of the deaf and dumb at Paris, exhibits a love of order which rises even to a passion, although all his faculties are extremely limited. If the most trifling article, a brush, for instance, be displaced, he immediately runs and replaces it. Pinel relates a very similar case. (Vol. 2, p. 289)

Victor (of Aveyron; ca 1788–1828; Figure 2) was a feral child. According to Paris physician Jean-Marc Gaspard Itard (1775–1838), his French parents neglected him and, from about the age of 4, he had, in fact, lived in the woods (Itard 1801, 1802). When captured, he managed to escape and then returned to the woods. He finally emerged in the small village of Saint-Sernin-sur-Rance and was subsequently sent to the Institut Nationale des Sourds-Muets, a facility for deaf-mutes, even though he might not have been deaf. Itard observed and worked with him for some five years, trying to teach him social manners and language, but came to the conclusion that the boy could never learn to speak.

Gall did not mention Itard by name, although he visited the institute and had a special interest in the case. Then again, Gall was often stingy in giving credit to others. French psychiatrist and reformer Philippe Pinel, cited in the quote above, was a notable exception. Gall reported that the "wild man" had a well or perhaps even overly developed "love of order," whereas many of his other faculties were underdeveloped.

Gall raised the following question: "Does social life give rise to factitious qualities or faculties?" (Vol. 1, p. 161). Man, like many species of animals, is a social organism, and some authors have argued that these propensities and tendencies are learned. Quoting Gall:

> But some think to prove that man is born without propensities and without faculties, and that he acquires these faculties merely by social life and by education; by citing the example of some individuals found astray in the woods, who, having received no education, have all the brutality of animals, and appear to be not only deprived of human faculties, but even of those of the least intelligent animals. (Vol. 1, p. 164)

Gall was unwilling to accept this conclusion. He pointed out that in most wild children brain organization is defective, and that many have heads that are too small or too large due to hydrocephalus and other disorders. "Ordinarily miserable creatures, of imperfect

Figure 2. Victor's portrait from the front cover of the report by Itard (1801).

organization" is one of the ways he described these children (Vol. 1, p. 164). Not only can they be a real burden to a family, but some poorly educated, lower-class people regarded them as bewitched, which is why some were abandoned in the woods.

Gall now elaborated on the behavioral and mental characteristics, and the shape of the head of the savage of Aveyron:

> The savage of Aveyron, placed in the deaf and dumb institution at Paris, is not different from those of whom I have just spoken. He is weak-minded to a great degree; his forehead is very little enlarged laterally, and very much compressed from above downward; his eyes are small and greatly sunken, his cerebellum [where Gall housed the organ for reproductive instinct] little developed. We were not able to convince ourselves that he had the sense of hearing; for, they could not in our presence render him attentive, either by calling him nor by sounding a glass behind his ears. His mode of existence is tranquil; his attitude and manner of sitting are decent; it is only remarked, that he is constantly balancing the upper part of his body and his head; he salutes by inclining his body, to the persons who arrive, and manifests his satisfaction when they depart. The sexual propensity does not seem to be active in him. He knows a few letters, and even points to the objects which the letters designate. In other respects, his favorite occupation is to restore to their former place any articles which have been displaced. Such is the result of the hopes which were formed of him, the efforts which have been made, and the patience and mildness which a benevolent woman has shown towards him. We may pronounce, with confidence, that these labors will never be crowned with any better success. The wild man found in the forests of Lithuania, who is cited by many authors as an example of the powerful influence of education, was certainly a similar being. (Vol. 1, pp. 165–166)

This quote shows that Gall had examined the boy. Although he mentioned the features of his skull, Gall did not associate them with specific behavioral propensities. But what seemed clear to him was that the boy's highest mental capacities were very limited. Consequently, and given his earlier experience with imbeciles, Gall did not believe that educational programs would prove worthwhile in his case.

But who was the wild man found in the forests of Lithuania, who Gall mentioned at the end of his quoted paragraph?[2] There was a report circulating in France at the time about a boy who was eight or nine years old when he was "discovered" in 1661. Gall did not give a source, but he could have learned of this case from an article in Louis Moréri's (1643–1680; 1674) *Le Grand Dictionaire Historique*. Moréri's one-volume edition of 1674 and posthumous two-volume edition of 1681 were expanded by others after his death, and they were often cited by physicians and philosophers interested in feral children. In this source, the boy was said to live in the forests with a group of bears, and for this reason he was called Joseph Ursin (for bear). After Joseph was captured, efforts were made to teach him normal human activities with little success.

This description is vague, and it is not clear what to believe from it (Benzaquén 2006). At the time, a number of cases of wild children were reported, and some found their way into the popular imagination and literature. In particular in France, Joseph's story was repeated many times and, as we have seen, it reached Gall, who doubted that the boy had been or ever could have been successfully cultivated back into civilization.

Returning to the "savage of Aveyron," Gall gave serious thought to what his and related cases revealed about the faculty he would call the "sense of locality":

> The sense of locality, making known the relations of space, I have been inclined to think that it might also be the sense of taste, of symmetry, and of order. It is certain that some persons are destitute of all spirit of order; while others, from their infancy, are pained at the sight of the slightest irregularity in the furniture, tables, &c. This sentiment sometimes amounts to a passion, even in idiots. I have already mentioned the *soi-disant* savage of Aveyron, in the Institution for the Deaf and Dumb at Paris, and I know many similar cases. (Vol. 4, p. 283)

This quote mentioning the deaf and dumb shows how Gall went about determining the nature of a faculty and a name for it. It might remind us of how Louis Leon Thurstone (1887–1955), Joy Paul Guilford (1897–1987), and other twentieth-century scientists tried to identify faculties that could account for intelligence with their test batteries.

Words, language, and God

Deaf and dumb individuals, as shown above, attracted the attention of philosophers and scientists interested in determining which aspects of mind might be inborn as contrasted with learned. Language played a special role in these disputes and in debates about whether we are unique. Hence, it is not surprising to find parts of these discussions in Gall's works.

Under "Of the Functions of the Senses in General," Gall began to address these issues, writing:

> But, no impression from without, no irritation from within, can become a sensation or an idea, without the concurrence of the brain. The faculty of perceiving impressions, of retaining and comparing ideas, and making application of them, is by no means in proportion to the senses either in men or animals, as is proved by the example of idiots and simpletons. (Vol. 1, p. 128)

Gall now brought up an argument formulated by French philosopher Claude Adrien Helvetius (1715–1771), who maintained, "In whatever manner we inquire of experience, she always answers, that the greater or less superiority of mind, is independent of the greater or less perfection of the organs of the senses" (Vol. 1, p. 129).

And Gall then moved on to the issue of whether words always first reflect some external observation and, by analogy, can be used to describe an abstract event. As put by Gall, "Who, then, will dare assert that the expressions, *strain, cold, warm, chill, palpitation, trembling, &c.*, have been designed to designate rather the qualities of external things, than those of internal sensations?" (Vol. 1, p. 130).

Gall's contention that language likely does not, in fact, originate from sensory experiences became even more compelling when he mentioned categories of words that are neither nouns nor verbs, And here he again turned to the deaf and dumb, bringing them into the picture he was painting.

> Whence come the words which do not precisely designate determinate ideas, but simply the mode of thinking; the prepositions, conjunctions, interjections, adverbs of interrogation and exclamation, &c, such as *hut, and, yet, notwithstanding, for, if, nevertheless, consequently, also, then, thus, alas, yes, no, &c.*?

> Do not the deaf and dumb, who possess reason, but who are deprived of the faculty of expressing themselves by articulate language, depict their internal sensations by gestures, which absolutely have nothing in common with the external world? If all our ideas come from the senses, what becomes of the general and purely intellectual ideas, whose signification is wholly independent of the material world? For example, "there is no effect without a cause;" "nothing can spring from nothing;" "matter can neither be increased or diminished;" "a quality, contrary to a subject, cannot belong to it; " "a thing cannot exist and not exist, at the same time." (Vol. 1, pp. 130–131)

As can be imagined, the most abstract idea in this philosophical dispute, especially during the 1700s and early 1800s, was the concept of God. How could an empiricist explain the experience of God? Is such a person forced to reject the notion of God? Toward the end of his books, Gall discussed the "Natural History of Man, in Relation to his Belief in God and his Propensity to Religion." There he explained:

> Every where and in all ages, man, urged by the feeling of his dependence upon every thing around him, is forced continually to acknowledge the limits of his strength, and to confess to himself, that his fate is controlled by a superior power. Hence the unanimous consent of all nations to adore a Supreme Being. (Vol. 5, p. 218)

And he continued:

> Such then is the sense of the Divinity, that there is not a single nation, however barbarous, however destitute of laws or of morals, it may be, which does not believe, that there are Gods. ... Men always have been led by an instinct, by a secret impulse, to acknowledge an omnipotent Being. (Vol. 5, p. 219)

Thus, Gall argued that, at all times and all over the world, people have acknowledged a Supreme Being and have engaged in a religious worship. He cited examples and involved the deaf and dumb in his arguments:

> It is still objected, that ideas relative to God and religion, never arise among the deaf and dumb; and hence it is concluded, that there is not in man any natural disposition to these ideas. But

can it be believed, that the man of the most cultivated mind could arrive at those ideas of God and religion, which we have, if he had not been brought up in these ideas? The faith of sectarians is the work of education, of arbitrary instruction, and the ideas, which the philosopher forms of God, are the fruit of the most elevated abstractions. We cannot expect either the one or the other from a deaf and dumb man, whose education cannot have been directed towards this point; but from what we see all rude people do, we might divine what the deaf and dumb would do, if living together in tribes; for, the want of hearing does not prevent the deaf and dumb from forming to themselves, the same idea of the external world, which other men form of it, and from drawing the same deductions from the events, which pass under their eyes. (Vol. 5, p. 225)

It is interesting to see how Gall discussed the notion of God, given that he was branded a materialist and a heretic. Of course, he repeatedly denied such charges, claiming, "*Dieu et cerveau, rien que Dieu et cerveau*" [God and brain, nothing else but God and brain]. True, he was not a deeply religious, church-going man. But he never rejected the idea of God. As he expressed it in his books, which were largely devoid of metaphysics, nature's wonderful architecture, with organisms perfectly designed to survive in their habitats, could only be the work of our Creator, of God (e.g., Vol. 4, p. 224).

Man would appear to be the only species capable of true, spoken language. But underlying spoken language is another form of communication, a "language of action." Here, Gall pointed specifically to the exchange of ideas between deaf and dumb individuals. They may lack language but they can nevertheless exchange ideas and feelings. Consequently, he argued:

If this language is not as generally perfect, as it might be, it is because we have greatly neglected it; it is too easily replaced by the language of words. But observe the deaf and dumb, before they have received any instruction; the exactitude and the readiness, with which they communicate to each other the emotions of their souls, feelings, sentiments, thoughts, and their intentions, will prove to you, that the language of action has many advantages over spoken languages. Do we not daily see, that numerous collections of people interpret without mistake the pantomimes of our plays? (Vol. 5. p. 296)

Thus, Gall contended that presumably imbecilic individuals could still communicate, even if they lacked or had only minimal verbal language. But putting this thought in context, it must be remembered that there were no strict criteria for labeling people as deaf and dumb at this time, making this literature unusually challenging for historians hoping to decipher it.

Special institutions

Gall attempted to develop a theoretical framework that would encompass all kinds of behaviors: taking care of the young, food preferences, sense of color, music, spatial orientation, humor, and more. Each of the faculties he settled on could be well developed, average, or poorly developed from one individual to the next. What others might consider abnormal or deviant was, for Gall, a consequence of one or more extremely developed or underdeveloped faculties or conditions (e.g., disease, religious study) that could cause extreme changes in the activity of specific faculties. In this context, and as noted earlier, he recognized that if the superior faculties were too weak to control our animal tendencies, this could result in conditions that could be disruptive to the individual, as well as to society.

Gall's approach to psychiatric patients (and criminals) was more humane than that of many of his contemporaries (Finger and Eling 2019, 363–390). This humane approach extended to the deaf and dumb. He favored special institutes for caring for them, institutes that might also be centers for learning more about what they might or might not achieve under various conditions:

> To soften brutal passions, and to dispose the people to honest enjoyments, moral, religious, and civil instruction will be imperative on all classes; even the malefactor will be judged worthy of compassion. Every where we shall see the institutions of benevolence multiplied; hospitals for the sick, for the insane, *for the deaf and dumb*, the blind, the incurable, old men, invalids, &c. The brothers and sisters of mercy will have the first claim to public esteem. We shall see asylums formed for lying-in women, for foundlings and orphans. Every where schools, academies, universities, museums, libraries will foster the arts, sciences, &c., for the purpose of increasing the happiness and ennobling the enjoyments of men. (Vol. 5, p. 166; italics added)

Conclusions

Gall was interested in studying the deaf and dumb when he began to construct his doctrine, and he never lost this interest. This is understandable, because the deaf and dumb represented an extreme of society. Much like criminals or great musicians, he believed they could shed light on the faculties of mind and the organization of the brain. Gall alluded to what he had gleaned from them when discussing the plurality of faculties, individual faculties, and the effects of the loss of superior faculties. He also pointed to them when discussing nature vs. nurture, the issue of whether all thinking must be based on sensations, and the question of whether important elements of language and speech could develop without the sense of hearing.

He pressed for specialized institutions that would be dedicated to the care and study of the deaf and dumb, although, unlike his novel *organologie*, which linked the mind and brain together in decidedly new ways, this was not a movement he initiated. But, although he suggested humanitarian reforms, he was not optimistic about making major strides with the majority of these children and adults, because he believed the deaf and dumb typically were severely impaired intellectually.

Gall collected skulls and plaster casts of interesting people and, in addition to his large book collection, they served as a sort of library for his research on the fundamental faculties of mind and their associated parts of the brain. He did have skulls and head casts from deaf and dumb individuals in his collections. Some seemed to suffer from hydrocephalus. Yet he encountered considerable variability within this group and was never able to find a reliable set of features characterizing the skulls of deaf and dumb children and adults.

What caught his attention, however, was how some deaf and dumb people could differ from others by displaying one or more other notable traits, making these individuals perfect candidates for his research program. In this context, he wrote about one deaf and dumb boy with a passion for order, noting that his other abilities were very limited. Similarly, he was interested in how well the sense of locality seemed to be developed in some deaf and dumb individuals, a particularly important faculty of mind for the "savages" struggling to survive in the woods.

Taken together, Gall's interests in the deaf and dumb shed considerable light on the organization of the mind and its dependence on the brain. And for historians, they provide a means to see how Gall was constructing and defending his ideas, and to appreciate more fully how he saw less fortunate humankind in natural ways.

Notes

1. When we compared the English translation to the French, we found the translation to be of high quality, with the translators remaining faithful to the original (see Finger and Eling 2019).
2. We thank Egle Sakalauskaite for her help in tracing reports on this Lithuanian wild child.

Disclosure statement

No potential conflict of interest was reported by the authors.

References

Benzaquén, A. S. 2006. *Encounters with wild children: Temptation and disappointment in the study of human nature.* Montreal: McGill-Queen's University Press.

De Sanctis, S. 1905. Types et degrés d'insuffisance mentale. *L'Année Psychologique* 12 (1):70–83. doi:10.3406/psy.1905.3709.

Finger, S., and P. Eling. 2019. *Franz Joseph Gall: Naturalist of the mind, visionary of the brain.* New York: Oxford University Press.

Fischbach, J. B. 1832. *Darstellung des k. k. Taubstummen-Instituts.* Wien: Doll.

Gall, F. J. 1822–25. *Sur les Fonctions du Cerveau et sur Celles de Chacune de ses Parties.* Vol. 6. Paris: J.-B. Baillière.

Gall, F. J. 1835. *On the functions of the brain and each of its parts: With Pbservations on the possibility of determining the instincts, propensities, and talents, or the moral and intellectual dispositions of men and animals, by the configuration of the brain and head.* Vol. 6. Ed. N. Capen. Trans. W. Lewis. Boston: Marsh, Capen and Lyon.

Gall, F. J., and J. G. Spurzheim. 1810–19. *Anatomie et Physiologie du Système Nerveux en Général, et du Cerveau en Particulier.* (4 vols with an accompanying Atlas). Paris: Schoell.

Itard, J. M. G. 1801. *De l'Éducation d'un Homme Sauvage, ou Des Premiers Développemens Physiques et Moraux du Jeune Sauvage de l'Aveyron.* Paris: Goujon Fils.

Itard, J. M. G. 1802. *An historical account of the discovery and education of a savage man: Or of the first developments, physical and moral, of the young savage caught in the woods near Aveyront in the year 1798.* London: Richard Phillips.

Moréri, L. 1674. *Le Grand Dictionaire Historique ou Le Mélange Curieux de l'Histoire Sacrée et Profane.* Paris: Les Libraires Associés.

Pinel, P. 1800. *Traité Médicophilosophique sur l'Aliénation Mentale ou la Manie.* Paris: Brosson.

Roget, P. M. 1824. Deaf and dumb. In *Encyclopaedia Britannica; or, A dictionary of arts, sciences, and miscellaneous literature, enlarged and improved.* Suppl. Vol. 3, 467–85. Edinburgh: Constable.

Stohl, A. 2000. *Der Narrenturm oder Die dunkle Seite der Wissenschaft.* Wien: Böhlau Verlag.

Gall's German enemies

Paul Eling and Stanley Finger

ABSTRACT

Franz Joseph Gall's (1758–1828) proposal for a new theory about how to represent the mental faculties is well known. He replaced the traditional perception-judgement-memory triad of abstract faculties with a set of 27 highly specific faculties, many of which humans share with animals. In addition, he argued that these faculties are dependent on specific cortical areas, these being his organs of mind. After several years of presenting his new views in Vienna, he was banned from lecturing for what he considered absurd reasons. The edict enticed him to make a scientific journey through the German states, both to present his ideas to targeted audiences and to collect more cases. This trip, started in 1805, was extended to include stops in Denmark, Holland, and Switzerland before finally ending in Paris in 1807. For the most part, Gall was received with great enthusiasm in what is now Germany, but there were some individuals who strongly opposed his anatomical discoveries and skull-based doctrine. In this article, we examine the concerns and arguments raised by Johann Gotlieb Walter in Berlin, Henrik Steffens in Halle, Jakob Fidelis Ackermann in Heidelberg, and Samuel Thomas Soemmerring in Munich, as well as how Gall responded to them.

The enemy in Vienna

In 1785, having finished his medical studies in Vienna, Franz Joseph Gall (1758–1828) opened a medical practice for well-to-do clients in the Austrian capital. Having always been a nature lover, he also spent considerable time studying individual and species differences. These two interests led him to write about a new way of viewing patients in a planned two-volume work, of which only the first volume was published (Gall 1791). This often-neglected work contained the first hints of what would become his revolutionary theory about the faculties of mind and the functional organization of the brain (Finger and Eling 2019).

Gall had been inspired by the ideas of the minister and philosopher Johann Gottlieb Herder (1744–1803) and, more specifically, his book *Ideen zur Philosophie der Geschichte der Menschheit* (1784–1791; see Finger and Eling 2019; Lesky 1967, 1970). A second source guiding his new way of thinking about individual differences, and the brain was of a girl identified only as Bianchi, who at age five exhibited a rare talent for beautifully singing songs she had heard just once or twice without any special training, yet who was ordinary in other ways (Eling, Finger, and Whitaker 2017). It occurred to Gall that having a special talent for music and not being special in other domains was incompatible with the traditional view of a general memory faculty. Rather, we must have special faculties for different kinds of memory. Thus, Gall reasoned that perception, judgment, and memory are not primary faculties, as had long been contended, but instead are secondary features of other, more specific, practical faculties, such as one for music and another for mathematics. This insight led him to develop a new psychology or philosophy of mind, although he chose to view it as physiology.

Gall was exceptionally skilled in anatomy, so much so that many of his contemporaries regarded him as an anatomist focused on the structure of the brain (Ackerknecht 1958). Almost everyone able to watch him dissect a brain admired his knowledge of its structures and how much he was able to reveal about the nerve pathways by dissecting from the bottom up, instead of working in the traditional top-down manner. His anatomical studies led him to the conclusion that the higher functions comprising the mind should not be associated with the ventricles or the underlying white matter. They helped focus his attention on the gray shell of the cortex, where the nerves appeared to end.

With these two fundamental assumptions—a multitude of practical faculties, and these faculties being seemingly dependent on the integrity of the cortical mantle—Gall took the logical next step. The different faculties of mind, he reasoned, must be dependent on separate territories, most involving the cerebral cortex. Here he drew from some boys he knew from his youth and afterward while in college. Those who excelled at memorizing verbal material had large, bulging eyes. He theorized that this was because of the growth of the brain just behind the orbits. And, if one mental faculty could be localized by cranial features, so could others.

Gall began to lecture about his new way of viewing the mind and brain in 1796. He did this from his fashionable Vienna home, in part because he did not have an academic position, but also because he kept his collection of skulls and casts from humans and various animals there. Skilled at lecturing and with new things to reveal, he attracted broad audiences from the region (e.g., physicians, students, clerics, and state officials) and visitors from other parts of Europe.

In 1798, he finally felt ready to publish his theoretical framework. Living in conservative Vienna and knowing his doctrine might stir controversy, he chose to publish his "credo" in the *Neue Teutsche Merkur*, a German periodical (Gall 1798). Here, he explained the basic assumptions of his theory and his methods, including why he favored studying the skulls of unusual or exceptional people (e.g., notorious criminals, the insane, and great musicians and artists). Nevertheless, he did not present or elaborate on his growing list of individual faculties of mind in his synopsis, which elicited a published response from Austrian censor Joseph von Retzer (1754–1824), who encouraged him to continue his research program.

Early in 1802, Gall received a brief note penned on Christmas Eve prohibiting him from further lecturing. The order had been approved by Francis II, the Holy Roman Emperor (1768–1835), who had been convinced Gall was promoting dangerous, destabilizing ideas. In brief, his theory linking the mind (read soul) to specific brain areas seemed materialistic, deterministic, and fatalistic. Not only was this an affront to religion but, by denying free will, common criminals could claim innocence, being merely victims of their physical brains. At the same time, the Emperor was also asking his ministers whether Gall had obtained permission to lecture, and if he might have been breaking the law in any way.

Gall disagreed with these accusations and, as would be typical for him, wrote a lengthy, assertive letter detailing why they should be discarded (see Finger and Eling 2019). He also requested a committee of knowledgeable men be appointed to evaluate the charges against him. To a large extent, the witnesses invited to express their opinions agreed with Gall that he had broken no laws and did not do away with free will. Nevertheless, an imperial order was issued in March 1802, stating that lecturing outside of the university was no longer allowed without permission.

Understanding the forces he was up against, convinced that the atmosphere was not about to change in Vienna, wanting to become better known, and seeing a need to obtain more feedback before publishing a book on his new doctrine, Gall made an important decision. He decided to leave Vienna voluntarily on a lecture and fact-finding tour through several German cities. He began his scientific journey through parts of what is now Germany early in 1805, not expecting that it would terminate in Paris in 1807 and that, instead of returning to Vienna within a year, he would never see the Austrian capital again (Mann 1984).

As revealed in the periodicals covering his journey, Gall met with mixed audiences. Whereas most physicians and lay people listened to him enthusiastically, and although he was showered with money and luxurious gifts from noblemen and women, there were also skeptics and some who rejected his anatomy, his skull-based doctrine, or both at the same time. Among other things, his rejection of metaphysical forces offended some deeply religious people and some philosophers; philosophies of soul still dominated thinking about the mind and higher-order behaviors. Moreover, Gall was not always polite to those who disagreed with him. Convinced he was always right and his critics always wrong, he had a short temper and often a disrespectful tongue.

With this in mind, let us examine how four of Gall's major critics responded to what he covered in his lectures and anatomical demonstrations in different German cities. These eminent men are Johann Gottlieb Walter (1734–1818) in Berlin, Henrik Steffens (1773–1845) in Halle, Jakob Fidelis Ackermann (1765–1815) in Heidelberg, and Samuel Thomas Soemmerring (1755–1830) in Munich. There were, of course, other critics, some (along with his supporters) leaving revealing comments on paper in their letters, essays, and books (see Finger and Eling 2019).

Walter in Berlin

Gall began his lecture tour in Berlin, where he encountered his first opponent, anatomist Johann Gotlieb Walter. Garlieb Merkel (1769–1850), a Lithuanian historian and writer then acting as editor of the *Freimüthige*, a Berlin newspaper, gave a detailed overview of the "combat" between Gall and Walter in his periodical (Merkel 1805a). Merkel mentioned Gall had received an overwhelmingly positive response to his lectures, Walter being the sole dissonant. Walter had spent a lifetime studying anatomy, starting in Berlin, earning his degree in Frankfurt an der Oder, and afterward returning to Berlin, where for years he reigned supreme (Figure 1). He considered himself an authority on the subject, and *the* authority in the region, having been appointed professor of anatomy in Berlin more than three decades earlier, in 1774 (Pagel 1896; Walter 1821).

Figure 1. Johann Gottlieb Walter (1734–1818) (with permission from Charité Universitätsmedizin Berlin Fächerverbund Anatomie").

After listening attentively and taking copious notes, he lashed out at Gall. Merkel wrote: "Behind him [Gall] stood the *Geheimrat* [Privy Councillor] Walter, who had made anatomical preparations for the museum." Merkel further noted how Walter "was often aroused, frequently his face expressed admiration, and sometimes he shook his head" (Merkel 1805b).

Walter (1805) published his criticisms in a 30-page pamphlet, titled *Etwas über Herrn Doctor Gall's Gehirn-Schädel-Lehre* (*Something About Mr. Doctor Gall's Brain-Skull Theory*). In it, he laid out his own opinions about the brain, which he maintained contradicted what Gall had been contending. Although we were unable to get a copy of Walter's rare pamphlet, its content can be inferred from Merkel's (1805a) extensive discussions about what it contained. His synopsis appeared across several issues of the *Freimüthige*.

Merkel began his review in the May 20 issue. After telling his readers how he did not like to perform this chore, he wrote that Walter was an anatomist with many years of experience, and therefore deserved to be respected. Nonetheless, Merkel went on to write, Walter's pamphlet was little more than a mixture of frequently rejected accusations and objections, of lewdness and bad jokes, of claims without justifications, and of statements that simply were not understandable. Because the small pamphlet was laid out so poorly, Merkel felt he had to comment on it on a page-by-page basis. His bias was decidedly negative but, as he put it, he was leaving final judgments to his readers.

An anonymous reviewer in the *Edinburgh Medical Journal* shared Merkel's opinion (Anonymous 1806). He covered not only Walter's pamphlet but also a book by Christian Heinrich Ernst Bischoff (1781–1861; 1805), which had an attached piece by Christoph Wilhelm Hufeland (1762–1836); the two esteemed physicians had also attended Gall's lecture-demonstrations in Berlin (see below). Bischoff and Hufeland focused on Gall's skull theory, as contrasted with Walter, whose gaze was more on his own specialty, the anatomy of the nervous system. The reviewer did not feel a need to devote a lot of space to Walter's book in his 12-page review. Although he covered the Bischoff–Hufeland book in considerable detail (they were intrigued by Gall's doctrine of mind and brain, but saw a need for more research), he devoted only a single paragraph to Walter's pamphlet, which was decidedly unfavorable (Anonymous 1806, 363).

The anonymous reviewer began by noting, "Professor Walter ... is very bitter against the author, commenting that 'The wrath of the venerable anatomist is sometimes quite laughable ...'" He went on to say things such as, "The pages of both these pamphlets are filled with dull attempts at wit," and he maintained that "Dr Gall is completely ignorant of anatomy ... and he [Walter] saw no such parts as were pretended to be shown." Nonetheless:

> The truth of the matter seems to be this: Gall's lectures tickled the fancy of the people of Berlin; he gave no less than six courses there: they became the universal topic of conversation in every society and in every place, from the palace, down to the post-house, and, like other much talked-of things, they became a subject of party-spirit. Professor Walter thought himself called upon to decide for the *Berlin publicum*, as he says in his title-page and preface; and he has signalized himself rather as a vain and vexed controversialist, than as a sound reasoner and formidable adversary. (Anonymous 1806, 363)

Walter not only attended Gall's general lecture series, which covered his organology and brought up some of his anatomy, he also accepted Gall's invitation to participate in his course involving autopsies and animal dissections, revealing his nervous system anatomy (Merkel 1805a). Gall separated his anatomical demonstrations from his more general lectures, as the former were designed for smaller and more specialized audiences, particularly for anatomists and physicians, and not for the squeamish. With the help of a few aspiring physicians, Walter drew up a list of 22 comments on what Gall tried to show in his specialized anatomy course. He sent it to Gall on April 22, along with a letter.

Walter's letter was quickly published in the *Freimüthige*, but not his list. In the published letter, he indicated that he was eager to learn more about Gall's ideas about the brain, and therefore had compiled a list of comments. He asked Gall whether he had understood his ideas correctly and whether he would now elaborate on selected points about the nerves, spinal cord, and brain.

As Gall saw it, Walter was blinded by archaic, old-fashioned ideas and opinions, and he pondered whether to respond to him or simply drop the matter. Some of his supporters questioned Walter's motives. As readers of a Dresden newspaper were informed, "Our primary physicians ... are convinced that Walter's essay is the result of excessive hurrying from the jealous scientist" (see Mann 1984). Others seemed to agree that Walter was envious and wanted to maintain his stature as the reigning authority on all things anatomical, so much so that he ignored what Gall was showing and seemed blind to what anyone with good eyes could see for themselves.

Although some encouraged him not to bother, Gall opted to respond. He finalized what he wanted to say while visiting nearby Potsdam on May 9. But when his written responses to each of the 22 statements were delivered to Walter, Berlin's famed anatomist refused to accept them. Walter asserted that he had presented his criticisms publicly and Gall was now obligated to do the same. Walter's comments, completed with Gall's responses, were subsequently published in the *Freimüthige* (Merkel 1805a).

Because Walter's list was based on Gall's anatomical demonstrations, his 22 items did not pertain to Gall's organology. Moreover, some of Walter's anatomical concerns were poorly worded, and many required a good knowledge of the then-current state of brain anatomy, especially the changes taking place between 1750 and 1800. For these reasons, we will not devote more space to Walter's 22 issues and Gall's responses to them. But before leaving Walter, we should state that we also examined a book published in 1821 by Friedrich Walter (1764–1826), the famed anatomist's son. This work dealt with the "Old Art of Painting," and it also described the life and works of his father. Here, the younger Walter presented his father's view (and, in fact, his own, as he was also a physician) on the anatomical structure of the brain, the nerve pathways, and the blood supply of the brain. In some 30 statements, he also elaborated on developmental changes, sensation and movement, the body, the soul, and life. What is interesting here is reminiscent of Sir Arthur Conan Doyle's *Hound of the Baskervilles*, one the most acclaimed of his Sherlock Holmes short stories. Like the malicious dog that failed to bark when expected, the younger Walter's book contained a notable omission. Notably, he did not include how his father had criticized Gall's research program, and especially his anatomy, some 16 years earlier.

Steffens in Halle

Gall encountered his second "enemy" in Halle. This man was *Naturphilosoph* Henrik Steffens, originally from Stavanger, Norway (Figure 2). He had studied theology and natural science in Copenhagen and became professor of philosophy at the University of Halle in 1804. He attended Gall's lectures there and lamented that Gall was making a mistake by ignoring the "higher sciences," meaning metaphysics.

In his autobiographical notes, Steffens nonetheless wrote about the strong impression Gall had on his listeners. He attributed some of his allure to his convincing presentation, and some to how he used extensive supporting materials and easily understood words:

> Gall entered the huge hall of the inn, surrounded by animal and human skulls. His lectures revealed his deep conviction, and he expressed himself with the ease of conversation. They impressed, and the comparison of the human skull with the animal skulls had something surprising. Thus, the skulls of infamous thieves were compared with those of [thieving] magpies. (Steffens 1842, 48)

Steffens felt compelled to add that the connection Gall was making between humans and animals was alarming, especially to a "deeper thinking man." More to the point, he was offended by them.

Consequently, he gave three lectures in Halle in response to Gall, which he then published in a 45-page pamphlet (Steffens 1805). In direct contrast to Walter, he limited himself to Gall's organology, having minimal knowledge about the brain. He wrote, "Gall, from his point of view, is entirely correct, but from my point of view about nature, he is incorrect in all aspects" (Steffens 1805, 8–9). And he continued by maintaining, "Gall's system is a hypothesis and like all other hypotheses an incorrect one" (1905, 19).

Steffens explained in his first lecture that one can distinguish features in animals just as can be done in humans. However, he continued, "it is the main law of natural philosophy to only look at things as they are, without reasoning, without an hypothesis, only in this

Figure 2. Henrik Steffens (1773– 1845; public domain).

way and without splitting the absolute harmony of the Whole that was not split by nature"
(Steffens 1805, 18). It logically follows from this philosophical premise that one cannot
study the brain without also looking at the entire body. And, according to Steffens, there
were no organs for the soul other than the entire body.

In his second lecture, Steffens examined opposing factors, bringing up the ancient idea
linking thinking to the brain and less noble functions to the heart, and so on. He linked
these older ideas to blood circulation and breathing, sleeping and wakefulness, and
orienting toward the outside world or not. He expanded on this theme by arguing that
there are principled differences between animals and humans, saying, "For sure, no organ
acts in an animal in the way it acts in a human being. An animal may see better, but it
never sees the harmony; it may hear better, but it never hears the relations among tones"
(Steffens 1805, 26). In addition, he maintained that each sense has two aspects: sponta-
neity, which is special for an action, and receptivity, which is universal. In combination,
they allow for sensible actions, but these cannot be separately observed or associated with
specific organs.

Steffens began his third lecture with the question: What is the world? His answer
implied that everything is connected: "No part of an organism has reality of its own, it
only becomes reality in the Wholeness" (Steffens 1805, 32). Along with his holistic
approach, he now elaborated on the principle difference between animals and human

beings. In brief, only humans possess free will. Consequently, Gall, who was studying lower animals and was drawing conclusions from them that pertained to brain organization and human behaviors, had to be wrong (1805, 37). Quoting Steffens, "But it is a fact that in man there is something that separates him from animal violence; we find in him higher organizations, freedom, truth, ethics" (1805, 40). Freedom is where there is integration with wholeness.

Thus, Steffens argued for a holistic approach toward explaining human behavior. He also contended that interactions with the world need to be considered simultaneously. Given these premises, Gall was misguided in his quest to identify separate organs responsible for specific faculties. And most importantly, humans differ from animals in that they have freedom: They are not slaves of "receptivity."

Der Freimüthige, which had been supporting Gall, came down hard on Steffens, with its writer describing his philosophy as "shallow" and an insult to one's intelligence. This anonymous journalist explained that Steffens was merely "touting phrases, which claim the imagination and vanity of the listener" (Mann 1984, 99). The anatomist Justus Christian von Loder (1753–1832), from Jena, also defended Gall, adding additional details about Steffen's lectures:

> The day after Gall had finished his lectures, Steffens immediately presented [a lecture] in the same room, after he had given free tickets to all the students who had heard Gall, and he also invited the professors. Indeed, he did this to have Goethe as a listener. But he [Goethe] was very wise as to avoid him and to return to Lauchstädt the evening before. I have not heard him [Steffens], but Wolf and others have been there two times *honoris causa*. Three times he lectured from 6:00–7:00. He illuminated the doctrine of the skull in a natural-philosophical manner and rejected it completely, even finding it immoral.

Further:

> The man [Steffens] presented himself, unrequested, as the teacher of the entire Halle. Only a so-called "Naturphilosoph" could have such arrogance and impertinence! He produced strange things, in a language that nobody understood, but our students were amazed and highly praised him. It is funny to hear this man speak in the most decisive tone about all kinds of things, of which he understands nothing at all. (Ebstein 1920, chap. 20.)

That a newly installed professor of philosophy would feel offended by Gall, who dispensed with all metaphysics, is hardly surprising. This was not the first time and would not be the last time philosophers would disagree vehemently with Gall. It should be noted that a major obstacle in the dispute between the moral philosophers and Gall had to do with the interpretation of the notion of a mental faculty (Spoerl 1936). Whereas philosophers had long interpreted a faculty as an abstract capacity of the mind or soul, such as perception or judgment, Gall was thinking more concretely about specific abilities for making music, learning a language, or taking care of one's young. To Gall, these were capacities that could be associated with specific "organs" in different parts of the brain. As he repeatedly pointed out, with focal brain injuries and many brain disorders, only some functions, and by no means all, would be affected.

In addition to these opposing views on faculties, Steffens was repeating the concerns expressed by others before him about Gall's new science being materialistic. As we have seen, this form of criticism began and was dismissed by Gall even before he left Vienna. Although Gall continued to address such concerns as he made his way through the

German states, he was never able to convince everyone that his science was not materialistic, and consequently did not do away with free will, as feared by some conservative clergymen and intransient moral philosophers.

Ackermann in Heidelberg

Walter was an outdated anatomist and Steffens a metaphysically oriented moral philosopher with limited knowledge of the structure and function of the nervous system. As such, their jabs and attacks could be dismissed as producing no lasting scars. Gall's next two opponents, however, were different. Both Jakob Fidelis Ackermann (Figure 3) and Samuel Thomas Soemmerring were well informed about the brain, and both were highly respected by their fellow physicians for their basic insights and influential writings. In fact, both men were involved in studies looking into differences between individuals with respect to behavioral characteristics and anatomical features, much like Gall himself. One might imagine how, had circumstances and personalities been different, they might have looked upon Gall, also a German by birth, as a brother perusing nature for the similar truths. Soemmerring was older than Ackermann, and he was his most important teacher. But we will start with Ackermann, because we have opted to follow Gall chronologically as he journeyed from city to city through the German states.

Considered one of the leading anatomists of his day, Ackermann was born in 1765 in Rudesheim, near Mainz. He studied in Cologne, Würzburg, and then Mainz under

Figure 3. Jakob Fidelis Ackermann (1765–1815; public domain).

Soemmerring, a giant in anatomy and anthropology. Soemmerring allowed him to use his extensive library and to make numerous preparations at his house, which became a center for the *Aufklärung* (Enlightenment) in the region. Soemmerring was best known for his work on the comparative anatomy of Europeans and Africans, and Ackermann focused on differences between men and women. Both topics were of considerable importance for Gall. So was Ackermann's contention that comparative anatomy by itself was insufficient to understand how the body functions—that one must also take physiology into account.

After finishing his dissertation in 1788, Ackermann embarked on a study trip, visiting Johann Peter Frank (1745–1821) in Pavia, as well as other leading figures in medicine. On his return, he took a position at the University of Mainz, which closed during the French occupation. In 1804, Ackermann then became professor of anatomy at Jena, and a year later accepted a professorship in Heidelberg. This was where he and Gall would battle, and where he would continue his reign until his death in 1815.

Ackermann first laid out his criticisms of Gall's doctrine in a book of almost 200 pages that was probably completed in September 1805. This was before the two men had their public dispute in Heidelberg. The latter took place in 1807, after Gall had returned to the German states following lectures and demonstrations in Denmark, the Netherlands, and Switzerland, but before he entered France. Ackermann's earlier book was titled *Die Gall'sche Hirn- Schedel- und Organenlehre vom Gesichtspunkte der Erfahrung aus beurtheilt und widerlegt* (*Gall's Brain, Skull and Organ Theory, Evaluated from an Empirical Point of View and Rejected*). In its first part, Ackermann dealt with Gall's brain anatomy; in the second, his skull theory; and in the third, his chosen organs, first more conceptually and then individually. He was organized and argued impressively, expressing his concerns in a large number of separate paragraphs that could be looked upon as sections.

Following Ackermann's plan, and therefore beginning with the anatomical section, we can see that he was not one for mincing words. Early on he wrote, "That especially the cortex is not the organ where the nervous system ends as Gall believes" (Ackermann 1806, 20), and that we err "when we trace the nervous system beginning from the spinal cord into the skull" (1806, 21). These principles, based on dissecting brains from the bottom up, which would be more revealing of the various nerve pathways, are now considered to be among Gall's more significant anatomical contributions, but Ackermann viewed them as heretical. Further reading suggests that many of Ackermann's objections in the anatomical section of his book stemmed from his erroneous belief that the brain does not arise and develop from the spinal cord but from the circulatory system.

Ackermann addressed an extremely intriguing subject later in this section, writing:

> Mr. Dr. Gall thinks that all organs of animal life are double, and this is mostly true, and has not been denied, although I would except the epiphysis (*Zirbeldrüse*), the aqueduct (*Trichter*) and the pituitary gland. But he believes that these double organs do not always serve the same function and that one can, as he expresses it, [act] as a reserve while the other is active, and he seems to be wrong in this. (Ackermann 1806, 40)

To the best of our knowledge, Gall's ideas on hemispheric asymmetry have not been seriously examined, and it is not our intention to do so here. But clearly, the matter of symmetry was raised by Gall, and it could explain why unilateral brain damage might, or might not, have certain effects. Ackermann was obviously a believer in hemispheric symmetry: the two sides of the brain having precisely the same basic functions.

But what about Gall's more controversial skull theory? Ackermann started his section on it in a rather blunt way, contending, "Gall did not know the structure of the brain, and he knew as little of the structure of the skull (Ackermann 1806, 53)." He then turned to the rationale behind Gall's craniology—the assumption that the surface of the skull truthfully reflects bumps and depressions of the underlying brain. This is the assumption of parallelism, which Gall was promoting as dogma. Ackermann warned that, although this might be correct in general, it might not hold for all parts of the skull and perhaps not for every individual. He gave two main reasons for deviations from this parallelism. The first involved the muscles that draw the bones upward or downward, and the second is the air that creeps into various areas in the brain and enlarges spaces among cells in the diploë. He then stated, "Dr. Gall's presentation of the formation of [skull] bones is entirely wrong" (1806, 55), in particular when he was maintaining that these bones "model themselves according to the form of the brain, that is, that they draw back when the brain grows, and sink, indeed thicken, if the brain mass decreases at some spot" (1806, 55).

Ackermann supported his statement that there is no tight correspondence between the surface of the brain and that of the skull by pointing to several things: his own observations regarding the thickness of skull bones at different places, variations among individuals even within a family, the skulls and brains of psychiatric patients and individuals committing suicide, and aged people, who, as Gall well knew, have thicker skulls that are not malleable and cannot accurately reflect later, underlying brain changes. Interestingly, although disputing the principle of parallelism of skull and cortex, Ackermann did not discuss the growing fashion of palpating skulls with the hope of feeling the individual bumps Gall was associating with specific talents and propensities.

The largest part (more than 100 of the 180 paragraphs) of Ackermann's attacks on Gall targeted his *Organlehre* (i.e., organology), starting with the very notion of an "organ." Ackermann claimed Gall defined his organs as the material instruments that the Creator gifted us, so we could experience the world and think in specific ways (Ackermann 1806, 92). An alternative definition used by Gall is that an organ is a part on which the mind acts in a particular activity, and therefore which is sensitive and organized for that specific influence (1806, 93). Ackermann was opposed to both definitions, first, because they do not cover all organs, for instance, the muscles and bones; and, second, because he felt they failed to explain anything. He maintained that, "An organ is the real representation of the force itself" and, "With the organ, also its function is presented" (1806, 93). In essence, Ackermann wanted to include the function (*Kraft*, or force) into the definition of an organ, whereas Gall was seemingly only referring to the anatomical or material precondition—that is, the instrument (*Anlage*).

A second topic addressed in this section was labeled functions of the brain. Ackermann claimed Gall was maintaining that the brain serves three kinds of functions: organic-vegetative, sensitive, and thinking (Ackermann 1806, 104). But the organic-vegetative function does not depend on the brain, and with regard to the sensitive function, Gall was mixing everything up by assuming that the nerve fibers leave and return to the brain. As Ackermann saw it, the activation of the brain is caused by peripheral nerves coming from the external sense organs. He implied that the process is "bottom up," and that there is no influence of the brain on the sense organs themselves.

Ackermann next turned to the notion of a general principle for sensation. By denying there is a general faculty for sensation, he argued, Gall showed how little he understood about the concept of "organism," which must be based on a perfect whole (Ackermann 1806, 114). Ackermann admitted that the localization of a general faculty for perception had yet to be discovered, but he pointed to how Soemmerring described the sensory fibers coursing up toward the ventricles.

He then moved on to discuss Gall's presumed multiplicity of organs and the evidence he provided for it. He was not willing to accept Gall's view that there are numerous organs, each with its own perceptual and memory capacities, and he formulated own his views, contending,

> the brain is given nothing but the substrate (the material) for thinking. Thought itself, the comparison of impressions, the union of the same, the emergence of the difference of ideas compared, and, finally, the repercussions upon the organs of the will, which depend upon the comparison, and the peripheral movement depend only on activity, on activation that takes place in the organ of the soul, be it in the lower (the colliculi) or in the higher (the marrow of the hemispheres). (Ackermann 1806, 123)

Having dismissed the various arguments Gall had formulated in support of multiple organs, and favoring a general faculty of mind, Ackermann now addressed Gall's claim that the nerves of the spinal cord go to the brain and fan out to the separate organs in the cortex. He argued that this was an anatomical and physiological impossibility. Moreover, he stated, it was not necessary to have a separate organ for each and every animal capacity.

He was now ready to present his critique of Gall's so-called special organs, here classified in three faculty categories: faculties that allow an individual to react to the environment, faculties that make an individual more familiar with the environment, and faculties associated with higher mental capacities. His verdict was that Gall failed to follow the rules of logic and had been deceiving himself. To cite just one example of Gall defying logic, why, Ackermann asked, did Gall favor the palm of the hand over his fingers for feeling the skull in search for bumps and their associated organs? After all, as was well known, the tips of the finger are much more sensitive than the palm (Ackermann 1806, 195).

Clearly, Ackermann's attack on Gall was much broader and more serious than the one Walter had launched in Berlin or that of Steffens in Halle. He targeted every aspect of Gall's research program and his conclusions, although he also made mistakes.

Thus Ackermann, in contrast to Gall, drew on the notion of a vital force that organizes and regulates the functions of all bodily organs, including the brain. He further maintained that the outer surface of the skull dod not faithfully reflect the surface of the brain. And as for Gall's definition of an organ of mind, he maintained it explained nothing, while warning others, yet again, that Gall was doing away with free will. In brief, Ackermann left nothing out of what amounted to an all-encompassing condemnation of Gall's methods, findings, doctrine, and conclusions in a book he published before Gall even arrived in the beautiful university town of Heidelberg.

Gall, as might be expected, was not one to overlook an onslaught, and certainly not one that was laid out in print for other physicians and the public at large. Hence, he chose to respond to every one of Ackermann's criticisms in the same way as he had done with Walter's accusations (Gall 1806). There is little need to reiterate all of Gall's comments, as most simply justified craniology and repeated the principles of his organology. The important point for us at this juncture is that, once again, Gall did not feel beaten or even injured by these criticisms. He would later write:

> Professor Ackermann of Heidelberg, whom my adversaries in Germany have adopted as their leader, and whom my adversaries in France have faithfully copied, has directed himself with a suspicious animosity against the innateness of the moral qualities and intellectual faculties. If these dispositions are innate, he said, we have done with moral liberty; our actions are inevitable, and malefactors of all kinds have gained their cause.
>
> All the objections of Ackermann turn around the same false definition of organ and I should be almost ashamed to regard them as worthy of the least attention, if they had not found so many partisans.
>
> As Professor Ackermann always continues to repeat these same objections, I am obliged to hold to the same answers. All his arguments have no other basis than this false definition: the organ is the true representative of the faculty. If the organ and the manifestation of its faculty were the same thing, and their co-existence were necessary, all the organs of animals and of man, those of automatic as well as those of animal life, would have to be continually and simultaneously in action, or an instant of cessation of the action would cause them to disappear. Where do we see any example of this in nature? Does a muscle disappear because it is inactive? Ackermann answers, that a muscle in motion is quite another muscle from that at rest. It would result from this reasoning that the same foot, according as it walks or remains immoveable, would be quite a different foot. (Gall 1835, Vol. 1, 233, 234, 239)

As implied, another fundamental point in the dispute had to do with the unity of the soul and whether there is a place from which all nervous activity is directed. In this context, Gall would later write:

> "The organization," says Prof Ackermann, "though divisible into several organs, yet offers one complete whole, in which all the organs depart from one point, and in which they must all re-unite." But, unhappily, he is obliged to concede, that the anatomy of the brain does not offer this principal point, where all the nerves of sense unite, and which transmit sensations to the organ of the soul. (Gall 1835, Vol. 1, 203)

Moreover,

> Van Swieten and Tiedemann have already remarked that a general point of union, where impressions of all sorts should arrive at once, would produce only confusion. Yet Professor Ackermann thinks, that such a union of the divergent nerves would be very possible, by means of an intermediate substance in which they should terminate; and as, according to his opinion, this might happen, he concludes peremptorily, that it is so. (Gall 1835, Vol. 1, 203)

Soon after Ackermann's book came out, Gall entered Heidelberg, and although he did not present his usual series of lectures, he agreed to a meeting at which he and his antagonist would present their arguments before an audience. An anonymous journalist for the *Baierische National-Zeitung* (*Bavarian National Newspaper*) covered the vitriolic sparring between the two men early in 1807, writing:

> During Dr. Gall's visit to Heidelberg, the habitual residence of one of his most important opponents, Professor Ackermann, there was a kind of public contest between these two famous anatomists. Although great proofs of dexterity were given, as usual nothing was decided. Dr. Gall began the dispute, presented his doctrine, and replied to some of Ackermann's objections. It was now Ackermann's turn; he appeared the following day. Like his adversary, he was equipped with all the evidence, with anatomical preparations, the only objects which Dr. Gall studied more closely. (trans. from Anonymous 1807, 535)

German neuropsychiatrist Paul Moebius (1853–1907) would later criticize Ackermann for his attack on Gall's position favoring localization of higher functions in the cortex, rather than in the center of the brain (Moebius 1905). Quoting Moebius, "What a horrible chatter German professors have formulated at that time against the bright presentations of Gall" (also see Ebstein 1924, 270).

Soemmerring in Munich

Having had his say in Heidelberg, Gall headed for Munich, where Samuel Thomas Soemmerring (Figure 4) was living and working. In 1778, at the age of 23, Soemmerring had described the organization of the cranial nerves in his doctoral work. He then became professor of anatomy in Kassel and afterward in Mainz. He had to give up his position and his post as dean of the medical faculty in 1795, when the French occupied Mainz, and he opened a private practice in Frankfurt. He subsequently rejected several academic invitations (e.g., Jena and St. Petersburg) before accepting one from the Academy of Science of Bavaria, which led him to Munich in 1804, where he also served as a court counselor.

Gall met Soemmerring in March 1807, and the venerated anatomist took notes on what they discussed. They covered Gall's discoveries and ideas, and Soemmerring's personal feelings about Gall. He did not, however, publish his notes while Gall was alive. They were not released until 1829, which was a year after Gall died and two years before

Figure 4. Samuel Thomas Soemmerring (1755–1830; public domain).

Soemmering's own death (Soemmerring 1829). Soemmering also left a diary with several entries bearing on Gall's visit to Munich (Mann 1983).

Soemmerring's diary entry for March 28, 1807 reveals what he thought about Gall and goes far toward explaining why he did not want to make his feelings public while Gall was still living.

> Dr. Gall comes with Dr. Spurzheim, hard-bodied, brazen, and coarse, and both treat me at first in far too denigrating a way, despite all assurances that they wanted to learn from me. I should see his manner of dissecting the brain. He absolutely does not to listen to my objections. I immediately expressed my doubts about his ideas of the unfolding of the brain. Frequently he is quite annoyed to find already in my work a great deal of what seemed like new discoveries to him. I assured him that I did not feel comfortable with his ideas. (Mann 1984, 174, translation ours)

Soemmerring remained the perfect, always polite gentleman in Gall's presence. He tried to engage with Gall in nonconfrontational ways, and dined with him and Spurzheim. When Gall started his course in Munich near the end of April, Soemmerring showed up in the auditorium of the Bayern Academia of Sciences along with about 80 other listeners. He also accompanied Gall on his visit to an insane asylum on May 11. And at the end the month, when Gall was preparing to leave, Soemmerring took part in a farewell party for him.

Although Gall found that "his" discoveries and ideas sometimes upset his well-mannered host, he believed Soemmerring really appreciated them. "He honored us all," Gall wrote, "as difficult as it was, and as much as it cost him to agree with all the new and irrefutable discoveries." He elaborated, adding, "often he could scarcely stand it; it was as if one should take a disgusting medicine, of whose effectiveness one is nevertheless convinced" (Mann 1984, 106).

Gall was mistaken, to say the least. Soemmerring felt that Gall did not provide a wealth of new neuroanatomical information—things he did not already know about neuroanatomy. In a reply to a letter from classical scholar and archeologist Christian Gottlob Heyne (1729–1812), head of the Göttingen State and University Library, Soemmerring brought up Gall's neuroanatomy and his cranioscopic ideas, and his stance contrasted with the views that would henceforth be expressed by the majority of Gall's critics. Most, especially those witnessing his dissections, would find Gall's neuroanatomy stronger than his cranioscopy. Perhaps because he was skilled and honored as an anatomist, Soemmerring wrote that he was not impressed by Gall's neuroanatomy and terminology. Nonetheless, he found Gall's more questionable cranioscopy intriguing.

> Gall's discovery of the organs, that is to say, the features and signs of certain capacities and mental powers on the skull, seem to me to be very important, and as far as I have examined here myself and have observed examinations, correct and true. But what I understand of his dissection of the brain, there seems to me nothing new, but only improper terminology. (Ebstein 1924, 285)

In his notes about Gall's lectures, Soemmerring was more to the point. Here he wrote that there could be little doubt about the 27 bumps on the skull that Gall had indicated (Soemmerring 1829). But he then added that Gall was unable to provide convincing evidence about the nature of the underlying organs, and that what he was claiming about how human functions could be related to those in animals also remained unproven.

Returning to Gall's anatomy, he devoted more space to how he felt about Gall's notions of the brain being unfolding tissue or being like membrane. He penned that Gall was presenting an old idea without new insights, and he wondered whether these thoughts were even consistent with Gall's idea of separate organs.

Thus, on one hand, Soemmerring seemed to endorse Gall's observation that there are bumps that could be seen and felt on skulls. But on the other, he was not convinced of the functions Gall associated with the organs underlying these bumps—and he was clearly unimpressed with Gall's neuroanatomical discoveries and insights.

For his part, Gall was less than pleased with many of the opinions Soemmerring was willing to share with him in person. And, being Gall, he could not contain his emotions. Instead, he expressed his strong feelings in ways that seemed brash and even offensive to Soemmerring, who remained his considerate and overly polite host (Mann 1983, 175).

Discussion and conclusions

From the start, Gall knew that his doctrine might be thought of as materialistic. "Hardly had I obtained any results from my researches," he would later write, "when I foresaw the objections touching materialism, fatalism and the irresistibility of actions" (Gall 1835, Vol. 1, 274.). These objections based on religious orthodoxy stimulated the Holy Roman Emperor to curtail his lectures in Vienna near the end of 1801.

As he would continue to do in his lectures elsewhere in Europe and in his two sets of books (Gall 1822–1825, 1810–1819), Gall argued that his approach to mind and brain was consistent with what revered philosophers and physicians had been long contending. He even cited the Greek philosophers, writing, "it has been held, even from the time of Alkmaon, that a favourable organization of the head and brain is an indispensable condition to the favourable manifestation of the mental powers" (Combe and Combe 1838, 315). Feeling it was important to bring Church history into the picture, Gall also mentioned mentioned how Saint Paul (c. 5–c. 67) and other pillars of the church, including Saint Augustine (384–430), understood that brain parts could, in fact, be associated with the human soul.

Quoting Gall,

> It leads no nearer to materialism when, instead of maintaining the dependence of the mental manifestations on the whole body (which has been demonstrated to be true by the foregoing facts and authorities), we limit the proposition to the brain, as being the special organ of the mental functions; a proposition on which also all physicians and philosophers are long since agreed. (Combe and Combe 1838, 316)

Furthermore,

> Having shown, by irresistible evidence, that the mental functions are essentially different from, and independent of one another, I draw the inference, that the whole brain cannot be regarded as a single organ, but that its entire mass is composed of as many distinct and independent organs, as there are different, independent, and particular mental qualities. Does this truth lead any nearer to materialism than the facts already stated? (Combe and Combe 1838, 316)

In other words, if it is not materialistic to think that mental processes must involve the brain, why is it now materialism to contend that many specific mental processes might depend on the integrity of many different parts of the brain? Throughout history, the basic idea of many faculties of mind had been sanctioned by the church. "The Jesuits, besides recognising the well-known distinctions between Memory, Judgment, and Imagination, taught likewise the existence of three different kinds of memory, memoriam realem, localem, verbalem, which have been established also by experience," Gall would contend (see Combe and Combe 1838, 318).

Gall continued his defense by mentioning highly regarded anatomists, physiologists, and clinicians who were not attacked by the clerics in this same context. Herman Boerhaave (1668–1738) and Gerard van Swieten (1700–1772), two physicians who helped shape Viennese medicine and who were greatly admired by Holy Roman Emperors, were on his list, and so were numerous others:

> [Even] Sömmering, in his Hirnlehre und Nervenlehre, 1791, § 83, says that it is not improbable that certain kinds of ideas arise in determinate parts of the brain; that certain mental functions are executed in determinate parts; in short, that these different powers appropriate to themselves different provinces of the brain. The same sentiments are expressed by Schelhammer, Willis, Viq D'Azyr, Glaser, Hocboe, Lanzisi, Morgangni, Schmit, Reil, Blumenbach, Cuvier, Platner, Tiedemann, Metzler, Herder, in short, by every celebrated and able anatomical, physiological, psychological, anthropological, and metaphysical author. Consequently, my fundamental principles have at all times been expounded by the greatest men, without any one having ever, on that account, become alarmed on the subject of materialism. (Combe and Combe 1838, 320)

At a hearing, as requested by Gall in his 1802 letter to the government of Lower Austria, various experts presented their opinions about these charges (for an English translation, see Combe and Combe 1838). In general, Gall's counter-arguments against the charges of materialism and fatalism were considered acceptable (Finger and Eling 2019). Nevertheless, as we now have shown, charges of materialism and fatalism against Gall continued to be raised by well-educated philosophers and esteemed anatomists, and not only the clergy or worried politicians, as he made his way through the German states.

In more recent times, Gall has been repeatedly portrayed in the popular press and in illustrations as a madman or a charlatan, sometimes even as a fraud bilking coins from unsuspecting, gullible people. Gall's tour through the German states and bordering countries argues strongly against these disparaging caricatures. Instead, Gall strove to be regarded as an astute student of nature, including human nature, and as a benefactor of humankind.

How Gall was received in Berlin, Jena, Heidelberg, Munich, and elsewhere during the first decade of the new century shows that he was highly regarded and appreciated by most medical and nonmedical people, even though he had some vocal critics. Importantly, although the latter disagreed with what he was claiming when dissecting human and animal brains, with his contention that the mental faculties are dependent on the integrity of specific parts of the cortex, and with his defense of craniology, these men did not consider him a charlatan, a fraud, or an imposter driven by avarice.

In this context, what each of the four critics we singled out had to say, as well as what they were defending, bears repeating. Walter was a senior anatomist with many outdated opinions about the brain and nervous system, who was motivated to show everyone in Berlin that he was still the authority on anatomy. Steffens, in contrast, was a moral philosopher disturbed by how Gall dismissed metaphysics, the subject he had just been hired to teach in Halle. His contention that Gall's doctrine was materialistic and fatalistic can be viewed as an attempt to bolster his image as a learned philosopher, both at his university and in the community.

Ackermann was a far more serious critic. As an esteemed anatomist, he brought up some notable flaws in Gall's thinking about nerve and brain anatomy, while also providing some reasons for rejecting his skull-based doctrine (e.g., that the outer surface of the skull does not reflect the detailed topology of the underlying brain parts). Additionally, his holistic philosophy and other thoughts about the mind resonated well in Heidelberg, where he reigned supreme and was not about to let his turf be challenged by anyone else. Finally, there was Soemmerring, another highly regarded anatomist who was put off by Gall's manners but, more importantly, by Gall's claims of priority for certain anatomical discoveries and insights, understandably because some were his own.

There are several commonalities here. First, all four of Gall's most visible opponents in the German states wanted to defend their positions and reputations—in effect, to maintain their established turfs against a brash challenger with new ideas. Second, several continued to promulgate the fear that Gall was promoting a soulless doctrine that was materialistic and fatalistic, despite how well he had been defending himself against such charges. And third, although disagreeing with Gall about certain features of his anatomy, his numerous faculties of mind, or his functionally distinct and independent cortical organs, none considered Gall a fraud or a charlatan. Even Ackermann, who criticized him on practically everything he was promoting, took the time and trouble to analyze Gall's contentions in exquisite detail, and to formulate his objections in a debate and for the ages in writing. Stated more positively, each of these revered and venerated men responded to Gall as a serious thinker, a talented anatomist, and an acute observer of nature. Had he been viewed otherwise, they would have ignored him or brushed him off as a person of no importance.

How Gall's enemies responded to him after he left Vienna in 1805 but before he resettled in Paris two years later speaks volumes about the man and what he was promoting at this time. Without question, he was brash and often lacking in social skills. But at the same time, he had many new ideas, some extremely insightful (e.g., specialized cortical areas), also others (e.g., his reliance on craniology) that would not withstand the test of time.

Importantly, what Gall's critics challenged during the opening decade of the nineteenth century was not his objectives or integrity. Instead, they focused on his methods, interpretations, and conclusions—including those that deviated from their own religious beliefs—and on his priority claims. Portraying Gall as a greedy charlatan preying on gullible minds, rather than as a serious naturalist, scientist, physician, and scholar, would be a later, unfortunate development, one that deviates from who he really was as student of nature and a researcher most of all craving praise and greater recognition from his peers.

Disclosure statement

No potential conflict of interest was reported by the authors.

References

Ackerknecht E. H. 1958. Contributions of Gall and the phrenologists to knowledge of brain function. In *The brain and its functions*, ed. F. N. L. Poynter, 144–203. Oxford: Oxford University Press.

Ackermann J. F. 1806. *Die Gall'sche Hirn- Schedel- und Organenlehre vom Gesichtspunkte der Erfahrung aus beurtheilt und widerlegt* [Gall's Brain-Skull- and Organ Theory on the basis of experience]. Heidelberg: Mohr und Zimmer.

Anonymous. 1806. Darstellung der Gallschen Gehirn- und Schädel-Lehre, von Dr C. H. Bischoff ... Hufeland ... Walther ..., 1805 [Account of Gall's Brain and Skull Theory, by Dr. C.H. Bischoff ... Hufeland ... Walther, 1805]. *Edinburgh Medical and Surgical Journal* 2: 354–66.

Anonymous. 1807. Miszellen [Miscellaneous]. *Bayerische National Zeitung*, May 14, 535.

Bischoff C. H. E. 1805. *Darstellung der Gall'schen Gehirn- und Schädel-Lehre; nebst Bemerkungen über diese Lehre von Christoph Wilh. Hufeland* [Account of Gall's Brain and Skull Theory; also Remarks on this theory from Christoph Wilh. Hufeland]. Berlin: Wittich.

Combe G., and A. Combe, eds. 1838. *On the functions of the Cerebellum by Gall, Vimont, and Broussais ... also answers to the objections urged against phrenology by Drs Roget, Rudolphi, Prichard, and Tiedemann.* Edinburgh: Maclachlan & Stewart.

Ebstein E. 1920. *Ärzte-Briefe aus vier Jahrhunderten* [Letters from Physicians from Four Centuries]. Berlin: Springer.

Ebstein E. 1924. Franz Joseph Gall im Kampf um seine Lehre: Auf Grund unbekannter Briefe an Bertuch usw. sowie im Urtheile seiner Zeitgenossen [Franz Joseph Gall in a struggle on his theory: Based on an unknown letter to Bertuch etc. as well as in the eyes of his contemporaries]. In *Essays on the history of medicine presented to Karl Sudhoff*, ed. C. Singer and E. Sigerist, 269–322. Oxford: Oxford University Press.

Eling P., S. Finger, and H. Whitaker. 2017. On the origins of organology: Franz Joseph Gall and a girl named Bianchi. *Cortex* 86:123–31. doi:10.1016/j.cortex.2016.11.010.

Finger S., and P. Eling. 2019. *Franz joseph Gall. Naturalist of the mind, visionary of the brain.* New York: Oxford University Press.

Gall F. J. 1791. *Philosophisch-medicinische Untersuchungen über Natur und Kunst im kranken und gesunden Zustande des Menschen.* Wien: Rudolph Graffer. [Also see http://www.deutschestextarchiv.de/book/view/gall_untersuchungen].

Gall F. J. 1798. Schreiben über seinen bereits geendigten Prodromus über die Verrichtungen des Gehirns der Menschen und der Thiere an Herrn Jos. Fr. von Retzer [Letter on his already finished prodromus on the activities of the brain of humans and animals to Mr. Jos. Fr. von Retzer]. *Neue Teutsche Merkur* 3:311–23.

Gall F. J. 1806. *Beantwortung der Ackermannschen Beurtheilung und Widerlegung der Gall's chen Hirn- Schedel- und Organen-Lehre vom Gesichtspuncte der Erfahrung. Hrsg. von einigen Schülern des Hrn. Dr. Gall, und von ihm selbst berichtigt* [Response to the Evaluation and Refutation of Gall's Brain-Skull- and Organ Theory on the basis of experience. Published by some Students of Mr. Dr. Gall, and Reported by Himself]. Halle: Neue Societäts und Kunsthandlung.

Gall F. J. 1822–1825. *Sur les Fonctions du Cerveau et sur Celles de Chacune de ses Parties* [On the Functions of the Brain and on Those of Each of its Parts], vol. 6. Paris: J.-B. Baillière.

Gall F. J. 1835. *On the functions of the brain and each of its parts: With observations on the possibility of determining the instincts, propensities, and talents, or the moral and intellectual dispositions of men and animals, by the configuration of the brain and head.* Ed. N. Capen, Trans. W. Lewis, vol. 6. Boston: Marsh, Capen and Lyon.

Gall F. J., and J. G. Spurzheim. 1810–1819. *Anatomie et Physiologie du Système Nerveux en Général, et du Cerveau en Particulier* [Anatomy and Physiology of the Nervous System in General and of the Brain in Particular], vol. 4 and an atlas. Paris: Schoell.

Herder J. G. 1784–1791. *Ideen zur Philosophie der Geschichte der Menschheit* [Ideas on the Philosophy of the History of Mankind]. Riga: Hartknoch.

Lesky E. 1967. Gall und Herder. *Clio Medica* 2:85–96.

Lesky E. 1970. Structure and function in Gall. *Bulletin of the History of Medicine* 44:297–314.

Mann G. 1983. Franz Joseph Gall (1758–1828) und Samuel Thomas Soemmerring; Kranioskopie und Gehirnforschung zur Goethezeit [Franz Joseph Gall (1758–1828) and Samuel Thomas Soemmerring; cranioscopy and brain research at the time of Goethe]. In *Samuel Thomas Soemmerring und die Gelehrten der Goethezeit*, ed. G. Mann and F. Dumont, 149–89. Stuttgart: Urban and Fischer.

Mann G. 1984. Franz Joseph Galls kranioskopische Reise durch Europa (1805–1807). Fundierung und Rechtfertigung neuer Wissenschaft [Franz Joseph Gall's Cranioscopic journey through Europe (1805–1807). founding and justification of a new science]. *Nachrichtenblatt Der Deutschen Gesellschaft Für Geschichte Der Medizin Naturwissenschaft Und Technik* 34:86–114.

Merkel G. 1805a. Dr. Gall und der Geheimrath Walter [Dr. Gall and the Privy Counsel Walter]. *Die Freimüthige* 98:390–391; 99, 393–395; 100,397–399; 101,401–403;103, 409–410.

Merkel G. 1805b. Dr. Gall's gelehrte Reise [Dr. Gall's scientific tour]. *Allgemeine Zeitung* 115:2.

Moebius P. J. 1905. *Franz Joseph Gall*. Leipzig: Barth.

Pagel J. 1896. Walter, Johann Gottlieb. In *Allgemeine Deutsche Biographie (ADB)*, Band 41, 26. Leipzig: Duncker & Humblot.

Soemmerring S. T. 1829. Meine Ansicht einiger Gallschen Lehrsätze [My opinion on some of Gall's assumptions]. *Göttingische. Gelehrten Anzeigen* 6–7:49–64.

Spoerl H. D. 1936. Faculties versus traits: The solution of Franz Joseph Gall. *Character and Personality* 4:216–31. doi:10.1111/j.1467-6494.1936.tb02124.x.

Steffens H. 1805. *Drei Vorlesungen über Hn. D. Gall's Organlehre* [Three Lectures on Mr. Dr. Gall's Organology]. Halle: Neue Societäts-Buch- und Kunsthandlung.

Steffens H. 1842. *Was ich erlebte* [What I experienced], vol. 6. Breslau: Max.

Walter F. A. 1821. *Alte Malerkunst und Johann Gottlieb Walter's Leben und Werke* [Old Art of Painting and Johann Gottlieb Walter's Life and Works]. Berlin: J. G. Hasselberg.

Walter J. G. 1805. *Etwas über Herrn Doctor Gall's Gehirn-Schädel-Lehre, Dem Berliner Publikum mitgetheilt* [Something on Mr. Dr. Gall's Brain-Skull Theory, presented to the public of Berlin]. Berlin: Wegener.

An early description of Crouzon syndrome in a manuscript written in 1828 by Franz Joseph Gall

Stephan Heinrich Nolte, Werner Hansen, Paul Eling, and Stanley Finger

ABSTRACT

Just a few weeks before his death in 1828, Franz Joseph Gall, the father of what others would later call phrenology, wrote a letter to an unknown person, presumably a fellow physician. The manuscript describes the case of girl, 19 months of age. The girl's skull showed marked deformations consistent with what would be called craniosynostosis or Crouzon('s) syndrome by physicians today. Gall related some clinical features of her case and suggested some treatment options. This case report is particularly interesting because it is almost 200 years old, predates Crouzon's description of the syndrome by 84 years, and shows that Gall was still involved with treating patients, even in his final year.

Franz Joseph Gall (1758–1828), whose name has been associated with phrenology, a term he never liked or used, shared several common traits with Franz Mesmer (1734–1815), who discovered what was called animal magnetism, and Samuel Hahnemann (1755–1843), the founder of homeopathy. All three men were physicians of German origin, studied for some time in Vienna, lived for years in Paris, and developed ideas that made them popular in some circles during their lifetimes. Although these men saw themselves as saviors or benefactors of humankind, each had vocal critics, especially among conservative forces and those not wishing to see their own ideas—and hence, standing—diminished in the eyes of their scientific and medical colleagues. Consequently, what Gall, Mesmer, and Hahnemann were proclaiming led to intense scientific and public debates, charges and counter-charges, and even name calling, with the opposition labeling all three men quacks, charlatans, or greedy frauds.

What should not be overlooked is that these men reflected the scientific and medical cultures of their times. Moreover, although all could and should rightly be recognized for presenting pseudoscientific doctrines, given their questionable methods and faulty reasoning, the fact remains that they also made important discoveries and/or tilled the soil for others to reexamine, reinterpret, and repackage some of their findings, insights, and ideas. In the case of Mesmer, for example, although his animal magnetism theory was misguided, his efforts were instrumental in setting the stage for hypnosis, and in showing the power of what amounted to placebo effects in medicine.

There is also much to be said for Gall, who was a superb neuroanatomist and an observant naturalist with a fertile mind. He was, in fact, the first physician to make a strong case in public for cortical localization of functions in the cortex, he drew needed attention to why individuals differ from one another, and he came forth with brain-based ideas that helped shape modern psychiatry and criminology, to name just a few of the outcomes of his research program, which nonetheless had glaring faults (see Finger and Eling 2019).

Gall's life and doctrine, and even his skull collections, have been examined in several lengthy articles and monographs (i.e., Ackerknecht and Vallois 1956; Finger and Eling 2019; Lesky 1979; Oehler-Klein 1990). Hence, only a brief synopsis as background to the letter he wrote in 1828 is needed.

Gall (Figure 1) was born in Tiefenbronn, in what is now southwest Germany. With an early interest in the wonders of nature, he first studied natural history and medicine at Strasbourg, but he completed his medical degree at Vienna. Staying in the Austrian capital, he opened a private practice and also pursued his scientific activities correlating mind, brain, and behavior to a large extent on exceptional cases, such as the men and women in the city's impressive new asylum and those in local jails. His basic ideas were that there are many innate faculties of mind (for music, mathematics, color, etc.), that each is associated with a distinct part of the brain, that the sizes of these organs will affect

GALL.

Figure 1. Franz Joseph Gall, Lithography by F. Gutsch, in Lewald A. 1835. *Europa, Chronik der gebildeten Welt*. Karlsruhe: Gutsch & Rupp (Property of S. H. N., reproduced with permission).

the developing skull's morphology, and that the skull's bumps and depressions can be correlated with behavioral propensities and character traits. The latter feature was especially important, as it opened his theory to anyone wishing to examine himself or herself, other people, and even animals.

Gall first presented his revolutionary doctrine linking mind, brain, and behavior to the public in writing in 1798. But he then spent years gathering more evidence for it and fine tuning its features before publishing his mature theory in his "great work." He settled on 27 faculties and associated cortical organs, the majority of which humans have in common with higher animals, locating the organs of those unique to humans in the front of the brain. This was his *Schädel-Lehre*, or *organologie* in French, the subject of two multi-volume sets of books he would publish in Paris following his lengthy but productive and largely rewarding Continental tour, which ended in 1807. Johan Gaspar Spurzheim (1776–1832) joined him as an assistant in Vienna in 1804 and accompanied Gall on his scientific tour.

The first of these sets came out serially between 1810 and 1819 and was titled *Anatomie et Physiologie du Système Nerveux en Général, et du Cerveau en Particulier* (Gall and Spurzheim 1810–1819–1819). Spurzheim was coauthor on the first two of the four volumes and on the atlas with its 100 beautiful copper plates. The other two volumes bore only Gall's name. The less-expensive edition, his *Sur les Fonctions du Cerveau*, had only Gall's name on the six volumes, which were finished in 1825 (Gall 1822–1825) and were translated into English 10 years later (Gall 1835). This set lacked the earlier detailed neuroanatomy and the expensive atlas, although Gall continued to cite specific illustrations from his earlier atlas in its pages.

Other than a brief trip to London, Gall spent the remainder of his life in Paris, always living in fashionable quarters. He also purchased a summer house in Montrouge, near Paris. He supported his rather lavish lifestyle mainly by lecturing and seeing patients, and less by what he received for his books, the first set of which had cost him a massive amount of money to bring to fruition.

Gall died on August 22, 1828, after a series of strokes. Following removal of his skull (now in the Musée de l'Homme) his body was buried in the Père Lachaise Cemetery. His tombstone was graced by a bust that was later beheaded, and this head has since vanished.

The newly found letter from Gall

The Gall letter describing what appears to be a case of Crouzon syndrome was part of a small collection given by the heirs of an anonymous collector to a German auction house to sell.[1] It was acquired by one of the authors (W. H.) in 2019, in a session for autograph collectors. There were no other items related to it in this collection, and we could find no evidence of an earlier letter or a response.

The three-page letter is handwritten in fluid French, although there are some orthographic flaws and problems with accents (Figure 2). Comparisons of the handwriting and examination of the signature make it clear that this letter was written by Gall, who, based on its date, wrote it just before he died.

---

[1] J. A. Stargardt, Autographenhandlung, Xantener Strasse 6, D-10707 Berlin.

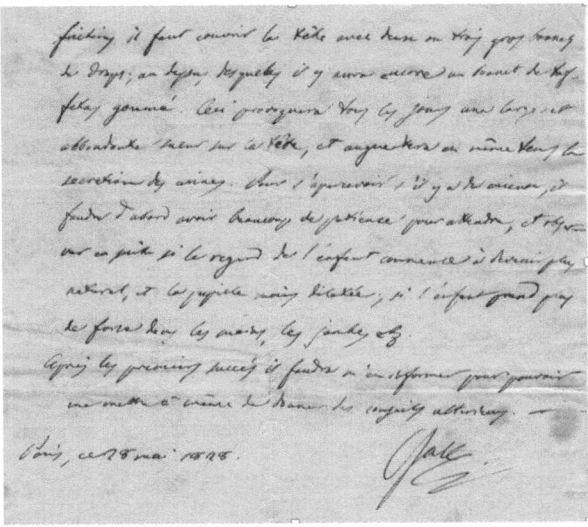

Figure 2. Page 3 of the manuscript with the signature of F. J. Gall, 1828 (Property of W. H., reproduced with permission).

The letter is almost certainly a communication from one physician to another, although it bears neither a name nor an address, making it hard to put it into a wider context. What is clear is that its subject is medical and also phrenological, which is why the case must have been of special interest to Gall, who attended to it even though he was failing physically at the time. In effect, it is a case report of a young girl with premature fusions of the cranial bones, described in the text as "soldered" sutures.

The girl is only 19 months of age, and the description is highly suggestive of craniosynostosis, or Crouzon syndrome, a disorder associated with certain facial and skull deformities. The early cranial fusions, giving the skull its characteristic features, can prevent the developing brain from growing properly, causing increased intracranial pressure. The result can be hydrocephalus internus. The two terms currently used for this disorder, craniosynostosis or Crouzon syndrome, were not a part of the medical lexicon at that time, and early-nineteenth-century physicians, Gall included, would likely have classified cases such as this one under the broad heading of hydrocephalus.

The text and its translation

In its original French, the text is as follows:

> Une fille de 19 mois, née avec une tête très-petite par derrière,
> de manière qu'à la première inspection on avait pu la prendre
> pour la tête d'un garçon, parce que l'occiput n'est pas bombé
> comme il a coutume d'être dans les têtes des filles. La partie fron
> tale, au contraire, est très bombée en avant; les tempes enfoncées;
> Les yeux gros, déprimés dans la figure; les pupilles extrêmement di-
> latées; le regard tellement incertain, qu'on a jamais pu se con-
> vaincre qu'elle fixe les objets. Du reste elle entend; elle a l'air
> quelquefois de sourire, comme si elle vous avait regardé, comme si
> elle vous avait vu. Les sutures de la tête sont parfaitement soudées.

Les extrémités supérieures et inférieures sont delicates et faibles, de ma-
nière qu´avec les mains elle ne saurait rien saisir; elle ne saurait
pas se tenir debout. Le corps et la figure sont tres bien nourris.
Quant à son intelligence, elle n`a jamais rien montré, si non
qu`elle parait prendre beaucoup de plaisir aux sons des instruments.
D´après ces données on peut conclure qu`elle a une conformation defec-
tueuse du cerveau, et plus spécialement qu`il y a un epanchement
ou une compression quelconque dans la region qui environne le nerf optique.

[Page 2] Quelque fois ces obstacles disparaissent par l`absorption qui a lieu
successivement, surtout en proportion que le cerveau se developpe,
et se consolide. Un tel changement a lieu ordinairement à l`ap-
proche de la puberté. Dans le but d`aider la nature dans
cette operation, il faut eviter tout ce qui pourrait affaiblir
l´enfant, il faut que l`enfant vive dans un air sec un peu vif,
à la campagne, eviter les endroit marecageux et entourer de
fumier. Il faut faire souvent à l`enfant des frictions seches.
Pour aider l'absorption du liquide epanché dans les cavités
du cerveau, il faut donner à l`enfant tous les jours la dose
prescrite des poudres suivantes:
Rp Calomel g. VI
Saccar. Alb. p. 3j
melez, et diviser le tout en 24 poudres, à prendre une
fois par jour dans du lait.
Il faudra aussi raser la tête à l`enfant, faire preparer un
liniment composé de jus de Genièvre et d´onguent mercuriel
parties egales. Avec ce liniment on frictionnera toute la
surface de la tête de l`enfant, une fois dans la matinée, ou
bien dans la journée; et on en prendra tout – au plus la
quantité de dix grains chaque fois: la friction sera continuée
tout doucement pendant un quart d`heure. Par dessus ces

[Page 3] frictions il faut couvrir la tête avec deux ou trois gros bonnets
de draps; au dessus desquels il y aura encore un bonnet de taffetas gommé. Ceci provoquera tous les jours une large et
abbondante sueur sur la tête, et augmentera en même temps la
secretion des urines. Pour s`apercevoir s´il y a du mieux, il
faudra d`abord avoir beaucoup de patience pour attendre, et obser-
ver en suite si le regard de l`enfant commence à devenir plus
naturel, et la pupille moins dilatée; si l`enfant prend plus
de force dans les mains, les jambes etc.
Après les premiers succès in faudra m`en informer pour pouvoir
me mettre à même de donner des conseils ultérieurs.
Paris, ce 28 mai 1828. Gall

The letter translates[2] as follows:

A 19-month-old girl, [was] born with a very small head posteriorly, in a way such that one would have taken it at first sight for a boy's head, because the occiput is not prominent,[3] as is usually the case with girls' heads. In contrast, the frontal part is very prominent with the temples sunk in, the large eyes impressed into the face, the extremely dilated pupils, and the gaze so uncertain that one could never decide if she could fixate on objects. For the rest, she hears and looks as if she smiles sometimes—as if she has looked at you, as if she had seen you. The sutures of the head are perfectly soldered.[4] The upper and lower extremities are very tender and weak, to the point that she cannot grasp anything with her hands and cannot remain upright. The body and the face are very well nourished. Concerning her intelligence,

---

[2]Our translation deviates from the grammar of the day to make what Gall was stating clearer.
[3]In French, *bombé* means domed or bulging.
[4]Gall's French word *soudée* implies more than just fused; hence, our use of the word "soldered."

she has never shown anything, other than that she seems to experience a lot of pleasure from the sounds of instruments.[5]

With these findings one can conclude that she has a defect in the configuration of the brain, and more specifically that there is a fluid accumulation or a compression of some sort in the region surrounding the optical nerve. These obstacles sometimes disappear over time by absorption, as the brain develops and consolidates. Such a change usually takes place when puberty approaches. With the aim to help nature in this endeavor, one must avoid everything that could weaken the child. The child must live in a countryside that is dry with fresh air and avoid marshes and muck. The child must often have dry rubbings. To help the absorption of the fluid distributed in the brain cavities, the prescribed doses of the following powders should be administered:

Rp Calomel g. VI[6]

Saccar. Alb. p. 3j[7]

Mix and divide all in 24 powders, to take one per day in milk.

The head of the child should be shaved, and a liniment composed of juniper juice and mercury ointment in equal parts should be prepared. Rub the entire surface of the child's head with this liniment once in the morning or during the day, using all of it or at least ten grains each time. The gentle rubbing must continue for a quarter of an hour. Apart from these massages, the head has to be covered with two or three thick cloth caps and, on top of this, another cap of rubbered taffeta.[8] This will provoke copious and abundant sweating of the head every day and will at the same time increase the urinary secretion. To see if it [she] is getting better, you will first need considerable patience to wait and see whether the child's gaze starts to become more natural and the pupil less dilated, whether the child becomes stronger in the hands, feet, etc.

You should keep me informed of your first results so I can provide more advice.

Paris, May 28th, 1828. Gall

Discussion

Individuals with congenital or early-in-life manifesting conditions with facial abnormalities, physical stunting, and intellectual disabilities were frequently called "cretins" during the nineteenth century, when Gall wrote this letter. This word derives from the French term *chrétien*, meaning a Christian, and it is still being defined as "one who is human despite deformities" (*Webster's Unabridged Dictionary of the English Language* 1996).

The medical definition of cretinism in former times covered all sorts of disabilities, especially if there were skull deformations. This is why German pathologist Rudolph Virchow (1821–1902; 1851) chose to distinguish different sorts of skull deformations along with what we would now consider to be hypothyroidism in a mid-century article that translates as "On Cretinism." Down syndrome (Down 1866) was the first condition to be differentiated as a distinct disorder under the umbrella term "cretinism." The term is now used less frequently in medical practice, and it has become more specific, signifying untreated congenital hypothyroidism.

---

[5]Musical instruments.
[6]Gran VI, six grains, 6 × 60 mg = 0.360 grams.
[7]3 sign for Drachme, which corresponds to 3.75 grams.
[8]Probably meaning a watertight rubber cap with taffetas inside.

Figure 3. Octave Crouzon (1874–1938). ©Bibliothèque de l'Académie nationale de médecine.

Virchow coined the term "craniostenosis" for a condition in which one or more of the sutures close too early. Physician, anatomist, and anthropologist Samuel Thomas Sömmering (1755–1830), whom Gall visited in Munich in 1807, identified the skull sutures as the site of cranial growth and concluded that premature suture fusion could result in cranial deformities. Scattered cases of this condition were reported in the literature, which remained fragmented until 1906, when Eugène Apert (1868–1940; 1906) first described syndromic craniosynostosis and coined the word "acrocephalosyndactyly"—a condition characterized by dysmorphic head and face shapes in conjunction with syndactyly of the hands and feet, giving them a webbed or conjoined appearance.

In 1912, Louis Edouard Octave Crouzon (1874–1938) described the features of another craniofacial syndrome without malformation of the extremities (Crouzon 1912).[9] Crouzon was born in Paris in 1874 (Beighton and Beighton 1987, 33; Roussy 1938; see Figure 3). He was proud of his modest origins, obtaining a scholarship from the town of Paris to enable him to study medicine. Among his academic teachers were Joseph Babinski (1857–1932) and

---

[9]Individuals studying human representations from Cook's Island; other artifacts; and, most recently, a pre-Columbian bowl have drawn attention to what they think might have been the same or a closely related syndrome in earlier cultures (see Deps and Charlier 2019).

Pierre Marie (1853–1940). Crouzon received his doctorate in 1904 and worked at the Hôtel-Dieu Hospital and the nursing school of the Salpêtrière in Paris, where he later became head of the Department of Neurology. Because of his organizational skills, he held official posts in the military health service in World War I and coordinated the reconstruction of the *assistance publique*, the French welfare system in the region of Paris. He was the general secretary of the Société Neurologique de Paris and editor of the *Revue Neurologique*, and he organized national and international conferences. It was only one year before his death that he received a chair at the medical faculty, the newly created chair for medico-social assistance.

Crouzon published numerous works on hereditary diseases, neuropathology, and social medicine. In 1912, he first described the syndrome of hereditary craniofacial dysostosis named after him (Crouzon 1912). But he was especially known for his interest in matters of public health: legal medicine, social insurance, and worker's pensions questions. In an obituary (Roussy 1938), his qualities were described as methodical, orderly, patient, firm, and clear. He knew what he wanted and why, he knew where to go to and whom to ask, when to speak out and when to remain quiet. He was armed for the battles of life. His friendly personality was described as warm, sensitive, kind, generous, and diligent, always prepared to help others, the poor, and the unfortunate.

Of the different forms of premature fusions, which are generally referred to as syndromal craniosynostosis, Crouzon's syndrome is now recognized as the most common

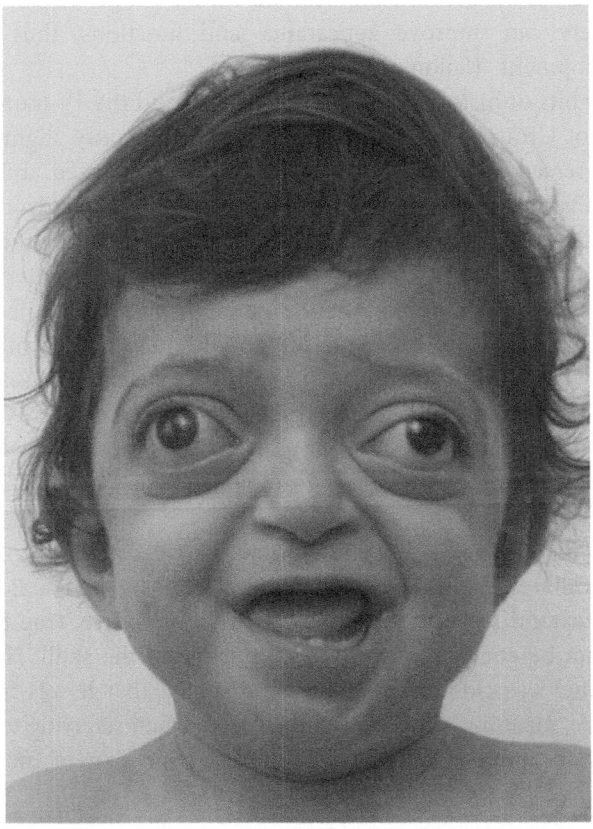

Figure 4. A toddler with Crouzon Syndrome.

(Dodge, Wood, and Kennedy 1959). It has a frequency of about 16 per 1,000,000 new-borns (Cohen and Kreiborg 1992) and is sometimes branded an orphan disease. It is characterized by an abnormal looking face with a flat occiput and wide-set, bulging eyes, hypertelorism (a large space between the eyes), and ocular protosis (the sudden protrusion of an eye from its socket).[10, 11] Some of these characteristics are detectable at birth, but the full clinical picture typically develops during the second and sometimes the third year of life, as Crouzon recognized (Figure 4).

The synostosis, meaning the fusion of adjacent bones, starts with the coronary suture and progresses to the lambdoid and sagittal sutures. Vision problems are frequent, including strabismus (problems aligning the eyes) and even lagophthalmus (inability to close the lids, due to eye protrusion). Optic atrophy is seen in about 20% of these patients. The ocular proptosis is primarily caused by retrusion of the lateral and inferior orbital margins with a very short orbital floor (Kreiborg and Cohen 2010). Intelligence is usually normal but, if there is increased intracranial pressure, intellectual disabilities can appear.

We now know that Crouzon's syndrome is caused by mutations in the fibroblast growth factor receptor 2 (FGFR2) gene (Reardon et al. 1994), either spontaneously or inherited in an autosomal dominant manner. Mutations of this gene stimulate immature cells to become bone cells during embryonic development, thus causing the bones of the skull to fuse prematurely. Signs and symptoms vary considerably in affected people, even in the same family. Treatment consists mainly of a neurosurgical procedure to keep the sutures open and to prevent complications from fluid circulation disturbances. Craniofacial surgery can improve appearance and functions, thus being helpful for psychosocial development (Solomon et al. 2011).

These facts and bits of history bring us back to Gall and the 19-month-old girl. At this age, the features of Crouzon's syndrome, although often only slightly visible at birth, have rapidly progressed, due to brain growth and the restricted ability of the skull to adapt to its expansion. Thus, the seminal features of the syndrome are now more prominent, especially the large, prominent eyes, the gaze, and other visual disorders (e.g., strabismus, amblyopia). The child described in Gall's letter seemed to show impairments in psychomotor development: She could not grab objects, stand, or walk (which normally begins before one year of age, although some children without disabilities might not walk until 15 months). Yet other malformations that characterize Alport's syndrome—such as syndactylia, which Gall almost assuredly would have mentioned—were not noted, and there is nothing in his letter about crawling or sitting. It is unclear whether this young girl was able to establish visual contact, but she seemed to hear and enjoy music.

Gall wrote that the girl had a brain malformation, consistent with his theory that the features of the skull reflect brain development. He postulated a liquid accumulation creating pressure around the optical nerve, which is, of course, true for internal hydro-cephalus but cannot be concluded from just the shape of the skull. He did not mention whether the fontanel was closed, but it can be assumed that it was from his use of the French word *soudée*, meaning soldered, tightly closed, when referring to the sutures. If the sutures remain open, fontanel bulging would be a clear sign of increased intracranial

---

[10]https://ghr.nlm.nih.gov/condition/crouzon-syndrome accessed 23.10.2019.
[11]www.orpha.net/consor/cgi-bin/OC_Exp.php?lng=en&Expert=207 accessed 23.10.19.

pressure. Gall speculated that the fluid might disappear by resorption as the brain continues to develop. As puberty signifies the time when the skull usually stops growing, he could well have been accurate in thinking that, if the child survived to puberty, there would be no further progression.

Gall's therapeutic advice is in line with how physicians were treating many disorders at this time, when there was great fear of miasmas or something harmful in the air, especially in low-lying, swampy areas. Physicians since the time of Hippocrates had recommended healthy locations, although they had no idea what was in the bad air (from which we have the term "malaria" for the mosquito-transmitted illness). Interestingly, when it came to selecting a site for the potentially curable insane, Gall, while still living in Vienna, recommended Heidelberg over another location, not just because of its university but because of its healthier climate (see Finger and Eling 2019, 370). Gentle rubbing with ointments would also have made sense, and we now know that massaging can help increase peripheral blood circulation, which could benefit weak toddlers.

For the resorption of the intracranial fluid, two evacuation measures were proposed, the first being the drug Calomel (Hg_2Cl_2). Mercuric chloride had been widely used in medicine from ancient times, but although toxic due to its low solubility in water, it is much less dangerous than other mercury-containing substances (Urdang 1948). In the early-nineteenth century, it was mainly used as a diuretic and purgative drug, and Gall recommended mixing it with sugar and giving the girl the powder once a day in milk. His prescribed dosage is low: 360 mg in 3.750 mg white sugar, divided into 24 portions, which equals 168 mg per dose.

Gall's second method to enhance resorption of the intracerebral fluid called for rubbing the girl's shaved head with a mixture of juniper juice and a mercury ointment. Both had known diuretic properties. As for the physical thermo-insolation with thick and rubber-coated cloth, the underlying idea was to induce sweating, to directly or indirectly reduce the amount of fluid in the body.

Regarding his prognosis, Gall predicted that, if those treatments worked, it would become apparent in the child's psychomotor development and sight. It is hard to tell how much confidence he had in these therapeutics, but he asked to be informed about how the girl was doing, even though he might have believed she would not live much longer.

When Peter Solomon Townsend visited him on July 5, six weeks after he wrote the letter, Gall was unable to rise, having hemiplegia of the left side of his mouth and head, and the right side of his body (Rosen 1951). He died on August 22, 1828.

Gall was a specialist when it came to cranial abnormalities. He had seen numerous other cases of cranial defects, and he had the skulls of many children with hydrocephalus in his collection. For this reason, and because the conditions giving rise to childhood cranial abnormalities were not being correlated with the finer features of the skulls at that time, he might not have been particularly intrigued by this case. But the fact that he wrote this detailed letter and asked for progress reports would suggest that it was of at least some interest to him, although perhaps more as a practicing physician than with regard to his organology.

We wondered whether cases of premature suture synostosis or syndromal craniosynostosis can be seen in the skull collections he left behind, which are now in museums. In 1824, he gifted what remained of his Vienna collection, almost 200 pieces, to Anton

Rollett (1778–1842), who opened a private museum in Baden bei Wien that still exists. As for his larger Paris collection, his widow gave it and his books to the state in exchange for a lifetime pension. First housed in the Musée National d'Histoire Naturelle, his skulls and casts were numbered, cataloged, and described in a catalog and a phrenological journal.

Gall's Paris list (Royer 1830) shows that he had some fetal and infant skulls with abnormalities, as well as many with smaller variations that he considered to be in the normal range. Item 137 is notable, because it is labeled, "Skull of a very young infant born with water in the head"; and No. 141 could, in fact, be a case of unilateral premature coronary synostosis, being, "A skull, of which the two halves are not symmetrical." Premature sagittal suture synostosis might be suspected in Case No. 159, which is described as, "Skull of a person who was both scrofulous and hydrocephalous. The longitudinal diameter is very large; yet the breadth is less than in ordinary skulls." Gall believed that the shape of this skull must have been altered by a disease of the brain and noted that this individual died in a lunatic asylum, but he also knew of several others "who enjoyed the possession of their reason." No. 250 is another infant born with hydrocephalous and very little brain, and it was the first case of this kind Gall observed. Thus, although there are some intriguing skulls in the collection, there is no skull that stands out as a clear-cut case of Crouzon syndrome.

Gall never gave up on cranioscopy, his primary method, yet the weak link in the edifice he was trying to construct. Nonetheless, the letter described in this article is more clinical than scientific, and there is little in it that bears on his theory of many independent faculties of mind, each dependent on a circumscribed cortical organ. The manuscript is written in sober, descriptive terms and hardly speculative or even controversial. This case report is important, in part because it had to be among the last ones, if not the last one, Gall would write, showing he was still very much a practicing physician as the curtain was coming down on his life. It is also important because it would seem to be the first known description by a physician of what others would call Crouzon syndrome in the next century.

Disclosure statement

No potential conflict of interest was reported by the authors.

References

Ackerknecht E. H., and H. V. Vallois. 1956. *Franz Joseph Gall, inventor of phrenology and his collection.* Madison: University of Wisconsin Press.

Apert E. 1906. De l'acrocéphalosyndactylie. *Bulletins et mémoires de la Société Médicale des Hôpitaux de Paris* 23:1310–30.

Beighton P., and G. Beighton. 1987. *The man behind the syndrome.* New York: Springer.

Cohen M. M., and S. Kreiborg. 1992. Birth prevalence studies of the Crouzon syndrome: Comparison of direct and indirect methods. *Clinical Genetics* 41:12–15.

Crouzon O. 1912. Dysostose cranio-faciale héréditaire. *Bulletins et mémoires de la Société Médicale des Hôpitaux de Paris* 33:545–55.

Deps P., and P. Charlier. 2019. A Crouzon syndrome from the classical period of Maya civilization? *Surgical and Radiologic Anatomy* 41:1525–27. doi:10.1007/s00276-019-02287-8.

Dodge H. W., M. W. Wood, and R. L. J. Kennedy. 1959. Craniofacial dysostosis: Crouzon's disease. *Pediatrics* 23:98–106.

Down J. 1866. Observations on an ethnic classification of idiots. *Clinical Lecture Reports, London Hospital, 3*, 259–62. https://web.archive.org/web/20060615010343

Finger S., and P. Eling. 2019. *Franz Joseph Gall. Naturalist of the mind, visionary of the brain.* New York: Oxford University Press.

Gall F. J. 1822–1825. *Sur les fonctions du cerveau et sur celles de chacune de ses parties.* 6 vols. Paris: J.-B. Ballière.

Gall F. J. 1835. *On the functions of the brain and each of its parts: With observations on the possibility of determining the instincts, propensities, and talents of the moral and intellectual dispositions of men and animals by the configuration of the brain and head.* 6 vols. Ed. N. Capen and Trans. W. Lewis. Boston, MA: March, Capen, and Lyon.

Gall F. J., and J. G. Spurzheim. 1810–1819. *Anatomie et physiologie du système nerveux en général, et due cerveau en particulier.* 4 vols. and atlas. Paris: Various publishers. [Gall was the sole author of vols. 3 and 4.]

Kreiborg S., and M. M. Cohen. 2010. Ocular manifestations of Apert and Crouzon syndromes: Qualitative and quantitative findings. *Journal of Craniofacial Surgery* 5:1354–57. doi:10.1097/SCS.0b013e3181ef2b53.

Lesky E. 1979. *Franz Joseph Gall, Naturforscher und Anthropologe.* Bern: Hans Huber.

Oehler-Klein S. 1990. *Die Schädellehre Franz Joseph Galls in Literatur und Kritik des 19. Jahrhunderts. Soemmering-Forschung VIII.* Stuttgart: Gustav Fischer.

Reardon W., R. M. Winter, P. Rutland, L. J. Pulleyn, B. M. Jones, and S. Malcolm. 1994. Mutations in the fibroblast growth factor receptor 2 gene cause Crouzon syndrome. *Nature Genetics* 8:98–103. doi:10.1038/ng0994-98.

Rosen G. 1951. An American doctor in Paris in 1828: Selections from the diary of Peter Solomon Townsend MD. *Journal of the History of Medicine and Related Sciences* 6:209–52. doi:10.1093/jhmas/VI.Spring.209.

Roussy G. 1938. Nécrologie Octave Crouzon. *La Presse Médicale* 86:1585–86.

Royer M. 1830. The phrenological journal and miscellany. Vol VI Andersen Edinburgh. Article III: Catalogue, numerical and descriptive, of heads of men and animals which composed the collection made by the late Dr. Gall. Presented by M. Royer, through Mr. Combe, to the Phrenological Society; and, by their permission, translated for insertion in the Phrenological journal. http://hdl.handle.net/2027/mdp.39015036634551

Solomon B. D., H. Collmann, W. Kress, and M. Muenke. 2011. Craniosynostosis: A historical overview. In *Craniosynostoses: Molecular genetics, principles of diagnosis and treatment*, ed. M. Muenke, W. Kress, H. Collmann, and B. D. Solomon, 1–7. Basel: Karger.

Urdang G. 1948. The early chemical and pharmaceutical history of Calomel. *Chymia* 1:93–108. doi:10.2307/27757117.

Virchow R. 1851. Über den Cretinismus, namentlich in Franken und über pathologische Schädelformen. *Verhandlungen der Physikalisch - Medizinischen Gesellschaft zu Würzburg* 2:230–70.

Webster's Unabridged Dictionary of the English Language. 1996. cretin. New York: Random House.

"My God, here is the skull of a murderer!" Physical appearance and violent crime

Jaco Berveling

ABSTRACT

Over the centuries, people have tried to determine character traits from a person's appearance, beginning with the physiognomic efforts of the Greek philosophers Socrates (ca. 470–399 BCE) and Aristotle (384–322 BCE) and still continuing today. In this quest, the discovery of criminal tendencies from someone's face always received special attention. This was also an important issue for physician Franz Joseph Gall (1758–1828). Gall maintained that a criminal's skull had a different shape than that of a law-abiding person. Phrenologists, as well as criminologists, including Cesare Lombroso (1835–1909), further propagated Gall's ideas and investigated countless heads of violent and petty criminals. This line of investigation led to much discussion and criticism. Were Gall, the phrenologists who followed him, and Lombroso sufficiently objective? Were these men really onto something, or were they led by prejudices? After Lombroso's time, physiognomy and cranioscopy were discredited. However, in the last decades, some researchers are again trying to find out whether people are indeed able to distinguish violent criminals from nonviolent criminals on the basis of their faces.

Introduction

It must have been quite a spectacle in 1824. French medical students had brought a skull for Franz Joseph Gall (1758–1828) to examine. Gall had a reputation for being able to "read" a person's character, based on palpating his or her skull. That's what the students wanted to see for themselves. They gathered expectantly around Gall, who looked at the skull and palpated it at all sides, and then began his analysis without a trace of hesitation:

> That's the head of an executed man; this man must no doubt have been carried into a crime by his unrestrained passions; … his character has been gloomy, and the desire to destroy has controlled him. His lusts, stimulated by loneliness and hardship, will have brought him to such a degree of furious excitement that all means of satisfying them were welcome, but above all the means of murder. (Anon. 1872, 90)

The students listened open-mouthed to his account. They had hoped to teach the arrogant doctor a lesson. The head they had acquired belonged to a criminal named Antoine Léger (1795–1824), who had recently been executed. He had kidnapped, raped, murdered, and eaten a 15-year-old girl. Everything Gall mentioned was consistent with the life of the criminal (Anon. 1825). To support his judgment, Gall pointed to the killer's low forehead as well as his strongly developed back of the head.

Gall's analysis of Léger has been described in a popular phrenological work that appeared in German in 1858 (Seidel 1858) and was later translated into Dutch (Anon. 1872). How reliable the description is cannot be determined with certainty. We do know from other sources, however, that Gall did indeed examine Léger's skull (Burgess 1858).

Gall's interest in the appearance of criminals fits into a long tradition. Greek philosophers already assumed a relationship between appearance and bad behavior, as did Italian scholar Giambattista della Porta (1535–1615) and Swiss minister Johann Kaspar Lavater (1741–1801). And even after Gall, countless numbers of nineteenth-century phrenologists, amateurs, and criminologists—the most notable being Cesare Lombroso (1835–1909) in Italy—tried to infer criminal tendencies from the shape of the cranium and facial features. Even in recent years, scientists have taken up the issue again.

The first physiognomists

Gall was not the last, but certainly also not the first, to establish a relationship between appearance and crime. Greek philosophers Socrates (ca. 470–399 BCE) and Aristotle (384–322 BCE) argued that faces in particular had much to say about an individual's character, and they acted accordingly. For instance, Socrates only allowed students to study philosophy if, by studying their faces, he was sure they were suitable for the study. And Aristotle advised that, if you were in doubt between two people, you should always choose the person with the best physical characteristics and stay away from people with bad facial features. In his *Historia Animalium* (History of Animals), Aristotle devoted several chapters to the art of physiognomy and how faces should be interpreted (Aristotle 1908; Finger and Eling 2019, 50–52). In *Physiognomica*, a treatise attributed to Aristotle, the major premises are that the character of animals is revealed in their form, and that humans resembling certain animals possess the character of these animals (Todorov 2017).

During the sixteenth century, Italian scholar Della Porta suggested that crime and appearance have something to do with each other. To find out more about this, he visited prisons. After studying executed criminals he concluded that people with small ears, bushy eyebrows, small noses, and big lips are more likely to commit crimes (Della Porta 1618).

Physiognomy never died, and during the late-eighteenth century it was bolstered by the writings of Swiss minister Johann Kaspar Lavater. In the period 1775–78, his masterpiece *Physiognomische Fragmente zur Beförderung der Menschenkenntnis und Menschenliebe* (Essays on Physiognomy, Designed to Promote the Knowledge and the Love of Mankind) was published (Lavater 1775-1778). In his *Fragmente*, Lavater made it clear that the appearance of a person reveals much about his or her inner self. He especially valued the face. He thought the shape of the head was important, but he also contended that you could read a lot from eyes, ears and noses.

Lavater looked at noses, ears, and eyes, but he insisted that a good physiognomist should also be guided by the "bones of man's body." When examining people, Lavater relied more on the hard parts (the bones) than the soft ones (skin and muscle tissue). After all, the hard parts are immutable. The skull he, believed, could be particularly informative. This was an idea Gall would develop, bringing the brain into the picture.

The Swiss minister wanted to provide his physiognomy with a scientific basis. This scientific basis came mainly from a huge assortment of portrait drawings, silhouettes, masks, and skulls. In his eyes, a face reader had to build a collection of the "most remarkable faces." Skulls and plaster prints of faces were important study materials. You could measure them from all sides and determine the proportions with a "mathematical precision." Lavater thought one should always measure in the same systematic way. To do this accurately, he devised a tool to allocate the outlines of the head. With this head measurer, the dimensions of human skulls could be precisely determined.

Lavater hoped to promote "human love" with his physiology. With the help of physiognomy, one could begin to understand oneself and others better, and this could bring people together.

Nonetheless, physiology would not only reveal the positive features of a person. It could also reveal an individual's negative characteristics, although this too was important. With the help of physiognomy, people would know who to trust and not trust. For example, Lavater studied the portrait and silhouette of the German criminal Johan Rutgerodt (1733–1775). Rutgerodt killed his handmaiden, with whom he had an extramarital relationship, and later his wife. Rutgerodt was executed in 1775 (Roos and Weege 1796), after which Lavater was sent his portrait and silhouette. At first Lavater thought Rutgerodt was a good man, but after some additional study he had to conclude that he must have been a barbarian, a "living Satan." Rutgerodt was a "fornicator" and "virgin killer" (Paape 1780, I, 81).

Lavater's *Physiognomische Fragmente* sold well. The book was translated into French, English, and Dutch and appeared in numerous editions. Lavater had numerous imitators, and some of them gave a practical twist to his physiognonomy. Thus "portable Lavaters" were created, including the Dutch *Phisiognomische almanach* (Physiognomical Almanac; Anon. 1782), *Le Lavater Portatif* (Anon. 1809), and *The Pocket Lavater* (Anon. 1817). These booklets were modest in size, so people could easily take them on their travels. When an owner shared a carriage with strangers, he could discreetly check what type of person he had accompanying him.

The pocket Lavaters usually contained about 30 portraits, with short explanations. Many of the people portrayed displayed all sorts of positive qualities, but there were also villains, impostors, and other unsavory characters. Portrait No. XXVI, for example (Figure 1), was a despicable character, a miser, and an impostor. His vices appeared from his mouth and eyes. They revealed that he was a hypocrite, without a shred of goodness or sensitivity (Anon. 1809).

Franz Joseph Gall, the first criminologist?

Like Lavater, Franz Joseph Gall was interested in skulls, but he was particularly interested in the functions of the brain. That is where individual differences between people originate, he argued, and one therefore had to understand and focus on it. Gall was a physician, and he felt Lavater's work had no scientific value, in part because his methods were poor but also because he did not understanding the faculties of mind and the organization of the brain (see Finger and Eling 2019).

Gall introduced his "organology" (one of many names given to his doctrine), his theory of brain functions, during the 1790s, laying out its fundamental tenets. In 1804, Johann Gaspar Spurzheim (1776– 1832) began to assist him with the finer details in his research program on the mind and brain. Gall did not consider the brain to be unitary. According to him, at least after he and Spurzheim made their way to Paris in 1807, the brain consisted of 27 independent, specialized organs, corresponding to specific tendencies, feelings, and abilities, also called "faculties." This was a groundbreaking insight and, numbers aside, it

Figure 1. Caractère nr XXVI, in Anon. (1809), *Le Lavater Portatif, ou précis de l'art de connaitre les hommes par les traits du visage* (Paris).

is fundamentally correct. Gall, however, had gone one step further. He believed that the skull followed the form of the brain. When a particular trait was strongly developed, it became visible through an elevation or bump on the head. So you could determine what abilities and character traits a person or an animal might have or have had from the skull. This was his "cranioscopy."

Among the 27 faculties, he distinguished some for appreciated social qualities, such as bonding and friendship, and goodness and compassion. But he also mentioned some less-desirable faculties, including "carnivorous instinct; disposition to murder (Würgsinn)" and "sense of property, propensity to steal (Hang zu Stehlen)."

Gall was particularly interested in the extremes—the geniuses and the insane—and also different types of criminals. Extreme behavior, like that of serial killers, was of interest for the development of his theory. Especially traits like the disposition to murder had to be clearly visible on the skull. The low forehead and the large head of the criminal Léger, for instance, pointed to a highly developed organ for "Würgsinn."

Finger and Eling (2019, 387) pointed out that Gall was quite interested in crime. First, he had a large number of skulls and casts of criminals in his collection. Second, he distinguished "disposition to murder" and "propensity to steal" as separate faculties. And third, he elaborated on the how and why of some of the murders criminals had committed, attempting to correlating their acts and whether they might be repeated with the states of their brains. For these reasons and for trying to approach the subject scientifically, some have called Gall the first criminologist (Savitz, Turner, and Dickman 1977).

Gall had no moral judgment about something like murder. Some animals are predators and kill to eat. People have this "animal" trait as well. The tendency to kill is present, but in most people it is not so strongly developed that it gets out of hand. The tendency to kill can be restrained by other faculties. And even when the appetite for murder is highly developed, it does not have to lead to killings and manslaughter. Gall recognized that these people might also find outlets in other behaviors. For example, a person with a propensity to kill could become a soldier, work in a slaughterhouse, or become an executioner (Finger and Eling 2019, 388).

Gall was especially interested in how "excessive activity" of an organ could lead to destructive and antisocial behaviors. The organ for the tendency to murder, for example, could be so strongly developed that it could not be restrained. Diseases, he knew, could also alter brain physiology, resulting in some more-animal-like organs overriding nobler faculties. Gall wrote in the English translation of his *Sur les fonctions du cerveau* (On the Functions of the Brain):

> You have the musician, the mechanic, the poet, all exclusive and ardent in their pursuits; but you also have the debauchee, the bravo, the robber, who, in certain cases, are passionate to such a degree, that the excessive activity of these propensities degenerates into actual madness, and deprives the individual of all power to restrain them. (Gall 1835, I, 253)

In some people, all restraints seem to disappear. For instance, Gall was fascinated by a Dutch serial killer. The killer, who played the violin at weddings, remained out of the hands of the authorities for a long time. However, some of his children's comments put police on his trail. Eventually, he confessed to the murders and told the police that he had killed people because he enjoyed bloodshed. "Carried before the magistrate," Gall wrote, "he confessed thirty-four distinct murders, and asserted that he had committed them without malice, and without any intention to rob, solely because he found extraordinary pleasure in them" (Gall 1835, I, 313).

Like Lavater, Gall developed his brain and skull doctrine largely by looking at and comparing heads (he also performed dissections, studied people with brain injuries, and so on). The skulls of the "debauchee," the "bravo," and the "robber" provided interesting study material. However, Gall was not a modern textbook example of objectivity. He leaned heavily on anecdotal evidence. When he found the same bulge on two skulls, he thought he was onto something.

One day, he noticed similarities between the skulls of two criminals:

> I was struck with the fact, that, though very differently formed in other respects, there was, in each, a prominence strongly swelling out immediately over the external opening of the ear. The same prominence I found also in some other skulls in my collection. It appeared to me not merely accidental, that, in these two murderers, the same cerebral parts should be so much developed, and the same region of the skull so prominently. (Gall 1835, IV, 51)

Gall could be quickly satisfied. When he had some evidence for a faculty and its organ, such as "disposition to murder," he could become so convinced that he would be dismissive of contradictory cases. This allowed him to present case after case supporting his contention, and to maintain a firm confidence in his craniology, which would not withstand the test of time.

Infamous murderers in Europe

Johan Caspar Spurzheim, Gall's assistant, went his own way around 1813, probably for several reasons, one of which was to promote himself as a deeper thinker. He renamed a large number of faculties Gall had distinguished (some with his help), deleted a few, and added some more. Spurzheim labeled his "new" skull doctrine "phrenology" (the science of the mind), a term he adopted from British physician Thomas Forster (1879–1860), who probably took it from American physician Benjamin Rush (1746–1813; Finger and Eling 2019).

Gall despised Spurzheim's changes, but his assistant's "phrenology" took on a life of its own. Spurzheim gave lectures in Great Britain and Scotland, where many phrenological societies were formed. European physicians and phrenologists—including Hubert Lauvergne (1797–1859), Carl Gustav Carus (1789– 1869), George Combe (1788– 1858), Frederick Bridges (???–1883), Giuseppe Maria De Rolandis (1793–1848), and Pieter de Riemer (1769–1831)—now went to work with the ideas of both Gall and Spurzheim.

Among other things, these men focused on the skulls of notorious criminals, including murderers and rapists. Death masks were also made after criminals had been executed. Many of those skulls and masks can still be viewed in museums and universities, such as the Henderson Trust collection at the University of Edinburgh. Edinburgh is home to masks of murderers such as John Linn, Joseph Fieschi, William Burke, and William Hare (Kaufman and McNeil 1989). Skulls of criminals were also collected and examined in France, Italy, England, Germany, Denmark, and the Netherlands, to name just six more of a far greater number of countries.

In France, not only was the head of the cannibal and murderer Antoine Léger interesting study material, but this was also true for the notorious murderer Pierre-François Lacenaire (1803–1836). Phrenologists examined him in prison, and a molding was made of his head (Figure 2). Lacenaire was executed in 1836, and four hours after he was guillotined, his head was already on the table at the *Société phrénologique* in Paris. The phrenologists of the

Figure 2. Plaster head representing executed French criminal Lacenaire, Science Museum, London.

Société initially had the idea that they were looking at a cast or model, but they soon found out that it was the fresh "truncated head" of the culprit (Anon. 1846a). After examining the skull, they concluded that Lacenaire had a mediocre mind, was vain, and had "the worst tendencies and most indomitable desires" (Anon. 1846b).

If a phrenologist wanted to know more about the "worst tendencies," then it was best to go to prisons, which is what French physician Hubert Lauvergne did when he entered the prison of Toulon. His book, *Les forçats considérés sous le rapport physiologique, moral et intellectuel, observés au bagne de Toulon* (The Convicts Considered from a Physiological, Moral and Intellectual Point of View, Observed in the Prison of Toulon), is considered the first semiscientific publication on criminals (Lauvergne 1841). He categorized criminals into thieves, murderers, rapists, and counterfeiters. Gall's faculties of "combativeness" and "destructiveness" were prominent in hardened killers. Lauvergne, for example, described the "cold" and "cunning" killer Pony. He had a stocky, crudely built body without grace, and "a classic criminal head." He possessed a relatively small skull, with bulges around the ears (Lauvergne 1841).

In Germany, philosopher and physician Carl Gustav Carus criticized Gall but still maintained a toned-down form of cranioscopy. Carus did not believe there were dozens of "faculties," as Gall maintained, but limited himself to three organs that left their traces on the skull, and strove for a *wissenschaftlich begründeten Cranioscopie* (scientifically founded cranioscopy; Carus 1841). He also studied the skulls of criminals, including female murderer Johanna Weichold, who was executed in 1837. Weichold—who, according to Carus, had lived a riotous life—poisoned her husband with arsenic. Carus created an image of her skull in his *Atlas der Cranioscopie* (Atlas of Cranioscopy; Carus 1843–1845). He contended that one could tell from her skull that she was an immoral human being. "As far as the form of the skull is concerned," he wrote, "it corresponds completely to the notion of a mean, low and ugly soul and is therefore a very remarkable proof of the importance of cranioscopy" (Carus 1843–1845).

In Scotland, phrenologists were interested not only in the murderers Linn, Fieschi, Burke, and Hare but also in David Haggart (1801–1821). Haggart's life was all about pickpocketing, theft, robbery, and murder. He brutally murdered a prison guard in 1820, in order to escape from prison, but he was arrested again in 1821. Before being executed in Edinburgh, he was visited by the "eminent craniologist" George Combe. Combe treated the killer with respect, gained his trust, and examined his skull. It is not surprising, of course, that the organs for combativeness (very large) and destructiveness (full) were strongly developed in Haggart (Anon. 1821). This was reflected in his behavioral history. During his childhood, Haggart had been constantly embroiled in scuffles, and he had even tried to drown someone and beat another to death.

Haggart recorded his life story in *The Life of David Haggart*. The book featured an appendix in which Combe gave a sketch of the "Natural Character of David Haggart, as indicated by his Cerebral Organization." Combe evaluated 34 faculties and concluded:

> The greatest errors have arisen from a great self-esteem, a large combativeness, a prodigious firmness, a great secretiveness, and a defective love of approbation. No others of the faculties appear to possess an undue degree of energy or deficiency. (Haggart 1821, 159)

Combe submitted his findings to Haggart, who responded in writing, admitting that Combe was right in almost every way.

In England, phrenologist Frederick Bridges was equally fascinated by criminals. In 1860, he published *Criminals, Crimes, and Their Governing Laws* (Bridges 1860). Bridges thought that palpating heads was not enough. He opined that the parts of the head had to be in proper relationship with each other. With criminals, the relationships were different than with honest people. With a "Phreno-physiometer" that he had developed, he was able to map the proportions of heads "en profile." The angle between the ear opening and top of the eyebrow he considered decisive. In murderers this angle was about 40°; in individuals who adhered to the law it was about 25° (Figure 3). Based on this angle, he felt able to indicate

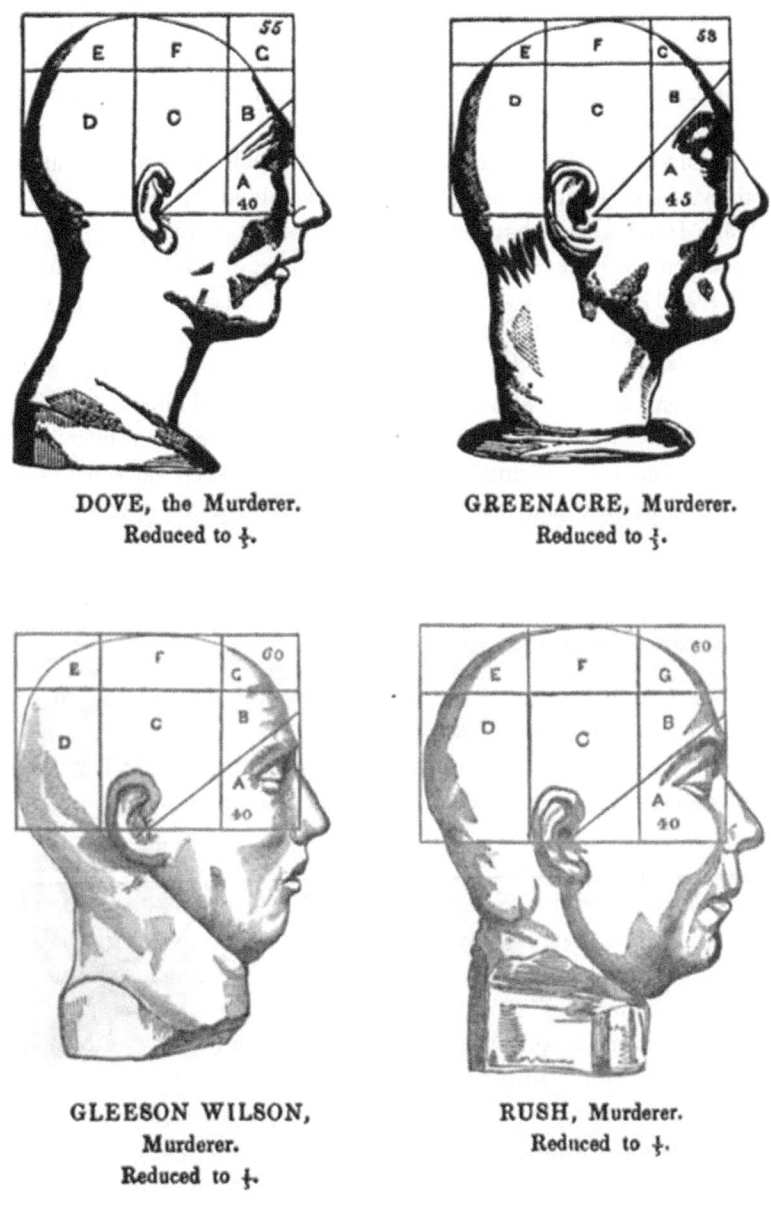

DOVE, the Murderer.
Reduced to ⅓.

GREENACRE, Murderer.
Reduced to ⅓.

GLEESON WILSON,
Murderer.
Reduced to ⅓.

RUSH, Murderer.
Reduced to ⅓.

Figure 3. Murderers in *Crimininals, Crimes, and Their Governing Laws* (Bridges 1860).

what kind of crime was committed: murder, theft, or scam. He discussed the appearance of several potential and impenitent criminals, including a 14-year-old boy with a "most dangerous type of head" and the killer James Spollin, whose story he treated extensively in another publication (Bridges 1858).

In Italy, phrenologists were particularly interested in the skull of Giorgio Orsolano (1803–1835). Orsolano was known as the "hyena of San Giorgo Canavese." He abused young girls, killed them, and chopped them up. He then processed them into sausages that he sold in his butcher shop. After he was executed, his skull was examined by Turin doctor and writer De Rolandis. De Rolandis found that the organs of cunning and destruction were overdeveloped, while Orsolano's organs for religion, benevolence, and wisdom were too small. In 1835, he published his findings in *the Journal de la Société phrénologique de Paris* (De Rolandis 1835).

In the Netherlands, thanks to the efforts of physician Pieter de Riemer, the skulls of the Zwartjesgoed Gang have been preserved (Anon. 1831). De Riemer wondered if Gall's cranioscopy was accurate, and he collected skulls of intellectuals and lunatics, as well as murderers, thieves, and arsonists to find out. He examined the skulls of Johan Hersberg (nicknamed "Jan Hanneke"), Johan Baptist (nicknamed "Grand Eye"), and Johannes van den Bos (nicknamed "Nolletjes Jan"). They were members of a notorious gang that engaged in theft, murder, and robbery between 1798 and 1806. Because of their dark skin, they were called *Zwartjesgoed* (meaning "black people").

After diligent detective work, the gang was caught in 1805 and tried in early 1806. The gruesome punishment consisted of being broken on a wheel and ultimately death. There was not much left of their bodies after their sentences were carried out, but their skulls remained undamaged. Gall visited De Riemer's museum in 1806, when he traveled through the Netherlands to give lectures and meet with scientists sharing his interests (Hekker and le Grand 2007).

Criticism of phrenological methods

The ideas of Gall, Spurzheim, and their followers were found interesting by many people, but there was also criticism. Many critics thought the phrenologists took imprecise measurements. For example, how was one supposed to determine the size and contents of a skull? Or the exact location of a cranial bump? After all, this was a precise work. In order to parry this criticism, all kinds of instruments were developed and used, including tape measures, calipers, a phreno-physio-meter, and other tools. With these tools, everything could be carefully checked by researchers. In 1828, Thomas Stone (d. 1854), junior president of the Royal Medical Society, checked countless skulls of criminals to see if the phrenologists were right. He measured and calculated, and his conclusion was scathing. The phrenologists were wrong on all fronts (Stone 1829).

Another point of criticism concerned the limited number of observations. The organ for a disposition to murder, for example, could not be determined on the basis of one or a few skulls. To meet this critique, there was an enormous appetite for gathering masks of living and dead people across Europe, and portraits and skulls of people with diverse backgrounds, including criminals. The more skulls the better, and the more extreme the crime, the more coveted the skull.

A third point of criticism concerned the potential biases of phrenologists when examining skulls of criminals. The temptation, of course, was great to find a highly developed organ for murder or theft. Around 1900, American phrenologist Henry C. Lavery (1870–1954) in Wisconsin realized that there was a need for a completely objective device (Kuhfeld 2020). He developed the "psycograph," a "mechanical" or "robot phrenologist"—a machine that could read skulls. No human mind or hand was involved. The machine looked like a hair-drying hood. The person would sit on a stool while a basket with dozens of metal antennae was placed on the head and fixed to the nose. When the machine was switched on, the head was supposedly scanned. Lavery claimed that the helmet measured different faculties from the skull, and within 90 seconds 160 different character traits could be evaluated. The device then printed out the result for 32 faculties (Risse 1976).

Galton and the compositional photo

English scholar Francis Galton (1822–1911) started working with photographs of criminals and tried to fuse these photos into one compositional photograph (Galton 1883). This, he believed, would reveal the appearance of the "average" criminal (Todorov 2017). Galton tried to refine his technique through trial and error for years, but ultimately he failed to produce the picture of a "typical criminal."

Lombroso and the criminal man

Italian criminologist Cesare Lombroso pursued the same course. Lombroso was appointed professor of psychiatry at the University of Turin in 1876. In the same year, the first edition of his main work *L'uomo delinquente* (Criminal Man) came out (Lombroso 1876). The book made him world famous, and it contains a mixture of physiognomy and phrenology.

Lombroso's eureka moment came when he was examining the skull of a criminal. He saw an anomaly and realized that criminals were "atavistic" creatures with the "ferocious instincts of primitive humanity and inferior animals" (Lombroso-Ferrero 1911, xiv–xv).

He discovered that the skulls of most criminals were unusually small or deformed. The faces of criminals also spoke volumes (Figure 4). Thieves, he said, had expressive faces, small eyes that never looked directly at you, thick eyebrows, flat noses, and a sloping forehead. Rapists had sparkling eyes, delicate features, and swollen lips and eyelids, and most were slender. Their faces deviated in every way from those of the average good citizen.

Lombroso also collected hundreds of skulls from deceased criminals and, with countless measurements and statistics, managed to create the appearance of objectivity (Gould 1996). Over time, however, it became clear that Lombroso's methods were poor and his findings were untenable.

Conclusions

Franz Joseph Gall was not the first to establish a relationship between appearance and frowned-upon behaviors or crimes. Greek philosophers, including Socrates and Aristotle, the Italian scholar Della Porta, Swiss minister Lavater, and numerous others preceded him.

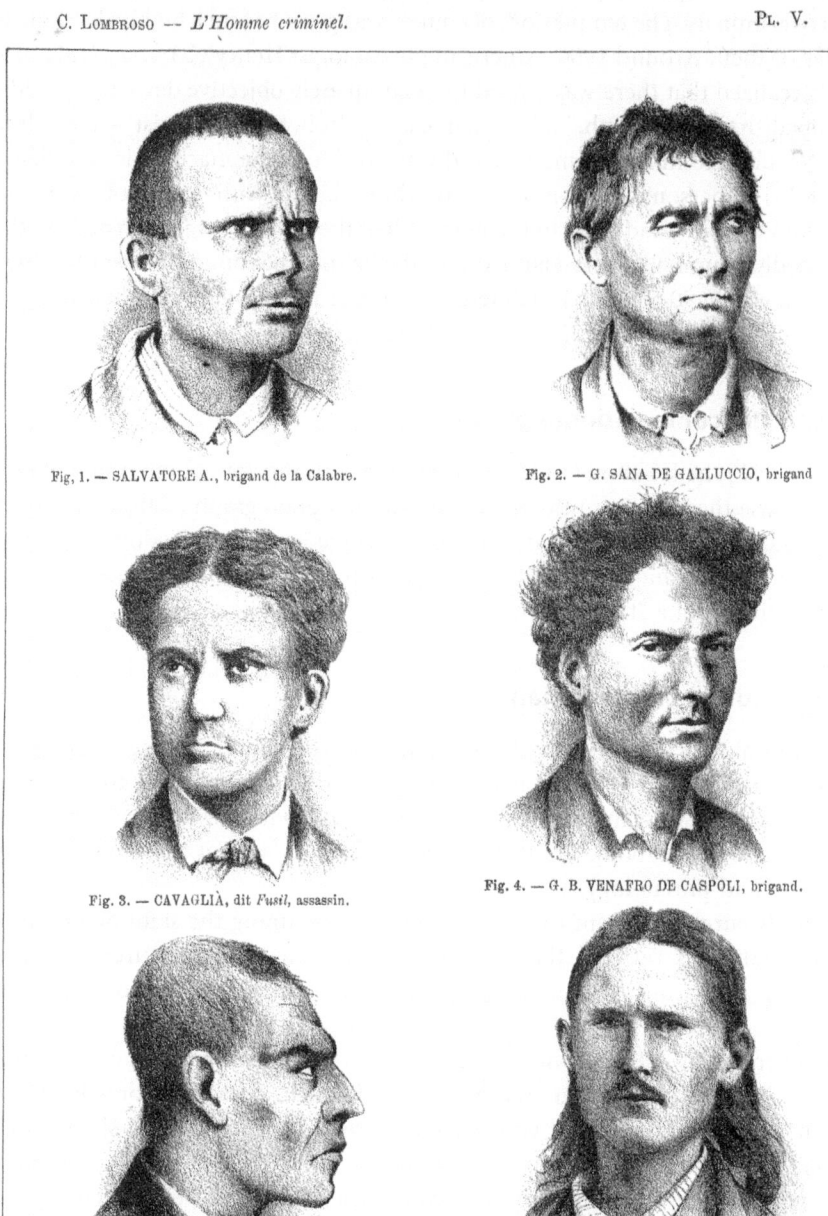

C. LOMBROSO — *L'Homme criminel.* PL. V.

Fig, 1. — SALVATORE A., brigand de la Calabre.

Fig. 2. — G. SANA DE GALLUCCIO, brigand

Fig. 3. — CAVAGLIÀ, dit *Fusil*, assassin.

Fig. 4. — G. B. VENAFRO DE CASPOLI, brigand.

Fig. 5. — O....., voleur napolitain.

Fig. 6. — CARBONE, chef-brigand.

Impr. Camilla et Bertolero.

TYPES DE CRIMINELS.

Figure 4. Types des criminels, Plate 5 of Cesar Lombroso's *L'Homme Criminel*, Wellcome Collection.

Gall was inspired by the ideas of these predecessors. For example, there is considerable overlap between Lavater's and Gall's ideas and how they were presented (Finger and Eling 2019, 64–68). Still, Gall had a unique view. Unlike Lavater, Gall was interested in the medical sciences and quoted more from the works of anatomists and physicians. He was primarily concerned with the brain, rather than the face, and how different faculties depended on anatomically distinct organs. He believed that the various faculties, such as the disposition to murder, and the extent to which they are developed could explain differences between individuals.

Gall, in turn, influenced a host of other "scientists," including his assistant Spurzheim and other phrenologists in Europe, such as Frenchman Hubert Lauvergne, the German Carus, the Scotsman George Combe, Frederick Bridges from England, De Rolandis from Italy, and Pieter de Riemer from the Netherlands. Cesare Lombroso also knew Gall's work. All of these men had a special interest in the relationship between crime and appearance, especially the construction of the face and skull. For that reason, the faces and skulls of the most notorious criminals were coveted and examined.

The interest in appearance and its relation to crime did not stop with Lombroso or with Galton, or even with Henry C. Lavery and his psychograph helmet in Wisconsin. Psychologists remained interested in physiognomy, and during the final quartile of the 1900s, several intriguing new studies were published. These publications showed that people, rightly or wrongly, were still reading a lot from faces. Questioning whether people are able to draw correct conclusions based on faces, a number of researchers have asked subjects to indentify who is a criminal from a face and the shape of a skull (Johnson et al. 2018; Royer 2018; Stillman, Maner, and Baumeister 2010; Valla, Ceci, and Williams 2011). In these studies, participants proved surprisingly adept at picking out the criminals. This certainly did not work in 100% of all cases, but it did occur more often than might be expected by chance. Needless to say, these newer studies have also had their detractors.

This historical overview teaches us three lessons. First, it appears that the fascination with a possible relationship between appearance and crime is not confined by time. One can draw a line from Aristotle, Della Porta, Lavater, Gall, and Lombroso to recent academic research.

Second, recent scientific research suggests that Gall and his followers might have been on to something when suggesting that there might be facial or skull features that violent criminals display above chance levels. Nonetheless, follow-up studies with better experimental designs and more thoughtful analyses will be needed to elucidate the differences subjects are attending to, if specific features can, in fact, be confirmed.

Third, it is clear that we cannot simply compare the approach of Della Porta, Lavater, Gall and Lombroso, in the sixteenth, eighteenth, and nineteenth centuries, respectively, with the laboratory methods of contemporary researchers. These four men relied mainly on anecdotal evidence. Gall, for example, ignored cases that challenged his skull doctrine. He also often knew in advance when he was presented with the skull or death mask of a criminal. Hence, he was focused on confirming what he expected. For example, he was once sent a coffin with skulls from a doctor who worked in a prison. As he unpacked it, he already had his conclusion ready: "My God, here is the skull of a murderer!" (Gall 1835, Vol. IV, 151).

Acknowledgments

For help and advice, I am grateful to Paul Eling and Stanley Finger.

Disclosure statement

No potential conflict of interest was reported by the author.

References

Anon. 1782. *Phisiognomische almanach voor de beminnaars der gelaatkunde, voor 't jaar 1783*. Utrecht: A. Stubbe.

Anon. 1809. *Le Lavater Portatif, ou précis de l'art de connaitre les hommes par les traits du visage*. Paris.

Anon. 1817. *The Pocket Lavater, or, the science of physiognomy: To which is added an inquiry into the analogy existing between brute and human physiognomy, from the Italian of Porta*. Van Winkle & Wyley.

Anon. 1821. Phrenological observations of the cerebral development of David Haggart. *Blackwood's Edinburgh Magazine* X (LIV):682–91.

Anon. 1825. 23. Versailles (Seine et Oise). - Cause d' Antoine Léger - Viol et homocide. In *Annuaire historique universel pour 1824*, ed. C. L. Lesur, 811–15. Chez A. Thoisnier-Desplaces, Libraire.

Anon. 1831. *Beredeneerde beschrijving van het Museum Anatomico-Physiologicum van P. de Riemer*, Vol. 1. Rotterdam: Wed. J. Allart.

Anon. 1846a. Lacenaire en de Bonaparte van het bagno. In *Nieuw Nederlandsch Magazijn, ter verspreiding van algemeene en nuttige kundigheden*, 397–99. Amsterdam: Gebroeders Diederich.

Anon. 1846b. Physiognomische gissingen aangaande den neus. In *Nieuw Nederlandsch Magazijn, ter verspreiding van algemeene en nuttige kundigheden*, 364–67. Amsterdam: Gebroeders Diederich.

Anon. 1872. *Kijkjes door venster, deur en dak in het binnenste van den mensch: De kunst om uit den schedel, het gelaat, de handen, houding, gang en gebaren, ja zelfs uit het schrift, het karakter van iemand te leeren kennen*. Zwolle: Van Hoogstraten & Gorter.

Aristotle. 1908. *The works of Aristotle*, Vol. 12. Oxford: Clarendon Press.

Bridges F. 1860. *Crimininals, crimes, and their governing laws*. London: George Philip and Son.

Bridges F. 1858. *Phreno-physiometrical characteristics of James Spollin: Who was tried for the murder of G.S. Little*. London: W. Horsell.

Burgess J. 1858. *The medical and legal relations of madness*. London: J. Churchill.

Carus C. G. 1841. *Grundzüge einer neuen und wissenschaftlich begründeten Cranioscopie (Schädellehre)*. Stuttgart: Balz'sche Buchhandlung.

Carus C. G. 1843–1845. *Atlas der Cranioscopie, oder Abbildungen der Schaedel- und Anlitzformen beruehmter oder sonst merkwuerdiger Personen*. Leipzig: AugustWeichardt.

De Rolandis G. M. April 1835. Lettre à M le docteur Fossati, sur un criminel convaincu de plusieurs viols, suivis de meurtre'. *Journal de la Société phrénologique de Paris* 3:244–47.

Della Porta G. 1618. *De humana physiognomonia*. Nicolaum Hoffmannum.

Finger S., and P. Eling. 2019. *Franz Joseph Gall. Naturalist of the mind, visionary of the brain*. New York: Oxford University Press.

Gall F. J. 1835. *On the functions of the brain and of each of its parts: With observations on the possibility of determining the instincts, propensities, and talents, or the moral and intellectual dispositions of men and animals, by the configuration of the brain and head*, Vols. 1-6. Boston: Marsh, Capen & Lyon.

Galton F. 1883. *Inquiries into human faculty and its development*. New York: Macmillan.

Gould S. J. 1996. *De mens gemeten*. Amsterdam: Contact.

Haggart D. 1821. *The life of David Haggart: Alias John Wilson, alias John Morison, alias Barney M'Coul, alias John M'Colgan, alias Daniel O'Brien, alias the Switcher*. W. & C. Tait.

Hekker B., and J. le Grand. 2007. Drie criminelen op de ontleedtafel. In *Universitaire collecties in Nederland: Nieuw licht op het academisch erfgoed*, ed. T. Monquil-Broersen, 59. Zwolle: Waanders Uitgevers.

Johnson H., M. Anderson, H. R. Westra, and H. Suter. 2018. Inferences on criminality based on appearance. *Butler Journal of Undergraduate Research* 4 (1):6.

Kaufman M. H., and R. McNeil. februari 25, 1989. Death masks and life masks at Edingburgh University. *British Medical Journal (BMJ)* 298:506–07. doi:10.1136/bmj.298.6672.506.

Kuhfeld E. R. 2020. *Phrenology and the psycograph*. http://washuu.net/Med-Lec/psycgraf.htm.

Lauvergne H. 1841. *Les forçats considérés sous le rapport physiologique, moral et intellectuel, observés au bagne de Toulon*. Paris: JB Baillière.

Lavater J. C. 1775-1778. *Physiognomische Fragmente zur Beförderung der Menschenkenntnis und Menschenliebe*. Leipzig und Wintherthur.

Lombroso C. 1876. *L' uomo delinquente*. Milano: Heopi.

Lombroso-Ferrero G. 1911. *Criminal man. According to the classification of Cesare Lombroso*. New York: The Knickerbocker Press.

Paape G. 1780. *Handleiding tot de physiognomiekunde*. Dordrecht: A. Blussé en zoon.

Risse G. B. 1976. Vocational guidance during the depression: Phrenology versus applied psychology. *Journal of the History of the Behavioral Sciences* 12 (2):130–40. doi:10.1002/1520-6696(197604)12:2<130::AID-JHBS2300120204>3.0.CO;2-U.

Roos G., and J. Weege. 1796. *Merkwaardige character- en levensschetsen van eenige ter dood veroordeelde persoonen*. Amsteldam.

Royer C. E. 2018. *Convictable faces: Attributions of future criminality from facial appearance*. Ithaca, NY: Cornell University Press.

Savitz L., S. H. Turner, and T. Dickman. 1977. The origin of scientific criminology: Franz Joseph Gall as the first criminologist. In *Theory in criminology*, ed. R. F. Meier, 41–56. Beverly Hills: Sage Publications.

Seidel F. 1858. *Einblicke durch Fenster, Thür und Dach in das Innerste des Menschen. Eine Quintessenz der Beobachtungen und Forschungen eines Lavater, Gall, Spurzheim, Roger, David u. A. m. über Physiognomik, Schädellehre, Mund, Zähne und Lippen, Haltung und Bewegung des Körpers, Stimme, Gang, Kleidung, Mimik, Deutung und Auslegung der Handschriften, Chiromantie*. Weimar: Bernhard Friedrich Voigt.

Stillman T. F., J. K. Maner, and R. F. Baumeister. 2010. A thin slice of violence: Distinguishing violent from nonviolent sex offenders at a glance. *Evolution and Human Behavior* 31 (4):298–303. doi:10.1016/j.evolhumbehav.2009.12.001.

Stone T. 1829. *Observations on the phrenological development of Burke, Hare, and other atrocious murderers: Measurements of the most notorious thieves confined in the Edinburgh Jail and Bridewell, and of various individuals, English, Scotch, and Irish, presenting an extensive series of facts subversive of phrenology*. Edinburgh: Robert Buchanan.

Todorov A. 2017. *Face value: The irresistible influence of first impressions*. Princeton, NJ: Princeton University Press.

Valla J. M., S. J. Ceci, and W. M. Williams. 2011. The accuracy of inferences about criminality based on facial appearance. *Journal of Social, Evolutionary, and Cultural Psychology* 5 (1):66. doi:10.1037/h0099274.

Spurzheim Reexamined

Johann Gaspar Spurzheim: The St. Paul of phrenology

John van Wyhe

ABSTRACT

Franz Joseph Gall's wayward discipline Johann Gaspar Spurzheim greatly modified Gall's original system and introduced it to the English-speaking world. Through an active program of itinerant lecturing, publishing and converting disciplines, Spurzheim made phrenology. He also developed a philosophy of following the laws of nature that was adopted and further promoted by his disciple, George Combe. Combe's book *The Constitution of Man* (1828) became one of the best-selling works of its genre in the nineteenth century. Thus Spurzheim, never particularly original, exercised an enormous influence on nineteenth-century culture.

Introduction

Johann Gaspar Spurzheim (1776–1832) occupies an unusual position in the history of science. He is considered both important and obscure. Although he is mentioned no less than 12 times in the new *Oxford Dictionary of National Biography*, there is no entry for him. His name has been given variously as Johann Gaspar, Johann Caspar, and just Caspar. In the *Dictionary of Scientific Biography* (1970–1976), his name is given as Johann Christoph Spurzheim. This erroneous name was first attributed to him in 1844 (Callisen 1844). The historian Roger Cooter wrote of Spurzheim in 1984:

> We know hardly more than that he was raised a Lutheran and that prior to studying medicine in Vienna at the turn of the century he was a student of divinity and philosophy at the University of Trêves ... due to the paucity of the right sort of biographical material. (Cooter 1984, 51)

A paucity there surely has been because Spurzheim was neither a Lutheran nor, it seems, a student of divinity and philosophy at Trêves (Trier). It has even been suggested that Spurzheim was a homosexual (Lynch 1985).

Despite his obscurity, Spurzheim can justifiably be called one of the most influential men of science of the nineteenth century. This is so because Spurzheim single-handedly transformed and transplanted phrenology to the English-speaking world. The social diffusion set in motion by Spurzheim's peripatetic lecturing, writing, and self-publicizing changed the face of nineteenth-century culture (van Wyhe 2004, 2007). Hundreds of societies were founded to discuss and promulgate his doctrine, thousands of books and articles were published expounding or condemning it, and millions of people came either to believe some or much of it—and it can safely be said that almost every adult human being in the Western world had heard of it. Few men of science are able to boast of such influence on the culture of their day or succeed in disseminating and perpetuating their ideas to such an extent. Yet Spurzheim was no great genius or even a very original thinker. The core of his phrenology was learned from his master and the founder of the doctrine, Viennese physician Franz Joseph Gall (1758–1828). Hence, it is hard to imagine that Spurzheim was anything other than a phrenologist.

However, his surviving 215 or so letters (from 1807–1832) make it clear that Spurzheim had other interests that help to explain why phrenology came to develop the way it did in the English-speaking world. He was an early anthropologist and theorized on laws of heredity, race, insanity, education, social organization, and natural laws of morality and nature. Spurzheim was not just a phrenological crank; he was a man of science fully enmeshed in the society of his time. As a learned gentleman, he moved almost seamlessly among elite men of science and medicine in France, Great Britain, Ireland, and America.

Spurzheim

Spurzheim was born on December 31, 1776, in Longuich, a small town nestled along the Moselle river in Germany close to the border of modern-day Luxembourg. He was baptized on the same day in the nearby baroque Catholic church of St. Laurentius as Caspar Sportzheim, son of Johann Sportzheim (1741–1787) and Anna Maria Bosenkeil (d. 1802), tenant farmers of the Maximiner abbey estate in Longuich, owned by the Benedictines. Southwest of the church lies the abbey complex of St. Maximin. Dating back at least to the twelfth century, the present buildings were rebuilt in 1714. Under the occupation of Napoleon's army from 1804, the Benedictine estates were secularized. In 1808, the farmland was divided into five portions and sold.

Spurzheim's godfather was given as Nikolaus Justen from Switzerland, standing in for R. D. Caspar Bosenkeil from Ochtendung.[1] Before standardized spelling, Caspar, Gaspar, and Kaspar were alternative spellings of the same name. Spurzheim used Gaspar when signing his books and letters. He was the fifth of seven children. His brother Willibrod became a watchmaker in Oedenburg in Hungary, Karl Theodor Heinrich became a master saddler in Vienna, Franz died (c. 1797) at Trier, and Joseph Lorenz worked as a saddler in Vienna. Spurzheim's sister, Theresia, married Nicolas Hermsdorf of Schweich (near Trier) and had a large family. She was widowed by 1834.

Spurzheim only spent his earliest years in the Maximin farmhouse in Longuich. Around the year 1797, the family moved to Trier. The last son, Franz Anton, was baptized in 1782 at St Michael in Trier. Most of those attending the church were servants and manual laborers of the Maximin cloister. Spurzheim's father died in 1787. His mother continued to live in the parish until her death in 1802 in the house of her brother, who was a clergyman at Selterns, near Limburg, in the Grand Duchy of Hessen.

In autumn of 1790, three years after the death of his father, Spurzheim entered the Gymnasium at Trier. Here he acquired the first rudiments of Greek and Latin. He was clearly intelligent, distinguished himself as one of the foremost students, and won several

---

[1] Kirchenbuch Nr. 1 der kath. Pfarrei St. Laurentius in Longuich, p. 70. Familienbuch Lonquich. Bistumsarchiv Trier, Abt. 77, Nr. 82.

prizes for scholarship.[2] By September 1796, he had completed all six classes. The French army occupied Trier in August 1794, and the university closed in 1798. As Spurzheim later wrote "At the beginning I was destined to become a clergyman, but when the French occupied my country I went to Vienna, 1799, in order to study medicine." (Spurzheim to George Combe, March 13, 1821, National Library of Scotland, MSS.7201–7515)

Two of Spurzheim's brothers had already settled in Vienna. Joseph Lorenz was a saddle boy who died in 1809 at the age of 29. The elder, Karl Theodor Heinrich, was a saddle master and wagon builder until his death there in 1838.[3] Spurzheim was received into the family of a Count Splangen[4] as tutor to the latter's two sons. Spurzheim passed his first *viva voce* in October 1803, the second in March 1804, and he was awarded his doctorate on August 7, 1804 (not 1813, as some sources state).

Count Splangen's physician was Franz Joseph Gall. Gall had begun, in the early 1790s, to create a new science of mind and brain that would eventually evolve, via Spurzheim's changes, into phrenology (see van Wyhe 2002, 2004). Gall offered public lectures on his new science from 1796 at his home in Vienna. Spurzheim began attending these lectures in 1800. In 1804, presumably after completing his doctorate, Spurzheim was employed by Gall as a dissectionist and assistant. Gall purportedly paid Spurzheim 1200 florins *per annum* (Moscati 1833). In later years Spurzheim attempted to overemphasize his creative role in Gall's science or phrenology. Rather than admitting that he was Gall's paid dissectionist, Spurzheim wrote, "I was simply a hearer of Dr Gall till 1804, at which period I was associated with him in his labours and my character of hearer ceased."[5] These words were repeated endlessly, usually verbatim, by phrenologists throughout the nineteenth and early twentieth centuries to emphasize the authoritative paternity of the man who was the source of their doctrines.

Gall's lectures and any future publications were banned by imperial decree in December 1801, ostensibly because Gall had no permission to lecture in his home, and his lectures were inappropriately materialistic, irreligious, and attended by ladies (Lesky 1981). In 1805, Gall was invited by his elderly father to visit him at the family home in Tiefenbronn in Swabia before he died. Gall took the opportunity to recommence lecturing on his favorite subject outside Austria. Spurzheim accompanied him as dissectionist and assistant. Gall's lecture tour was repeatedly extended as nobility, physicians, and *literati* in city after city invited him to lecture on his new science of the brain and mind, which Gall called *Schädellehre* (doctrine of the skull). Gall was showered with money and gifts and became a great celebrity. As merely the dissectionist and demonstrator of the famous Dr. Gall, Spurzheim was almost never mentioned in the countless books and newspaper accounts of Gall's lectures across Europe (Blöde 1805; van Wyhe 2002).

After years of traveling, Gall and Spurzheim arrived in Paris in 1807. Here their reception was particularly warm and publication of the doctrine seemed a possibility, so they stayed. It is from this period that the first known Spurzheim letters began. They show us that at this time Spurzheim (Figure 1) was still thrilled to be close to the center of

---

[2]Trierisches Wochenblatt 1791, Nr 40, 1792, Nr 39, 1793, Nr 38, 1795, Nr 39 1796, Nr 38.

[3]His son, Karl Spurzheim (1809–1872), became a well-known alienist.

[4]This name is given in almost all accounts of Spurzheim's life, but no record of anyone by this name has been found. It must be a misspelling. Spangen is a possible alternative.

[5]Come (1824, 11). This quotation is from Spurzheim's *Essai Philosophique sur la Nature morale et intellectuelle de l'Homme*, Appendix p. 213. This introduction by Come also emphasizes strongly Spurzheim's common laboring character; Gall and Spurzheim were "constantly together, and their researches were conducted in common" (p. 12).

Figure 1. A portrait of Spurzheim drawn by his wife and engraved by W. H. Lizars. Frontispiece to Mackenzie, *Illustrations of Phrenology* (1820).

attention and fame, Gall, and that he did not yet claim copaternity of the science. Gall recommenced lecturing in November 1807.

The break with Gall

The separation of Gall and Spurzheim has been a source of speculation since the early nineteenth century. Here a new account is given taking account of all surviving evidence. In Paris, Spurzheim was 31 and began to grow ambitious to be more than just the assistant of Dr. Gall. He spoke of setting out on his own. Gall wrote to friends at the time,

> The sole reason the name Spurzheim is on my work (the *Mémoire*), and will be written on the large one is that he knows my doctrine completely, has already contributed much to its perfection, and will propagate it further after my death. I was never jealous of the praise of men, and the doctrine must persist because it is true and useful. (Neuburger 1917, 24)

The joint *Mémoire* was presented to the French Institute on March 14, 1808. Five men of science, led by the great comparative anatomist Georges Cuvier (1769–1832), presented a report in response on April 18.[6] Cuvier found the cerebral anatomy largely unoriginal, although the least flawed of the work. Cuvier could see no grounds for deriving

---

[6]The report was probably written solely by Cuvier. See Outram (1984). See also Cuvier's 1822 account of the controversy with Gall in Flourens (1856, 189). See also the valuable discussion in Clarke and Jacyna (1987, 43).

psychological claims from the anatomy of the brain. Gall was essentially rejected and informed that his work was useless for science. He was not elected to the *Académie des Sciences* (see Gall 1821). Historian Dorinda Outram wrote that for Cuvier, the 1808 controversy with Gall meant consolidating Cuvier's own authority to define proper science (Outram 1984, 124). With profitable lectures to give and a wealthy elite clientele as a physician (which his fame assured him), Gall remained in Paris, becoming a naturalized citizen in 1819.

A settled period of lecturing and the preparation of publications began. Gall's first treatise on his science was published as *Anatomie et physiologie du système nerveux en général, et du cerveau en particulier* (1810, 1812). Spurzheim, it seems, furnished only the references and directed the completion of engravings (Gall 1818, xvii). In later years Spurzheim claimed as much as he could: "all anatomical discoveries made after 1804 are the result of my labours" and, "My discoveries form its (*Anatomie et physiologie*) principal object" (Spurzheim 1826, xiv). These distortions have remained largely unrecognized and unremarked.[7] In later years, Spurzheim's acolytes maintained the story of Gall misappropriating Spurzheim's work (Gibbon 1878, 211).

In June 1813, Spurzheim left Gall. Henceforward they pursued their interests independently, and it is not known that they ever met again. The reasons for and context of the separation of Gall and Spurzheim remain obscure. There may have been a sudden angry break, although surviving letters show that the two continued to correspond at least until 1816. It is clear that Spurzheim continued to consider Gall as a patron who could confer benefits on him.

Clearly the two men differed as to how the doctrine should be developed and applied. Gall must have rejected Spurzheim's speculations and inclinations. Living with Gall, Spurzheim could not publish his own ideas (or Gall's) or, it seems, contract any profitable financial arrangements. Many years later, Spurzheim wrote to his Scottish disciple George Combe (1788–1858):

> I was since told that Madame Gall opposed to my visiting her husband [on his deathbed]. — Let it be observed en passant that I suspect the same person having greatly contributed to Gall's first attacking me.[8]

Spurzheim left Paris in June 1813 and traveled to Vienna, perhaps because he had nowhere else to stay apart from with his brother, the saddler. It is from Vienna, in September 1813, that the lengthiest part of the surviving correspondence begins. The letters were written to a young French woman named Honorine Pothier (c. 1790–1830).[9] Pothier was the widowed daughter of a wealthy Parisian family with two sons and one daughter. Spurzheim and Pothier hoped to marry, but because Spurzheim had no fortune, it was considered impossible. Spurzheim set out to earn money as Gall had done when traveling about Europe in 1805–1807.

In January 1814, Spurzheim left Vienna and traveled to Britain. Spurzheim had studied English, at Gall's suggestion, so he was prepared to lecture on new ground. Spurzheim was

---

[7]The exceptions are John Elliotson's lengthy footnotes providing full citation comparisons and quotations from his own conversations with the elderly Gall.

[8]Spurzheim to G. Combe, August 23, 1828, NL Scotland.

[9]Pothier's maiden name is given at the Countway Library and by writers that have cited the letters (e.g., Walsh and Kaufman) as Perier. However, a close reading of the letters indicates that Perier was Honorine's brother-in-law; he was married to one of her two sisters. Although Perier was therefore part of her family, it was not her maiden name.

unknown apart from his name sharing the title page of Gall's works since 1808. Almost nothing was known about the doctrine in Britain or Spurzheim's role in the collaboration with Gall. It was a unique opportunity to shape his own name and image. Rather than lecturing on "craniology" or "the system of Dr. Gall" or some such name as Gall's doctrine had been known in British periodicals, Spurzheim chose to claim copaternity for the whole doctrine. His first "respectable medical and amateur" London listeners (at three Guineas each) heard lectures on "the physiognomical system of Drs. Gall and Spurzheim" (*Philosophical Magazine*, 44, Dec. 1814, 470).[10]

Spurzheim would always afterward maintain that the science was as much his as Gall's. Spurzheim never revealed the fact that he was only Gall's paid dissectionist and assistant, and that Gall alone had lectured in Europe. Spurzheim also claimed that he had cowritten Gall's works, which bore both names on the title page, and it was claimed that he later plagiarized Gall's works (which were not translated into English until after Spurzheim's death). Gall later complained:

> Of the one hundred and twenty pages ... one hundred and twelve are copied from my own works. ... He will say that he was right to do this, because he is supposed to be the collaborator. At least he could have indicated the source of his riches ... others have already accused him of plagiarism. It is at the very least, quite ingenious to make up books by means of scissor snips. (Hollander 1920, 340–343; see, e.g., John Elliotson's (1840) lengthy footnotes providing full citation comparisons and quotations from his own conversations with the elderly Gall.)

As "phrenology" consisted of the followers of Spurzheim's teachings, there was never any doubt for them that he was an authority equal with Gall, and often his superior. Ever since 1814, Spurzheim had been described in the English-speaking world as a cofounder, associate, collaborator, or partner of Gall; and phrenology, according to the *Oxford English Dictionary*, was "the theory originated by Gall and Spurzheim". These references were often the exact phrases used by Spurzheim (see Combe 1824; Poole 1824; Capen 1836; Gibbon 1878, vol. 1, 151, 211). Spurzheim sometimes advised his disciple George Combe not to give too much credit to Gall: "Hitherto in your journal you ascribe more to Gall than he deserved. ... Gall stood still since 1813 when I left him" (Spurzheim to Combe [before 15] March 1830. NL Scotland). Clearly, Spurzheim was ambitious, status conscious, and not entirely honest.

Spurzheim made considerable alterations to Gall's original system. He arranged the faculties of the mind into a hierarchical taxonomy of orders and *genera* ascending from faculties common to man and lower animals, such as *Philoprogenitiveness*, the love of offspring (starting at the back of the head and moving forward), to "the moral sentiments," some of which were shared by man and animals, and other sentiments proper to man, such as veneration. He added new faculties to Gall's 27, resulting in 33. Spurzheim also changed the names of several faculties. He also asserted that the faculties had proper and improper functions. Finally, he added the traditional theory of the four temperaments or humors (lymphatic, sanguine, bilious, and nervous) to the system.

---

[10]Spurzheim's first four publications carried Gall's name in the title in an effort to appear equal with Gall as codiscoverer. Only after controversy had made Spurzheim well-known did he cease to make use of Gall's name.

Spurzheim's lectures

Spurzheim's first lectures were delivered at his London lodgings, the first floor of 11 Rathbone Place near Oxford Street, which had a room "large enough for fifty auditors" (Spurzheim to H. Pothier, June 12, 1814, Countway B MS C 22.1 fd. 1). He also invited the editors of newspapers to attend *gratis* knowing that a mention in print, by bringing more auditors to subsequent lectures, would more than compensate for a few free tickets. From London, in September 1814, he wrote, "I have already a good number of prosolites who speak for me, and engage other persons to attend my lectures" (Spurzheim to H. Pothier, September 1, 1814, Countway B MS C 22.1 fd. 1). He took steps to ensure the propagation of his doctrine. "I … make acquaintances with different persons who can be useful in inserting different articles in scientific journals" (Spurzheim to H. Pothier, September 20, [1814], Countway B MS C 22.1 fd. 1).

Spurzheim described his lectures as "demonstrative lectures." Like Gall, he lectured without notes and offered dissections when he could procure fresh brains of humans or other animals, and he always made use of skulls, busts, models, and diagrams. Spurzheim would tell his audience about a particular faculty of the mind and then offer some examples of its corresponding bump from his large collection of skulls and casts or point to a chart to make an anatomical point clear. Auditors were able to buy charts or books with diagrams from Spurzheim or a bookseller.

Spurzheim's first book, *The Physiognomical System of Drs. Gall and Spurzheim*, was published in 1815. He then set off on his lecture tour of Great Britain. He traveled first to Bath and then Bristol. In September he traveled to Dublin and later to Cork, Limerick, Coleraine, and Waterford. While in Ireland, he first saw the devastating review of his book in the *Edinburgh Review* (Gordon 1815). The anonymous author was a young and ambitious Edinburgh anatomy lecturer named John Gordon (1786–1818) whose work focused on the brain. Spurzheim was an unwelcome interloper who threatened to overturn accepted dissection techniques and certainly was receiving more attention than Gordon. The review was ferocious and soon notorious. The *Edinburgh Review* was one of the most widely read and respected periodicals in Britain. The review both hurt and helped Spurzheim. Sales of his book suddenly slumped. Although Spurzheim was widely thought to be a charlatan or a fool, at least now almost every literate person in Britain had heard of him and his doctrine.

By April 1816, he was in Liverpool. He traveled through Manchester, Lancaster, and Carlisle before reaching Edinburgh in July 1816. What ensued was a masterstroke for Spurzheim. Taking a letter of introduction, he called on Dr. Gordon, who was said to be away. Spurzheim tried again until Gordon received him but pretended not to recognize his name. Spurzheim then gave a public demonstration of his cerebral dissection techniques with a copy of the *Edinburgh Review* open beside him. He demonstrated to a large audience, in Gordon's own lecture theater at Surgeons' Square, that the structures and characteristics described in his book could be demonstrated despite the mocking denial in the *Edinburgh Review*. Many who had previously laughed at Spurzheim began to think he had been unfairly misrepresented.

Outside the anatomy theater, Spurzheim continued to promote his system. He had met an Edinburgh lawyer named Dundas on the coach to Edinburgh, and perhaps this gave him access to this community of educated Edinburgh society. It was at the home of the

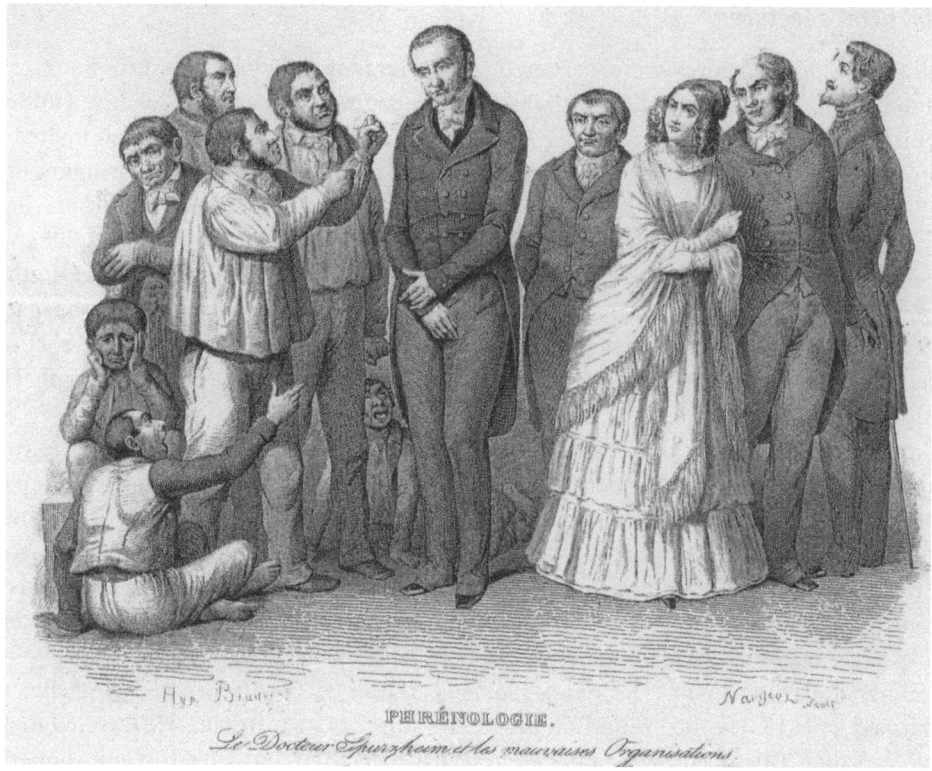

Figure 2. Hippolyte Bruyères, *La Phrénologie: Le Geste et la Physionomie* (Paris: Aubert, 1847, facing p. 513). Bruyères was Spurzheim's son-in-law. He claimed that this posture was particularly characteristic of Spurzheim.

solicitor James Brownlee that Spurzheim was introduced to George Combe. Combe would become the leading phrenologist in Britain and, after Spurzheim's death, the world. After six months in Edinburgh, Spurzheim had turned the city from the focus of his greatest embarrassment into the phrenological capital of the world. In 1820, a number of gentlemen converted by Spurzheim founded a society devoted to his science, which had then come to be known as phrenology.

In July 1817, Spurzheim returned to Paris with the intention of settling there. He married Honorine Pothier in 1818, and they lived in the Rue de Richelieu, Paris. In the remains of The Henderson Trust Collection at the University of Edinburgh Medical School are several large diagrams apparently sketched by Pothier and used by Spurzheim in his lectures. Also in the collection of the Wellcome Institute, London, is a pencil drawing of Spurzheim from 1825 by Pothier.

In 1818, Spurzheim published—in French—his work on insanity (Spurzheim 1818). This was followed in 1820 by his *Essai philosophique sur la nature morale et intellectuelle de l'homme* (*A Philosophical Catechism of the Natural Laws of Man*). In the following year he received the title of doctor from the University of Paris in order to practice medicine there. In 1818–1819, the Paris *Athénée* hired Spurzheim to lecture on the "nature of moral and intellectual man applied to social institutions" (Staum 1995). His work with the insane is depicted in a drawing by his son-in-law.

During these years in Paris we have almost no evidence of him, but he was probably reading writers such as Cabanis, Baron d'Holbach, and Constantin François de Volney. Cabanis argued that changing people's environment could perfect them and make them equal, and this would then be passed on as had purportedly been done with animals. We can gather that Spurzheim read these authors because it was from their works that his philosophy of natural laws was largely derived (Figure 2).

This philosophy is first evident in the manuscript of his *Philosophical Catechism of the Natural Laws of Man* (in French). Spurzheim proclaimed divinely ordained, invariable, inherent, and regular natural laws to which "all inanimate and all living beings are subject" (Spurzheim 1825, 12). He presented three classes of laws to which man was specifically subject: vegetative, intellectual, and moral. These laws exerted a mutual influence, it was evil to break them—and suffering and unhappiness were the result. "All suffer alike who infringe, as all without exception prosper who obey, the natural laws" (Spurzheim 1825, 22). Spurzheim claimed to work his way toward his conclusions from the properties of matter to ask: "*What is the grand object of the philosophy of man? ...* to show the necessity of man's ... submission to the laws which nature imposes" (Spurzheim 1825, 6–7; emphasis in the original). The role nature played in Spurzheim's text was the same role it came to play in the phrenology controversies in which he had taken part— that of the ultimate authority of appeal (van Wyhe 2003). The emphasis on the necessity of man's submission was an indirect attack on the authority of revelation and an assertion of Spurzheim's "undoubtedly egotistical" authority as a great philosopher (Cooter 1984, 51).

Spurzheim's natural laws were clearly derived from the writings of the French *idéologue* and revolutionary writer Constantin François de Volney (1757–1820), whose *La Loi Naturelle* (*The Law of Nature*; 1793) was also written as a catechism. Volney's other work, *Les Ruines, ou méditations sur les révolutions des empires* (*The Ruins, or Meditation on the Revolutions of Empires*; 1791), probably also influenced Spurzheim. For Spurzheim, the moral was whatever was in harmony with all the innate faculties of man (as revealed partly by phrenology). Spurzheim announced that, were his philosophy to be adopted, "Morality would become an exact science" (Spurzheim 1825, 20). So Spurzheim's philosophical program shared the scientific pretensions of his phrenological life while extending his claims to authority even further. Some of the scientifically necessary alterations of behavior included those with hereditary diseases or defects should not have children (an injunction to become more familiar in the later eugenics movement), neither should close relations; extremes of behavior were also to be avoided, such as unbridled accumulation of wealth in business, and abuse of privilege—especially inherited privilege. The only privilege that was acceptable was for those who invent or make useful discoveries (such as Spurzheim himself). Only privately did Spurzheim confess: "Nature is my greatest authority. I reject every interpretation [of the Bible] which is not conformable to the eternal and universal laws of Creation" (To George Combe, December 5, 1826, NL Scotland).

Spurzheim's *Catechism* was translated into English in Edinburgh in 1822 by the Scottish medical writer and member of the Edinburgh Phrenological Society, Robert Willis (1799–1878).[11] George Combe later claimed that he read the French manuscript

---

[11]Willis also translated Spurzheim, *Anatomy of the Brain*, 1826.

in 1824.[12] The *Catechism* was printed in 1826, but, following Combe's advice, it was withheld and not distributed. Fearing the condemnation of its secular naturalism which Combe foretold, Spurzheim did not publish the small volume until June 1828—the very month when Combe published his *Constitution of Man*. Curiously, however, *Catechism* bore the date 1825 on its title page and its title was changed to *A Sketch of the Natural Laws of Man*.[13] Combe's work became one of the bestselling philosophical works of the century; in print until 1899, it sold at least 300,000 copies. Its influence is difficult to exaggerate (see van Wyhe 2004). Yet it was borrowed almost wholesale from Spurzheim.

Spurzheim also gave two three-month lecture courses each year on his version of phrenology in Paris. From 1822, the Jesuits forced the government to restrict public lectures to those that were licensed. In 1824, Spurzheim was refused a license to instruct a class larger than 20. As he later wrote to George Combe, "It was necessary for me either to give up lecturing on Phrenology and to stick to the practice of Medicine in Paris or to give up my practice in Paris and to lecture in England" (Spurzheim to G. Combe, December 5, 1825, NL Scotland). Spurzheim returned to England in March of 1825. Much had changed since his departure in 1817. His doctrines had been further spread by his acolytes, and societies and journals had been founded to propagate them further.

His wife died after a long illness in London in December 1829. Thereafter Spurzheim continued to travel and offer lectures in Paris, London, and Ireland. His most ambitious destination resulted from an invitation to lecture in the United States. He arrived in Boston in June 1832. There, he continued to visit public health institutions and perform public brain dissections and lecture on phrenology. He fell ill and died of fever in Boston on November 10, 1832. His American hosts were so impressed by him that he was buried with honors, and a new phrenological society was founded in December 1832 to carry on the discussion and dissemination of his doctrines. Despite his short sojourn in the United States, Spurzheim had succeeded in establishing the phrenological movement there (see Carmichael 1833; Walsh 1972).

Acknowledgments

For help and advice during the research over many years that lay behind this article, I am grateful to Nils Bütner, Anne Secord, Jim Secord, Shelley Innes, Marc Renneville, Thora Bleckwedel, Antranig Basman, Clare Brown, Carly Douglas, Lee Perry, Chris Mycock, Paul Eling, Stanely Finger, and Vassiliki Michou.

Disclosure statement

No potential conflict of interest was reported by the author.

---

[12]See the preface to most editions of Combe's *Constitution of Man*.
[13]Spurzheim, *A Sketch of the Natural Laws of Man* (1825; actually 1828; Boston ed. 1833, 5th American ed. Boston, 1839; another ed. was New York, 1840). Revised and republished (Glasgow, 1844, pp. 34); *The Natural Laws of Man: A Philosophical Catechism* (7th ed. Manchester, London; c. 1870). See Cooter, *Phrenology in the British Isles* (1989, pp. 312–313).

References

Blöde K. A. 1805. *Dr. F. J. Gall's Lehre über die Verrichtungen des Gehirns, nach dessen Dresden gehaltenen Vorlesungen in einer fasslichen Ordnung mit gewissenhafter Treue dargestellt. Mit einer dreyfachen Abbildung eines von Gall bezeichneten Schädels.* Dresden: Arnold. 2nd supplemented and improved in 1806.

Callisen A. C. P., ed. 1844. *Medicinisches Schriftsteller-lexicon der jetzt lebenden Verfasser.* Copenhagen.

Capen N. 1836. *A Biography of the author.* In *Spurzheim, Phrenology, in connexion with the study of physiognomy.* 3rd ed. American ed. Boston, MA: Marsh, Capen, & Lyon.

Carmichael A. 1833. *A memoir of the life and philosophy of Spurzheim.* Dublin: Wakeman.

Clarke E., and L. S. Jacyna. 1987. Nineteenth-century origins of neuroscientific concepts. *Medical History* 32(2): 211–13.

[Combe G]. 1824. Preliminary dissertation on the progress and application of phrenology. *Transactions of the Phrenological Society* 1:1–62.

Combe G. 1828. *The Constitution of Man Considered in Relation to External Objects.* Edinburgh: Anderson.

Cooter R. 1984. *The Cultural Meaning of Popular Science: Phrenology and the Organization of Consent in Nineteenth-Century Britain.* Cambridge: Cambridge University Press.

Cooter R. 1989. *Phrenology in the British Isles: An Annotated, Historical Bibliography and Index.* Metuchen, NJ, & London: Scarecrow Press.

Elliotson J. 1840. *Human physiology incorporating much of the elementary part of the institutiones physiologicae of J. F. Blumenbach.* 5th ed. London: Longman, Orme, Browne, Green, and Longmans.

Flourens P. 1856. *Recueil des éloges historiques lus dans les sciences publiques de l'Académie des sciences.* Vol. 2. Paris: Garnier.

Gall F. J. 1818. *Anatomie et Physiologie du system nerveaux in general et du cerveau en particulier.* Vol. 3. Paris: d'Hautel.

Gall F. J. 1821. *Demande pour être nommé membre de l'Académie Royale des Sciences.* Paris.

Gall, F. J., and J. G. Spurzheim. 1810–1819. *Anatomie et physiologie du système nerveux en général, et du cerveau en particulier.* Paris: F. Schoell.

Gibbon C. 1878. *The life of George Combe, author of "The constitution of man.* Vol. 2. London: Macmillan.

[Gordon J]. 1815. The doctrines of Gall and Spurzheim. *Edinburgh Review* 25 (June):227–68.

Hollander, B. 1920. *In search of the soul: And the mechanism of thought, emotion, and conduct.* London: K. Paul, Trench, Trubner & Co.

Lesky E. 1981. Der angeklagte Gall. *Gesnerus: Revue trimestrielle publiée par la société d'histoire de la médecine et des sciences naturelles* 38:301–11.

Lynch M. 1985. "Here is adhesiveness": From friendship to homosexuality. *Victorian Studies* 29:67–96.

Moscati, F. M. 1833. Biographical paper on the character and phrenological organization of Dr. Spurzheim. *Lancet* 18 (20 Jan.):399–403.

Neuburger M. 1917. Briefe Galls an Andreas und Nannette Streicher. *Archiv für Geschichte der Medizin* 10:3–70.

Outram D. 1984. *Georges Cuvier: Vocation, Science and Authority in Post-Revolutionary France.* Manchester: Manchester University Press.

Philosophical Magazine, 44, Dec. 1814, p. 470.

Poole R. 1824. View of some of Dr. Spurzheim's lectures as delivered at Edinburgh in the Winter of 1816. *Transactions of the Phrenological Society* 25:89–130.

Special Collections Section of the Francis A. Countway Library of Medicine. Boston, MA: Boston Medical Library, B MS c 22.

Spurzheim J. G. 1818. *Observations sur la folie ou Sur les dérangements des fonctions morales et intellectuelles de l'homme.* Paris: Treuttel et Würtz.

Spurzheim J. G. 1825. A sketch of the natural laws of man [actually 1828], Boston edition, 1833, 5th American edition Boston, 1839; another edition New York, 1840. Revised and republished Glasgow. 1844.

Spurzheim J. G. 1826. *A philosophical catechism of the natural laws of man.* London: Charles Knight.

Spurzheim J. G.. c.1870. *The natural laws of man: A philosophical catechism.* 7th ed. Manchester, London.

Staum M. 1995. Physiognomy and phrenology at the Paris Athénée. *Journal of the History of Ideas* 56 (3):443–62. doi:10.2307/2710035.

Trierisches Wochenblatt. 1791. Nr 40, 1792, Nr 39, 1793, Nr 38, 1795, Nr 39 1796, Nr 38.

van Wyhe J. 2002. The authority of human nature: The Schädellehre of Franz Joseph Gall. *British Journal for the History of Science* March 35:17–42. doi:10.1017/S0007087401004599.

van Wyhe J. 2003. Science vs. religion or my gods are better than your gods? The controversies over Combe's *Constitution of Man* 1826-60. *Intellectual News.* Vol. 11/12, 24–31.

van Wyhe J. 2004. *Phrenology and the origins of Victorian scientific naturalism.* Farnham, UK: Ashgate.

van Wyhe J. 2007. The diffusion of phrenology through public lecturing. In *Science in the marketplace: Nineteenth-century sites and experiences,* ed. A. Fyfe and B. Lightman, 60–96. Chicago: Chicago University Press.

Walsh A. 1972. The American tour of Dr Spurzheim. *Journal for the History of Medicine and Allied Science* 27:187–206. doi:10.1093/jhmas/XXVII.2.187.

The split between Gall and Spurzheim (1813–1818)

Harry Whitaker and Gonia Jarema

ABSTRACT
An acerbic footnote in Volume 3 (1818) of the five-volume great work of Franz Joseph Gall and Johann Gaspar Spurzheim, *Anatomy and Physiology of the Nervous System in General and of the Brain in Particular with Observations on the Possibility of Understanding the Many Moral and Intellectual Dispositions of Man and Animals by the Configuration of Their Heads*, marked the end of the collaboration between Gall, the founder of *organologie*, and Spurzheim, promoter of phrenology. We discuss the background of this note and the nature of the rift that marked the end of Gall and Spurzheim's collaboration.

Mentoring has a special role in science, humanities, the arts, and religion. In the first decades of the twenty-first century, academic medicine has paid increasing attention to that relationship, both successful and unsuccessful, between mentors and mentees or disciples. Recent studies of mentor–disciple relationships are of two general types, the first of which could be characterized as genealogical, typified by the classic study of Boring and Boring (1948) that examined the question of "who taught whom" with the data summarized in a tree diagram. This model was copied for German psychologists by Wesley (1965) and later for French psychologists by Wesley and Hurtig (1969). The second type focuses on the interpersonal interactions in the time leading up to or during a mentoring process (Sudzina & Knowles, 1993; Sambunjak, Straus, & Marusic, 2006; Strauss, Chatur, & Taylor, 2009). These latter surveys discuss both successful and failed mentor–disciple relationships. Evidently, the second approach has not captured the same interest among historians of medicine, psychiatry, or behavioral sciences. Although there is frequent mention of mentors and their students in the major journals in these fields (*Journal of the History of Medicine, History of Psychiatry, Journal of the History of the Behavioral Sciences*, and, of course, *Journal of the History of Neuroscience*), most articles document the fact in the manner of Boring and Boring (1948) rather than explore the nature of the relationship, for example, the recent paper by Bentivoglio, Vercelli, and Filogamo (2006). Mentor–disciple interactions may be complicated by pride, prestige, jealousy, and, of course, money; to some degree all of these play a role in a classic example of a failed relationship that of the founder of *organologie*, Franz Joseph Gall (1758–1828), and his student and collaborator, Johann Gaspar Spurzheim (1776–1832).

Gall synthesized craniology from physiognomy and from contemporary knowledge of brain function (e.g., Unzer, as well as Prochaska, who was one of his teachers), utilizing an empiricism associated with the development of eighteenth-century science. Spurzheim first met Gall in 1800 when he began attending Gall's lectures in Vienna; in 1804, he became Gall's paid assistant (Van Wyhe, 2004) and, by 1808, Spurzheim was a formal collaborator whose name appeared on their publications until 1813 (Gall & Spurzheim, 1808, 1810–1813). Between 1808 and 1813, it is reasonable to argue that Gall had the major influence on the development of *organologie*; however, it is equally reasonable to argue that Spurzheim's contribution to that development was well beyond that of merely a paid assistant. To defend that assumption, consider the following: With the exception of the three books in 1808, 1810, and 1813, neither Gall nor Spurzheim formally collaborated with anyone else; when put into historical context, we take note of the fact that coauthorship of scholarly work was much less common in the early-nineteenth century than it is today. The plausible conclusion is that Spurzheim's contribution was significant enough to Gall that he added his name to the title page. After 1813 until his death in 1832, Spurzheim's singly-authored books influenced the spread and popularity of what he first called physiognomy then phrenology, in the English-speaking countries of England, Scotland, and the United States. Gall published nine more books after 1813, all singly authored.

During the preparation of their great work on the anatomy and physiology of the nervous system, Gall and Spurzheim parted company. As noted above, Volumes 1 and 2 of this work (1810–1813) had been printed with both names on the title page. From the first footnote in Volume 2, p. 147, we learned that "Dr. Spurzheim, having left Paris to teach the doctrine of the functions of the brain in England and to continue to gather new facts there, is no longer working on the writing of this work." According to Van Wyhe (2004), "the reasons for their separation remain unclear" (p. 27). As Temkin (1947) concluded, "[S]trictly speaking, therefore, Spurzheim's collaboration in the main work extends only over Vol. 1 and pp. 1–146 of Vol. 2" (footnote 24, p. 280). Although the biographical nature of the footnote is uncommon, there is no hint of the animosity that Gall was to develop towards his former protégé until one encounters Footnote 2, Sections ix–xix of Volume 3 (Gall, 1818–1819c) subtitled "Remarks on the Work Titled: Observations on Phrenology, or the Knowledge of Moral and Intellectual Man, Founded Upon the Functions of the Nervous System by G. Spurzheim, D.M." (hereafter "Remarks").

Our focus here is an analysis of Footnote 2, Gall's1818"Remarks" with attention to how he perceived the science of craniology or organology, how he perceived Spurzheim to have abandoned these principles, and how and why Gall chose to attack his former student, associate, and collaborator. In the "Remarks," we find two criticisms, the first a potentially interesting scientific critique of Spurzheim's alterations to Gall's system, and the second, mixed in with the scientific critique, is a personal attack on Spurzheim himself. Gall lashes out rather viciously, his anger expressed in several direct and quite virulent attacks on Spurzheim's intellect and character; several sarcastic remarks are clearly intended to belittle Spurzheim and there is an effective rhetorical device designed to disparage Spurzheim's work by posing a series of questions underscoring what Gall perceived to be Spurzheim's faulty reasoning.

Keeping the temporal context in mind, remember that Spurzheim had already published four books *in English* between 1815 and 1817, three of which mentioned Gall's name in their title. For example, in 1815, he wrote *The Physiognomical System of Drs. Gall and Spurzheim*, published in London. Since Gall could read English, we find it notable that it was not until Spurzheim's 1818book *in French* that he reacted. Why? To begin, we surmise that Gall had decided that the Parisian milieu was his demesne; the previous English works by Spurzheim did not invade this domain but the French one did. We also note that the title of Spurzheim's 1815 book included Gall's name and referred to their project as physiognomy; however, in the 1818 French book[1] Spurzheim did not mention Gall's name in the title and he also referred to the project as phrenology, a term that Gall did not like — Was he rubbing salt in the wound?

Another clue is found on the first page of the "Remarks." Gall categorizes Spurzheim with other "auditors" who had published synopses of his doctrine since 1802 (Froriep, Bloede, Bischoff, Demangeon, Nacquart, et al.). Gall professes to have been unconcerned about them; however, he then says:

> At the moment when the third volume was about to be published Mr. Spurzheim[2] found it appropriate to place himself in the class of a great number of my auditors and to publish a very incomplete treatise[3] on my doctrine. He claims in several passages the pretense to have introduced views that are more philosophical than those of the first author who, according to the expressions used in the newspapers by the friends of Mr. Spurzheim, had left his child in the cradle. (Gall, 1818–1819c, x)

This last sarcastic remark refers to a newspaper article in *La Quotidienne, 9 décembre 1818*, that suggested Gall had not developed his ideas very well in comparison to Spurzheim, as seen in the 1818 French book, mentioned above. Thus, not only did Spurzheim invade Gall's demesne but he received favorable publicity at Gall's expense for doing so, ostensibly adding insult to potential financial injury. The last line of Gall's second footnote is a vitriolic gem:

> Mr. Spurzheim's work is 361 pages long, of which he has copied 246 pages from me. He says he has the right, because he was supposed to be my collaborator in the first volume.... But he knows that he was only charged with the task of furnishing literary notes. At least he should have indicated the source of his riches. He no longer had the same right to my works on the organ of the soul and on the plurality of organs. Others have already accused him of plagiarism; it is at the least very ingenious to have made a book by cutting with scissors. (Gall, 1818–1819c, x)

The sarcasm in suggesting that Spurzheim did not write his book but assembled it mechanically with scissors and paste needs no comment other than to applaud the quality of the metaphor; we also note the subtler sarcasm of claiming that Spurzheim did not contribute to the content of the by-now-famous coauthored works but merely tracked down library references. We see no reason to take this at face value; why would a secretarial assistant be given coauthorship?

Gall criticizes Spurzheim on philosophical grounds for failing to adequately justify his "phrenological" faculties, at one point making what we regard as an astute observation that "everything you see on the surface is not necessarily a faculty"; in other words, there are bumps and then there are *bumps*. We take this as both scientifically valid as well as perceptive, albeit it remains debatable whether Spurzheim attempted to follow such

---

[1] *Observations sur la phraenologie, ou la connaissance de l'homme moral et intellectuel, fondée sur les fonctions du système nerveux.*
[2] We note that it was "Dr. Spurzheim" in the first footnote of 1813.
[3] No reference to the three English books.

scientific rigor. Noting that some of the faculties, for example, *amativity*, do not apply equally to man *and* to animals, Gall argues that the name of the faculty should reflect whether or not it is a universal primitive or only a human one. He then claims that (a) 23 of the faculties proposed by Spurzheim, "except for changing the name, conform to my discoveries," whereas (b) all of the new ones have not yet been proven by sufficient and precise empirical observations. As is well known, Gall originally proposed 27 *faculties* as far back as 1805 and, continuing until his death in 1828, he never published more than the original 27. On the other hand, first author Whitaker has personally seen a painted snuff box preserved in the Gall museum in Tiefenbronn, Austria, on which 35 faculties are written in ink in Gall's own handwriting. This judgment is based on comparing the handwriting to that of a letter written by Gall, found in the leaves of a book in Theodor Meynert's laboratory library in Vienna. The snuff box probably dates from Gall's Vienna period, prior to the European tour that terminated in Paris. According to Stanley Finger (personal communication, 2016), Gall discussed over 30 faculties while in Vienna; Finger found reference to 33 in 1801 and 32 in early 1805; during 1805, he proposed 26 and finally settled on 27 at the end of 1805. The indecision over the number of faculties occurred during Spurzheim's tenure as an "auditor" in 1800–1804 and as a famulus in 1804–1808.

Consequently, Spurzheim knew well that the number of faculties had been frequently changed up to 1805 and, therefore, it is realistic to conclude that he is unlikely to have considered 27 faculties a particularly special number. He began adding faculties to the list from his first sole-authored book in 1815; by the 1818 book, he had a total of 31 (Whitaker, 2000) and shortly thereafter had increased the number to 35. In the 1818 book in French, Spurzheim had not only added four new faculties but he also replaced four of Gall's original ones, therewith, we surmise, rubbing more salt in the wound. Although implicit, we think it might have been the case that Gall would in fact have accepted Spurzheim's eight new faculties, if (a) he thought there were adequate evidence for them and, presumably, (b) had Spurzheim kept them in English texts. In the "Remarks," Gall rhetorically asks, since "for now, the 8 [new ones] are mere conjectures, why doesn't Mr. Spurzheim wait until experience supports them?" (p. xii). Claiming that his own order of presentation of the faculties is more natural than Spurzheim's, Gall suggests that:

> [w]hat he calls putting philosophy into the physiology of the brain, something I never had the ambition to do,[4] forces one to make unnatural divisions[5] between the faculties, for example jumping from *destruction* to *construction*; one could believe that this monstrosity was created on purpose to make it almost impossible to study [the faculties] empirically...
> The triumph of this new division of the faculties of the soul is when Mr. Spurzheim puts among the intellectual faculties, the faculties of touch, taste, odor, hearing and vision, the five external senses. Spurzheim knows perfectly well that the intellectual faculties exist independently of the five senses. (Gall, 1818–1819c, xii)

On this point, anyone would have to concur, regardless of what one thinks of phrenology or craniology. Gall continues:

---

[4]Gall was not a native speaker of French; he may have intended "philosophie" to mean a proposition unsupported by empirical evidence.
[5]Van Wyhe (2004) notes that the appeal to nature was popular gambit used by phrenologists to defend their system.

Mr. Spurzheim, either because he wishes to flirt with ignorance or because he was intimidated by the facetiousness of a certain journalist, rejects moral evil[6] as an innate faculty ... Spurzheim thinks that all faculties are in themselves good and endowed for a beneficial goal... but, there are propensities to vice; without them, how would evil exist within us? (Gall, 1818–1819c, xiii)

With that remark, Gall has perhaps given us an early hint of sociobiology. This final point by Gall is, to our knowledge, the first to suggest one of the more telling criticisms of later-nineteenth-century phrenology, particularly as it was to develop in the United Kingdom and United States, that is, what one might call the "fortune-teller's bias": All the innate faculties are good and thus each person has only good characteristics.

In summary, Gall's critique makes no effort to veil his emotional reaction; clearly, he was angry and the sarcastic style reflects that fact. But it also fails in rigor and logic; in many places, it is less an argument for a science of organology than it is a defense of the territory Gall had staked out in Paris. Van Wyhe (2004) makes the same point when discussing John Gordon's attack on Spurzheim's competence as an anatomist. The overt argument is (a) that Spurzheim changed the natural structure of the faculties and how they relate to each other, (b) that Spurzheim shifted the faculties to a unidimensionality of *goodness*, and (c) that Spurzheim ignored the requirement for empirical evidence to support each faculty. Even the overt arguments are objectively weak: Gall admitted that Spurzheim did little more than a name change to many of the faculties, and that he, Gall, would probably accept the eight new faculties proposed by Spurzheim if evidence could be adduced in their favor. What constituted evidence according to Gall, Spurzheim, and their followers is of course questionable.

We now turn to what Spurzheim had to say. In 1826, he published a book that he claims to have translated from an unpublished French manuscript. In the preface, he responds to Gall's 1818 attack. Spurzheim first reminds the reader that Gall developed his doctrine from 1796, when he occasionally gave lectures at his home in Vienna, through 1800, when Spurzheim first became acquainted with Gall's ideas, up to 1804, when Spurzheim began to collaborate with him,[7] up to 1813, when Spurzheim claims that he left in order to pursue his own research. The last claim is fully supported by Gall's first footnote to Volume 2 of the great work (see above). As an aside, we note that the date on the title page of Volume 2 of Gall and Spurzheim, which contains the brief footnote mentioning Spurzheim's departure, is 1812, not 1813; the discrepancy in dates may very well be due to the fact that the five volumes of the great work were published in separate signatures over a period of several years, a common publishing practice in the early-nineteenth century. In the 1826 book, Spurzheim also notes that reports by those who had attended Gall's lectures right from the beginning contained both favorable and unfavorable criticisms; this is also true. Spurzheim constructed his role with Gall (1804–1813) not only as a collaborator but as something of an equal; we agree that this is debatable, noting that no unequivocal evidence for or against this interpretation has yet been forthcoming. Although Spurzheim does acknowledge that Gall was the originator of the doctrine, he avers that each of them supplied his own ideas, each contributed to the coauthored books, which date from the 1808 memoir to the *Académie des sciences* in Paris. This memoir was Gall's pitch to be inducted into the *Académie*; it was

---

[6]Our translation of "mal moral."

[7]An exaggeration, for at that time, Spurzheim was a paid assistant not a collaborator.

subsequently rejected. Spurzheim explicitly takes individual credit for the hypothesis of the folding of the brain during development, an idea similar to current thinking in developmental neuroscience. Spurzheim truthfully admits to copying from Volume 1 of the great work in the present 1826 book because "he has a right to it" as coauthor; he makes a particular point about his contribution to the anatomical plates in the great work (eventually published as Tome V but notably without Spurzheim's name on it) and he also candidly notes that this work was "continued by Gall, singly after the middle of the second volume (xiv)," supporting Temkin's conclusion stated earlier (see above). Spurzheim's preface concludes with his own attack on several other writers for having appropriated *their* (Gall and Spurzheim's) work without proper acknowledgment. For example, Cloquet (1827) evidently copied Gall and Spurzheim's anatomical plates, and Serres (1824), who after first criticizing Gall and Spurzheim on some points, then audaciously borrowed other points without acknowledgment. As an aside, Serres (1824) won a prize from the *Académie des sciences*, an obvious sore point with Spurzheim. Since Spurzheim's accusation falls outside the focus of this article, we have not compared the works of Serres (1824) and Cloquet (1827) against that of Gall and Spurzheim's great work and, thus, cannot verify his accusation.

By 1826, Gall's health had deteriorated; he suffered his first stroke in April of that year. He had finished and published the six volumes of his last work in 1825. Spurzheim on the other hand was enjoying great success in England, Scotland, and, shortly to come, the United States. Habitually, most references to phrenology at this time, both positive and negative, referred to Gall and/or Gall and Spurzheim, thus it is quite plausible that Spurzheim had something to lose if he were to dissociate himself from Gall. On the other hand, we know from an 1824 remark by Fossati (1857) that Gall's animosity toward Spurzheim was unchanged at least 6 years after the published attack in Footnote 2, and indirect, secondary references suggest that the two never reconciled before Gall's death in 1828 (Van Wyhe, 2004). Nevertheless, we must point out that, remarkably, in Gall's last work (1825), he quoted Spurzheim's own cases as empirical support of the craniological, not phrenological faculties—estranged but perhaps never completely separated? Van Wyhe (2004) discusses a different perspective on the split:

> In Paris in 1813, half-way through the second volume of *Anatomie et Physiologie* Gall and Spurzheim parted forever. The reasons for their separation remain unclear. Spurzheim claimed he left Gall because the latter had taken all the credit for Spurzheim's discoveries. Gall remarked in writing simply that Spurzheim had gone to England to teach the physiology of the brain and to gather new facts. A different explanation is also attributed to Gall. John Elliotson, materialist phrenologist, mesmerist, professor of clinical medicine at London University, and critic of Spurzheim, related Gall's story: "one day, Dr. S said he himself was going alone to England and he actually left Gall in a week, it turning out that he had been learning English with this view in Gall's house, without Gall's knowledge for six months". Elliotson's ally, the phrenologist F. M. Moscati, who also knew Gall, later corroborated Elliotson's rendition and added that Gall claimed in 1824 that *he* had separated from Spurzheim over "some pecuniary disagreement." (pp. 27–29, emphasis in original)

If this were true, Elliotson's report of what Gall said would indeed add another element to consider in analyzing the split between Gall and Spurzheim. However, in a footnote on page 29, Van Wyhe (2004) contradicts the story by Elliotson:

As early as April 1807, while Gall and Spurzheim were in Munich, Soemmerring wrote in his diary: "I advised [Gall] to be somewhat subtler in England — and to be firmly saddled in anatomy ... I am to admonish Spurzh, who is to learn English, to be diligent." (as quoted in Mann, "Franz Joseph Gall [1758–1828] und Samuel Thomas Soemmerring: Kranioskopie und Gehirnforschung zur Goethezeit," 1983, 178)

In our view, the evidence favors the interpretation that Gall would very likely have left Spurzheim to his own devices after his departure in 1813 if (a) Spurzheim had not returned to Paris to stake a claim in Gall's territory, which of course threatened Gall's income from his lectures and if (b) the French press had not criticized Gall by comparison to Spurzheim. The separation was in 1813, Gall's attack was five years later; during the interim period, Spurzheim had published four books on the same subject without any protests from Gall. Whether or not Spurzheim actually contributed discoveries of his own to their joint publications is unknown; the fact that his name appeared on the first two volumes of the "great work" and as well on the earlier 1808 work alone would certainly give Spurzheim legitimate claim to their contents if not credence to his role in their generation.

Acknowledgments

The authors gratefully acknowledge the helpful comments of Stanley Finger, Rhonda Boshears, Paul Eling, and Ola Selnes.

References

Bentivoglio M, Vercelli A, Filogamo G (2006): Giuseppe Levi: Mentor of three Nobel laureates. *Journal of the History of the Neurosciences* 15: 358–368.

Boring MD, Boring EG (1948): Masters and pupils among the American psychologists. *The American Journal of Psychology* 61: 527–534.

Cloquet J (1827): *Manual of Descriptive Anatomy of the Human Body, Illustrated by Two Hundred and Forty Lithographic Plates*, Godman JD, trans. Boston, Rutgers Press.

Fossati GA (1857): Gall. In: Hoefe F, ed., *Nouvelle Biographie Générale*. Paris, Firmin, Didot & Freres, Vol. XIX, pp. 271–283.

Gall FJ (1818): Remarques sur l'ouvrage intitulé: Observations sur la phraenologie, ou la connois-sance de l'homme moral et intellectuel, fondée sur les fonctions du système nerveux par G. Spurzheim, D.M. In: Gall FJ, ed., *Anatomie et physiologie du système nerveux en général et du cerveau en particulier avec des observations sur la possibilité de reconnoître plusieurs dispositions intellectuelles et morales de l'homme et des animaux par la configuration de leurs têtes*. Paris, Libraire Grecque-Latine-Allemande, Tome III, pp. ix–xix.

Gall FJ (1825): *On the Functions of the Brain and Those of Each of Its Parts* (6 vols.). Paris, J.-B. Baillière.

Gall FJ, Spurzheim JG (1808): *Recherches sur le système nerveux en general, et sur celui du cerveau en particulier; mémoire présenté a l'institut de france, le 14 mars 1808; suivi d'observations sur le rapport qui en a été fait a cette compagnie par ses commissaires, avec une planche*. Paris, F. Schoell, Rue des Fossés – S. – Germain –L'Auxerrois, No. 29.

Gall FJ, Spurzheim JG (1810–1813): *Anatomie et physiologie du système nerveux en général et du cerveau en particulier avec des observations sur la possibilité de reconnoître plusieurs dispositions intellectuelles et morales de l'homme et des animaux par la configuration de leurs têtes*. Paris, F. Schoell, Rue des Fossés – S. – Germain –L'Auxerrois, Tomes I et II, No. 29.

Sambunjak D, Straus SE, Marusic A (2006): Mentoring in academic medicine: A systematic review. *Journal of the American Medical Association* 296: 1103–1115.

Serres ERA (1824): *Anatomie comparée du cerveau dans les quatre classes des animaux vertébrés, appliquée à la physiologie e à la pathologie du système nerveux*. Paris, Chez Gabon et Compagnie, Libraires, Tome Premier.

Spurzheim JG (1815): *The Physiognomical System of Drs. Gall and Spurzheim: Founded on an Anatomical and Physiological Examination of the Nervous System in General, and of the Brain in Particular; and indicating the Dispositions and Manifestations of the Mind*. London, Baldwin, Cradock, and Joy.

Spurzheim JG (1818): *Observations sur la phraenologie, ou la connaissance de l'homme moral et intellectuel, fondée sur les fonctions du système nerveux*. Paris, Treuttel et Würtz.

Spurzheim JG (1826): *The Anatomy of the Brain with a General View of the Nervous System* (translated from an unpublished French manuscript). London, S. Highley.

Strauss SE, Chatur F, Taylor M (2009): Issues in the mentor-mentee relationship in academic medicine: A qualitative study. *Academic Medicine* 84: 135–139.

Sudzina M, Knowles JG (1993): Personal, professional and contextual circumstances of student teachers who "fail": Setting a course for understanding failure in teacher education. *Journal of Teacher Education* 44: 254–262.

Temkin O (1947): Gall and the phrenological movement. *Bulletin of the History of Medicine* 21: 275–321.

Van Wyhe J (2004): *Phrenology and the Origins of Victorian Scientific Naturalism*. Burlington, Ashgate.

Wesley F (1965): Masters and pupils among the German psychologists. *Journal of the History of the Behavioral Sciences* 1: 252–258.

Wesley F, Hurtig M (1969): Masters and pupils among French psychologists. *Journal of the History of the Behavioral Sciences* 5: 320–325.

Whitaker HA (2000): Phrenology. In: Kazdin AE, ed., *Encyclopedia of Psychology*. Washington, DC, American Psychological Association, Vol. 6, pp. 188–191.

International Connections

The reception of Gall's organology in early-nineteenth-century Vilnius

Eglė Sakalauskaitė-Juodeikienė, Paul Eling, and Stanley Finger

ABSTRACT
Much has been written about the development and reception of Franz Joseph Gall's (1758–1828) ideas in Western Europe. There has been little coverage, however, of how his *Schädellehre* or organology was received in Eastern Europe. With this in mind, we examined the transmission and acceptance/rejection of Gall's doctrine in Vilnius (now Lithuania). We shall focus on what two prominent professors at Vilnius University felt about organology. The first of these men was Andrew Sniadecki (1768–1838), who published an article on Gall's system in the journal *Dziennik Wileński* in 1805. The second is his contemporary, Joseph Frank (1771–1842), who wrote about the doctrine in his memoirs and published an article on phrenology in the journal *Bibliotheca Italiana* in 1839. Both Frank and Sniadecki had previously worked in Vienna's hospitals, where they became acquainted with Gall and his system, but they formed different opinions. Sniadecki explained the doctrine not only to students and doctors but also to the general public in Vilnius, believing the new science had merit. Frank, in contrast, attempted to prove the futility of cranioscopy. Briefer mention will be made of the assessments of Johann Peter Frank (1745–1821) and Ludwig Heinrich Bojanus (1776–1827), two other physicians who overlapped Gall in Vienna and went to Vilnius afterward. Additionally, we shall bring up how a rich collection of human skulls was used for teaching purposes at Vilnius University, and how students were encouraged to mark the organs on crania using Gall's system. Though organology in Vilnius, as in many other places, was always controversial, it was taught at the university, accepted by many medical professionals, and discussed by an inquisitive public.

Introduction

In 1798, Franz Joseph Gall (1758–1828), who had remained in Vienna after obtaining his medical degree, wrote a formal letter to the public censor, Joseph Friedrich Freiherr von Retzer (1754–1824), outlining his nascent theory of organology (Gall, 1798). Gall contended that there are many independent faculties of mind and that small territories of the cortex provide the necessary instruments for our propensities, ways of thinking, and character traits.

Color versions of one or more of the figures in the article can be found online at www.tandfonline.com/njhn.
© 2017 Taylor & Francis

These territories, he explained, could be discerned by correlating cranial features with exceptional behaviors or deficiencies: his so-called system of bumps. Some of his contemporaries agreed with his notions, while others considered his ideas about the mind to be materialistic, scandalous, or unsubstantiated and antireligious (e.g., Walther, 1802; Bischoff, 1805).

Much has been written about the development, dissemination, and reception of Gall's doctrine, especially after his landmark, five-volume *Anatomie et Physiologie du Système Nerveux en Général, et du Cerveau en Particulier* [Anatomy and Physiology of the Nervous System in General and of the Brain in Particular] was published in Paris during the second decade of the nineteenth century (Gall & Spurzheim, 1810–1819).[1] Nonetheless, the more widely disseminated reviews have been almost (if not entirely) about what transpired in Western Europe and the United States (e.g., Temkin, 1947; Davies, 1955; Ackerknecht, 1958; Young, 1970), especially his organology's transformation into "phrenology"[2] and the debates it generated in Great Britain (e.g., Cooter, 1984; Van Wyhe, 2002, 2004, 2007). How the doctrine fared in Eastern Europe has for too long gone virtually unnoticed, although it too should be a part of its history (see Strojnowski, 1965).

The aim of this article is to examine the reception of organology in Vilnius, the capital city of the Grand Duchy of Lithuania until 1795 and the center of Vilnius Governorate of the Russian Empire during the nineteenth century. In particular, our focus will be on Vilnius University, which was one of the most important institutions in the eastern region of scientific Europe during the first half of the nineteenth century. To this end, after some opening comments about Vilnius and its university, we shall analyze an article by physician Andrew Sniadecki (1768–1838), his *Krótki Wykład Systematu Galla z przyłączeniem niektórych uwag nad iego Nauką* [Short Lecture on the System of Gall with some Comments about his Science]. This paper was written in Polish and was published in 1805 in the journal *Dziennik Wileński* [Vilnius Daily]. We then shall turn to Joseph Frank's (1771–1842) *Mémoires Biographiques de Jean-Pierre Frank et de Joseph Frank son fils* [Biographical Memoirs of Johann Peter Frank and his son Joseph Frank], written in French, and the younger Frank's article, *Della frenologia* [On Phrenology], written in Italian and published in *Bibliotheca Italiana* in 1839. Briefer mention will be made of Joseph's father's, Johann Peter Frank (1745–1821), assessment, and that of Ludwig Heinrich Bojanus (1776–1827), yet another physician overlapping Gall in Vienna before heading north to Vilnius. Some concluding remarks about Gall, his system, and "phrenology" in Lithuania and Eastern Europe will follow.

Vilnius University

Vilnius University was founded in 1579 by Jesuits in the multinational[3] and multiconfessional city of the same name in the Catholic Grand Duchy of Lithuania (for its history, see Bendžius et al., 1977; Bumblauskas et al., 2004). It is one of the oldest and most respected

[1]Johann Gaspar Spurzheim (1776–1832) was the second author on the first two volumes and the atlas. A less expensive edition (without the atlas) bearing only Gall's name appeared between 1822–1825, and this edition was translated into English in 1835 (Gall, 1822–1825, 1835).
[2]Spurzheim popularized the term phrenology, which Gall detested, but Spurzheim did not coin it when he first used it in print in 1818 (Spurzheim, 1818). Its origin can be traced to Philadelphia physician Benjamin Rush (1745–1813), who employed it in two medical school lectures published in 1811 (Rush, 1811a, 1811b; Noel & Carlson, 1970). Four years later, it was used by British physician Thomas Forster (1789–1860), a friend of Spurzheim (Forster, 1815).
[3]Even in the beginning of the nineteenth century, Vilnius remained a multinational city: "More than 35 000 people, of whom 22 000 Catholics, 600 Greeks, 500 Lutherans, 100 Protestant Reformers, 11 000 Jews, and 60 Mohammedans, lived in Vilnius" in 1804, can be read in the memoirs of Joseph Frank (Frankas, 2013, p. 51).

universities in Central and Eastern Europe, its founders were strongly influenced by Renaissance, Reformation, and Counterreformation notions and ideas, and it was modeled after the Jesuit College in Rome and originally had two faculties: philosophy and theology.

After the abolition of the Society of Jesus in 1773, with the papal bull of Clement XIV (1705–1774), Vilnius University came under the jurisdiction of the Educational Commission of the Polish-Lithuanian Commonwealth, which promoted a new curriculum of civic education and secularism. The Faculty of Medicine was founded during this era, in 1781. Special attention was paid to the promotion of natural sciences in accordance with the ideals of the Age of Enlightenment.

On May 3, 1791, the Constitution of Polish-Lithuanian Commonwealth was approved. Its ideas were then defended against the invading Russian Army during the uprising of 1794. The final partition of the Polish-Lithuanian Commonwealth was carried out a year later. Even after the abolition of the Polish-Lithuanian Commonwealth and the annexation of the Grand Duchy of Lithuania by the Russian Empire, Vilnius University was able to maintain the same rapid pace of intellectual life, continuing to stimulate the formulation of new ideas in the natural sciences.

Little can be stated with certainty about courses involving the nervous system and its diseases, or about how systems of medicine were taught in Vilnius, before the nineteenth century. What is known is that Jakob Marquart (1583–1658), a doctor of theology, had introduced the basics of human anatomy to philosophy students during the seventeenth century using Andreas Vesalius' (1514–1564) *De humani corporis fabrica* (Bogušis, 1997). Also, after the Faculty of Medicine was founded in 1781, Stephanus Bisius (Stephanus Bisio, Bisis Frexonariensis; 1724–1790), a physician of Italian descent, began to lecture on anatomy and physiology, including not only osteology, splanchnology, angiology, and myology but also what would become neurology, this happening between 1781 and 1787 (Biziulevičius, 1997). Bisius, in fact, might have been the first physician in Vilnius to publish an original study on nervous and mental diseases. This work was titled *Responsum Stephani Bisii Philosophiae et Medicinae Doctoris ad Amicum Philosophum De Melancholia, Mania et Plica Polonica sciscitantem* [A Reply of the Doctor of Philosophy and Medicine Stephanus Bisius to the Question from the Philosophical Society about Melancholia, Mania and Plica Polonica][4] (Bisius, 1772). Additionally (as revealed in the curriculum of 1783–1784), Professor of Surgery Jacques Briôtet (1746–1819) lectured on external head wounds, concussions, subdural and intracerebral extravasations, and trepanation; Professor of Anatomy Joannes Andreas Loebenwein (1758–1820) (as announced in the curriculum of 1787) lectured on the theory of *sensibilitas* [sensibility]; and an Irish Professor of Therapy, John O'Connor (1760–1801), lectured through the academic year of 1800–1801 on the inflammatory diseases, including "inflammations of cerebral membranes" (Biziulevičius, 1997).

By the early-nineteenth century, Vilnius University had become the largest in the Russian Empire, based on student numbers and university departments (Bumblauskas et al., 2004). It remained a center of scientific thought and political freedom until it was

---

[4]*Plica polonica*, a tuft of matted, felted and filthy hair, is a phenomenon that was often considered an affliction confined exclusively to Poland and Lithuania. It had been thought to be a punishment from God, which could not be disposed of simply by cutting off one's hair, as this could lead to serious complications and even the patient's death (see Klajumaitė, 2013). *Plica polonica* was believed to be associated with number of other pathologies: diseases of brain, spinal cord, nerves, diseases of bones, cartilages, tendons, muscles, membranes, blood vessels, heart, lung, skin and other visceral organs (see Kaczkowski, 1821).

closed by Russian imperial authorities in 1832, following the suppression of a Polish and Lithuanian uprising during the previous year. It was during this era, which was marked by an increased interest in new biological theories and clinical medicine, that Dr. Joseph Frank, Professor of Special Therapy and Clinical Medicine, and Dr. Andrew Sniadecki (Jędrzej Śniadecki, Andrzej Śniadecki), then Professor of Natural Sciences, as well as Joseph's father Johann Peter and Ludwig Heinrich Bojanus, lectured at the university.

Franz Joseph Gall: His ideas and early followers

During the first half of the eighteenth century, philosophers, physicians, and scientists, following in the footsteps of René Descartes (1596–1650), were still searching for the elusive seat of the soul. Although Descartes' pineal gland theory was no longer taken seriously, these men were not yet attending to the cerebral cortex. Rather, they were still more drawn to subcortical structures. More attention started to be paid to the cerebrum later in the century, although the clergy, influential philosophers, and physicians continued to view the largest part of the brain holistically, in part because they considered the soul an indivisible entity. Johann Gottfried Zinn (1727–1759), one of Albrecht von Haller's (1708–1777) most talented students and collaborators, was not alone when he wrote, "*animae sedem per omne cerebrum esse extensam*" [the soul resides throughout the brain] (Zinn, 1749; for more on this zeitgeist, see Haller, 1762; Neuburger, 1897; Finger, 1994; Karenberg, 2009). No one was venturing to propose anything like cortical localization of higher functions during this era, other than Swedish mystic Emanuel Swedenborg (1688–1772), whose insights seemed unknown to his scientific contemporaries, and whose best writings on the subject remained unpublished until late in the nineteenth century (Swedenborg, 1740–1741, 1882–1887, 1938–1940; also see Akert & Hammond, 1962; Finger, 1994; Norrsell, 2007).

As noted, Gall introduced his *Schädellehre* or organology, or what others would soon begin citing as his phrenological doctrine, in his published 1798 letter to the censor, written while still in Vienna. Driven to study nature empirically, including human nature, he was not one to delve into metaphysics and deal with an immaterial soul. Instead, he focused on parts of the brain critical for thinking and behaving, as revealed by brain damage, development, and cross-species comparisons. Prior to the start of the new century, he was already concluding that there are many faculties of the mind, each dependent on a specific territory, with most, but not all, involving discrete, small parcels of the cerebral cortex. What he was promoting was a major break with the past, and it is generally viewed as the starting point for the longstanding tradition of localizing higher functions in the cerebral cortex (for reviews, see Critchley, 1965; Young, 1970; Finger, 1994; Eling & Finger, 2015).[5]

Nevertheless, correlating functions with structures proved challenging to Gall, because he could not get or preserve the human brains he needed for making comparisons. Because he knew that the growing brain helped shape the cranium, he concluded that unusually large and active cortical regions would cause bumps to form on the overlying,

---

[5]Gall's doctrine would guide Jean-Baptiste Bouillaud (1796–1881) and Paul Broca (1824–1880) to their breakthroughs in clinical-pathological studies on speech, which would show more convincingly that the human cortex is not functionally uniform. The new localizationists, however, would reject Gall's skull-based localizations.

still-maturing cranium, which could be detected throughout life by sight and with palpation. Consequently, by 1798 he was already engaged in amassing a great number of skulls and casts of skulls of people with known talents and propensities (e.g., vanity, music, worshipping, killing), forming what amounted to a working library for his research program. The majority of his human skulls came from executed criminals and the insane, not from gifted musicians, mathematicians, and mechanics, whose prized skulls were far harder to obtain (see Hagner, 2003). He also collected animal skulls and was always looking for dogs, horses, and other animals that behaved in exceptional ways. Gall clearly favored cranioscopy over all other methods (e.g., dissections, studying the effects of brain mutilations in people or animals), proclaiming his science could be confirmed by anyone willing to observe nature with an open mind. Along with its scope and utility, its openness to more than the scientific or medical community was one of the most attractive parts of his doctrine, although also its most controversial feature.

Gall would eventually settle on 27 faculties of mind, but the number, names, and locations of his faculties varied while he was in Vienna. The same held true while he was making his way through the German states and other parts of Western Europe during his 1805–1807 lecture tour, prior to settling in France, where he would finalize his system and publish his long-awaited books. Most of our faculties of mind, he emphasized, such as love of offspring and memory for locations, could be observed in both humans and animals, although some, like wit and religion, are unique to humans. He associated these highest, most human faculties with territories in the front of the cerebrum, an exceptional large region in adult men and women, but one that is poorly developed in animals.

Between 1796 and until 1805, Gall presented his new ideas about the brain and behavior to the Viennese public and foreign visitors in lectures, demonstrations, and private meetings at his home, where he housed his large collection of skulls, casts, and other supporting materials. Although he did not have an academic appointment, he also discussed his ideas with officials, colleagues, and students at Vienna's schools and prisons, as well as at its impressive new hospital with its imposing, tower-like insane asylum.

The origins of Vienna's General Hospital go back to 1693, when Emperor Leopold I (1640–1705) arranged for the establishment of the large hospital. In 1697, the first ward was finished, and it housed 1,042 persons. The hospital grew, and, in 1784 under Emperor Joseph II (1741–1790), a new state-of-the-art hospital opened that was constructed according to the plan of the Hôtel-Dieu from Paris. It had an attached maternity ward, orphanage, and the aforementioned lunatic asylum (the *Narrenturm* or Fool's Tower), which was the first building of its kind solely for the accommodation of mental patients. Additionally, the *k. k. medizinisch-chirurgische Josephs-Akademie*, or Josephinum, a modern academy for army surgeons, was opened in 1785 (see Wien Geschichte Wiki, n.d., *Altes Allgemeines Krankenhaus*). These renovations and advances, in connection with the new medical curriculum emphasizing bedside teaching (as developed by Gerard van Swieten [1700–1772] of the Medical Faculty), helped draw students and recently graduated physicians from all over Europe to Vienna.

Ludwig Heinrich Bojanus was one of the first foreign physicians to report on Gall's lectures. After finishing his studies in medicine at the University of Jena (1797), Bojanus practiced in the General Hospital of Vienna between 1797–1798, while also attending Gall's public lectures. Then, in 1801, he began a tour through several European countries (to prepare for founding a veterinary institute). While in Paris, he lectured to the members of the *Académie de Médicine* about Gall's system, publishing his information in the *Magasin Encyclopédique, an. VIII*, the year from the Napoleonic *Calendrier Républicain* (or *Révolutionnaire*) *Français* that would correspond to 1801. An English translation appeared a year later in the *Philosophical Magazine* (Bojanus, 1801, 1802). In this 16-page article, Bojanus described the 33 organs that Gall had mentioned in his lectures, while rejecting the idea that Gall's thinking "leads immediately to materialism." He stated, however, that it will now be up to the author of the new doctrine "to furnish us with further details on the subject" and "to convince us, in an incontrovertible manner, of the truth of his system" (Bojanus, 1802, p. 138).

Bojanus became Professor of Zoology at Vilnius University in 1806, and several years later became its Professor of Comparative Anatomy. He was also a founder of modern veterinary medicine and a pre-Lamarckian pioneer of evolutionism in Vilnius (see Otto, 1831; Fedorowicz, 1958; Biziulevičius 1997; Viliūnas, 2014). Nonetheless, we have not found any documents showing that Bojanus gave lectures on Gall's organology at Vilnius University, although it is certainly possible that he did so.

Andrew Sniadecki and Joseph Frank overlapped Bojanus in Vienna. Sniadecki practiced in the city's hospitals and the new General Hospital of Vienna from 1795–1796. As for Joseph Frank, he served as its Chief Physician from 1796–1802 (J. Frank, 1848; Railienė, 2005).[6] Like Bojanus, both Sniadecki and Frank became well acquainted with Gall's new ideas, and, as we shall now show, both expressed strong opinions about his doctrine after leaving Vienna for Vilnius.

Andrew Sniadecki in support of Gall's theory

Andrew Sniadecki (see Fig. 1) was born near the small town of Žnin, in the Polish-Lithuanian Commonwealth and was a prominent physician and scholar (for biographical details, see Railienė, 2004, 2005). Sniadecki graduated from the University of Pavia in 1793 and spent 1794 furthering his studies at the University of Edinburgh. He then traveled to Vienna, where he served as physician in the city's hospitals, including the General Hospital of Vienna from 1795–1796.[7] After leaving Vienna, Sniadecki worked as a general practitioner in Wołyń, a province of Polish-Lithuanian Commonwealth. In 1797, he went to Vilnius, where he accepted a position as Professor of Natural Sciences and became head of the newly established Department of Chemistry at the university. He would hold this post for more than two decades from 1797 until 1822.

Sniadecki lectured with great enthusiasm, and his presentations were attended by students, citizens, and "even ladies" (Railienė, 2004).[8] Moreover, by acting as an editor for the science and culture journal, *Dziennik Wileński*, he became known to even more people. He was personally convinced that science should be accessible to a wide audience, and this was one reason why he did many things, including publishing articles on "practical" chemistry.

---

[6]Joseph Frank wrote about his practice in General Hospital of Vienna: "Four senior physicians worked at the hospital; each of them treated a hundred and fifty, and in winter time – hundred and eighty patients" (Frankas, 2015, p. 299).
[7]According to Birutė Railienė (2005), Sniadecki stayed at Professor Johann Peter Frank's apartment during this period.
[8]This might seem unusual, but it is worth remembering that women also attended Gall's lectures in Vienna, Germany, and France.

Figure 1. Andrew Sniadecki (Jędrzej Śniadecki, 1768–1838). Painted by Aleksander Sleńdziński (1803–1878). © Lithuanian Art Museum. Reproduced by permission of Lithuanian Art Museum. Permission to reuse must be obtained from the rightsholder.

Sniadecki became the first chairman of the Vilnius Medical Society, which Joseph Frank established in 1805 (Triponienė, 2012). Joseph Frank knew Sniadecki from studies at the University of Pavia and from medical practice in the General Hospital of Vienna. After Frank arrived in Vilnius in 1804, he provided this brief portrait of his talented medical colleague:

> Andrew Sniadecki lectured on chemistry, and has published a textbook in Polish. There was not a more eloquent professor, but he experimented little. Sniadecki was busy with his medical practice from morning 'til evening, and had not much time for laboratory works. (Frankas, 2013, p. 59)

One of Sniadecki's most important treatises was *Teorya Jestestw Organicznych* [Theory of the Organic Entities], the first volume of which was published in Warsaw in 1804 (Sniadecki, 1804). This treatise is often considered the first major book on biochemistry in Polish, and it dealt with a wide range of subjects, including the mind. Sniadecki defined organic and inorganic entities, introduced the notion of psychic processes and stated that the origin of mental activity involves a specific form of metabolism in the nervous system. He also claimed that, even though the human brain consists of separate parts, it operates as an entity (also see Railienė, 2004).

Sniadecki's article, *Krótki Wykład Systematu Galla z przyłączeniem niektórych uwag nad iego Nauką* [Short Lecture on the System of Gall with some Comments about his Science] is of greater significance for historians of organology/phrenology (Sniadecki, 1805). This 27-page article was written in Polish, the dominant language in Vilnius at the time,[9] and was published in the first volume of *Dziennik Wileński* in 1805 (see Fig. 2). Sniadecki had concluded that Gall's theory is important and was worth communicating to the public. He mentioned that Gall had not yet published his system in a book, and that his only publication on his doctrine was his 7-year-old letter to Baron Retzer in the *Neuer Deutscher Mercur*. He continued:

> Whatever the fate of his theory will be, it definitely shows the features of genius and exceptional courage, and involves many important and interesting facts. Therefore, considering that this is worth our public attention, we expect to do a favor to our readers and provide a short description of Gall's theory, adding some observations. (Sniadecki, 1805, p. 16)

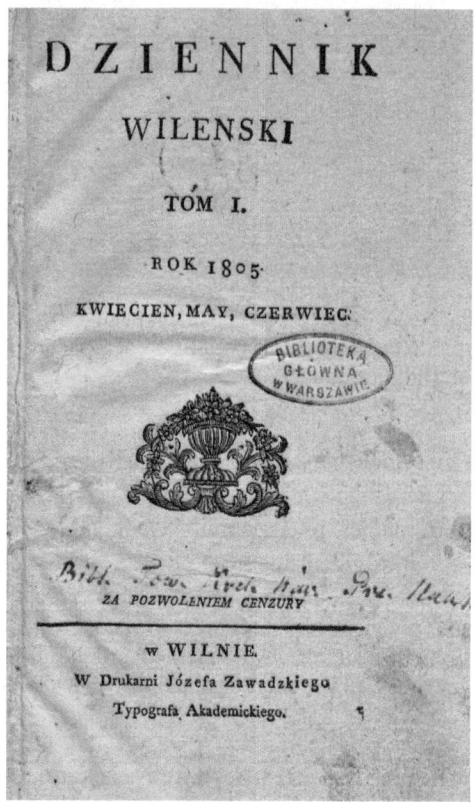

Figure 2. The title page of the journal *Dziennik Wileński* (1805), in which the article on the system of Gall by Andrew Sniadecki was published. © Vilnius University Library, Rare Book Department. Reproduced by permission of Vilnius University Library, Rare Book Department. Permission to reuse must be obtained from the rightsholder.

---

[9]In the middle of the nineteenth century, there were only about 35% Lithuanian-speaking inhabitants in Vilnius Governorate of the Russian Empire (Bairašauskaitė, Medišauskienė, & Miknys, 2011, p. 108)

He then introduced Gall's main principles to the readers. He wrote that, according to Gall, human talents and inclinations are innate and not dependent on education; talents differ from inclinations; and every basic talent and inclination can be associated with a different area of the brain. He further explained that, the stronger the talent or inclination is, the larger the area it occupies, and the external form of the head conforms to the brain, because the cranial surface is shaped by the growing brain (Sniadecki, 1805, pp. 18–23).

Carefully following Gall's teachings, Sniadecki emphasized how important it is to observe human heads and skulls. He stated that one should touch heads with the full palm, that the person doing this should pay attention to each bump and depression, and that it is necessary to explore the heads of persons with exceptional talents and abilities, as well as those who lack them, for example, people with certain insanities or criminal propensities. Additionally, one should investigate the heads of animals with known traits and apply these observations to humans (Sniadecki, 1805, pp. 24–25). In this context, he further noted:

> All less noble parts of the brain necessary to support life or to transmit the nerves to more intelligent structures lie at the bottom of the skull and are common to almost all animal species. However, the more animals evolve, the more the brain is spread to the sides and above, so that the most outstanding part is the human forehead, the noblest part of the brain. (Sniakecki, 1805, p. 26)

He then described Gall's two organs devoted to life itself. The first is the organ of the life force, and the second is what he called the organ of binding to life. The former is localized between the end of the spinal cord and *medulla oblongata*, and the second near the *foramen magnum* or in the *corpus callosum*. Neither of these subcortical organs will appear among the 27 faculties Gall would later list, all but one of which would involve the cerebral cortex (the exception being the reproductive drive, the first organ on Gall's final list, which he associated with the cerebellar cortex). These two primitive subcortical organs remained a part of Gall's system in 1805. One of the first substantial accounts written at the time of Gall's post-Vienna European tour was a 10-page report by Justus Arne Mann or Arnemann (1763–1806), a Professor of Medicine in Göttingen and later a Hamburg practitioner. It appeared in the March 1805 issue of the *Medical and Physical Journal* and was based in part on transcriptions of the lectures Gall had given in Berlin during the spring of that year. Arnemann (1805) referred to these two organs as the organ of life and the organ for preserving life. Bojanus, Ludwig Friedrich von Froriep (1779–1847), Gerardus Vrolik (1775–1859), and Christian Heinrich Ernst Bischoff (1781–1861) also mentioned the medullary organ of the life force. As for the organ of binding to life, it can also be found in Froriep, where it is mentioned as *Organ des Lebenserhaltungstriebes* [organ of the instinct to stay alive]. This drive would be included in what Gall would later characterize as the instinct of self-defense.

Sniadecki next mentioned the organ of lustfulness seated in the cerebellum. As noted, the cerebellum would be where Gall would continue to localize the site of the organ for the reproductive instinct, the most basic of his 27 faculties.

Sniadecki now turned to Gall's cerebral cortical organs, starting with the organ of love and attachment of offspring, and then mentioning organs for friendship, love, loyalty and goodwill, stubbornness, disposition to murder, cautiousness, cunning, propensity to steal, and memory, which is indicated by the bones that form the bottom of the eye vaults and consists of some separate organs: memory of things, persons, locality, words, languages, numbers, and tonality or organ of music. He also mentioned an organ underlying a talent for painting and perceiving colors and one for crafts.

Gall's remaining faculties, Sniadecki continued, are characteristic of humans alone, and they occupy the front and the top of the head, thus producing prominent foreheads in people with the corresponding highest traits. They are the organs of attentiveness, reasoning, wit, goodness, generosity, enthusiasm and imagination ("this organ makes persons become poets"), stability of decisions, pride, and love of the truth (see Fig. 3). When describing the latter faculty, Sniadecki commented:

> The place of the last faculty is unknown, and that is not strange, because people usually like flattery and fear to hear the truth. Gall and his followers will have to examine many skulls before finding this faculty. (Sniadecki, 1805, p. 37)

Clearly, Gall was just beginning to develop his system while Sniadecki was with him in Vienna in 1795 and 1796, and more than a decade would pass before he would arrive at his final system. This can be evidenced by reading the reports of others attending his lecture-demonstrations in Vienna and then in Germany. For example, this can be seen in the reports presented by Bojanus (1802), Froriep (1802), Walther (1802), Doornik (1804), Vrolik (1804), and others, which overlap Sniadecki's to a large extent, while also showing differences in the names and the number of organs described, reflecting Gall's continuing uncertainty about the organs making up the human mind. In this context, it is notable that Sniadecki described an organ of attentiveness, another organ that failed to make Gall's final list. Froriep called it the *Organ der Beobachtung* [organ of perception], Philipp Franz von Walther (1782–1849) listed it as the *Organe des Beobachtungsgeistes* [organ of sense of perception], Vrolik had a *Werktuig ter Waarneming dienende* [organ serving perception], and Jacob Elisa Doornik (1777–1837) mentioned the *Het werktuig van het Opmerkend*

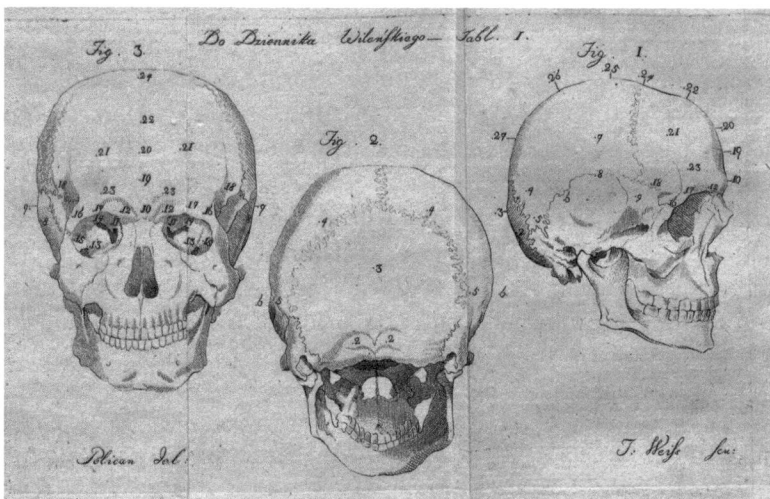

Figure 3. The illustration of Gall's cortical organs in Andrew Sniadecki's article (*Dziennik Wileński*, 1805). © Vilnius University Library, Rare Book Department. Reproduced by permission of Vilnius University Library, Rare Book Department. Permission to reuse must be obtained from the rightsholder.

Vermogen [organ of the attending faculty]. Finally, Sniadecki mentioned an organ of love for the truth, and this organ is also mentioned by Froriep as the *Organ der Wahrheitsliebe* and by Doornik as the *Het werktuig der Waarheidsmin* [organs of the love of truth]. It too failed to survive.

There are a number of faculties on Gall's final list that did not "exist" when Sniadecki learned about his organs of the mind. The faculty of comparative sagacity, the faculty of imitation and mimicry, and the faculty of God and religion seem to be among Gall's later discoveries or transformations of his earlier faculties. Doornik mentions the *Het werktuig der Godsdienstigheid en der Godsdienstdweperij* [organ of religiousness and religion fanaticism]. Vrolik lists the *Werktuig voor Godvrugt* [organ of devotion] and the *Werktuig voor den Talent van Schouwspelers* [organ of the talent of actor], which is compatible with Gall's faculty of imitation and mimicry. Vrolik also lists the organ *Voor wijsgerige Bespiegelíngen of bovennatuurkundíg Onderzoek ten voorbeeld strekke* [organ for philosophical contemplation and metaphysical investigation, serving as an example]. The latter seems to relate to Gall's organ of comparative sagacity.

In Table 1, we present both the 28 organs Sniadecki listed in 1805, based on his time with Gall during the 1790s but possibly also drawing from later lists, and the 27 faculties Gall would eventually settle on in his texts, starting with *Anatomie et Physiologie* of 1810–1819. These lists show that his set of faculties seemed to be stabilizing around 1804 but would still have to be adjusted before being finalized. Complicating matters, there could also have been differences in how his listeners were interpreting what he was saying, since there was no "authorized" list and because he used different names at different times for some of his faculties. These factors make it difficult to determine the basis of Sniadecki's list, and it is not hard to understand why Gall was expressing an eagerness to receive feedback from others that might help him finalize his still-emerging system.

After this introduction to Gall's system, Sniadecki provided some commentary and opinions about it. He stated that he agrees with Gall that our talents and inclinations are innate, since "daily experience and history teach us that the mind and the heart are definitely hereditary, like the features or deficiencies of the organic entities" (Sniadecki, 1805, p. 38). Nevertheless, Gall's idea that talents differ from inclinations, and that every talent and inclination must be housed in a different brain area, left him contemplating these more controversial contentions. "What is mind without imagination and memory?" he asked. He also focused on the unified whole, concluding the following: "Therefore every faculty of mind should be perceived as united, associated with each other, and at the same time located in different brain areas" (p. 39). And he wondered: "If someone has no organ of wit, but has good memory, quick perception, proper thinking, [and an] active and cheerful temperament, wouldn't that person be a good humorist?" (p. 40).

Sniadecki also thought that there were some discrepancies between the psychological and anatomical views of the mind and functions of the brain within Gall's system:

> Before every talent is placed in a different part of the brain, it is necessary to show that the brain is indeed composed of these different regions. Anatomists divide the brain into separate parts, but they do not teach us the functions of these parts. Gall, on the other hand, localizes the most important and noblest organs in that part of the brain that was perceived as a solid and integral substance, a substance that is unable to perform such different functions due to its structural integrity. (Sniadecki, 1805, p. 40)

Table 1. Summary of the 27 faculties, as described in the 1835 translation of Franz Joseph Gall main works (Gall, 1935) and the 28 organs mentioned by Andrew Sniadecki in his description of Gall's system in 1805.

Gall's Faculties	Sniadecki's Organs
I. Instinct of generation, of reproduction; instinct of propagation, etc.	Organ of lustfulness (lubieżności), Fig. 3, N. 2
II. Love of offspring	Organ of love and attachment of offspring (przywiązania do dzieci i rodziców), Fig. 3, N. 3
III. Attachment. Friendship	Organ of friendship, love, loyalty and goodwill (przyjaźni, miłości, wierności i uprzejmości), Fig. 3, N. 4 Organ of life force (mocy życia), Fig. 3, N. 1 Organ of binding to life (przywiązania do życia)
IV. Instinct of self-defence, disposition to quarrel, courage (Muth, Raufsinn)	Organ of disposition to muder (zabóyczy), Fig. 3, N. 6
V. Carnivorous instinct; disposition to murder (Wurgsinn)	Organ of cunning (chytrości), Fig. 3, N. 8
VI. Cunning, trick, tact (List, Schlauheit, Kluheit)	Organ of propensity to steal (kradzieży), Fig. 3, N. 9
VII. Sense of property, instinct of providing, covetousness, propensity to steal (Eigenthumssinn, Hang zu Stehlen)	Organ of generosity (szczodrobliwości), Fig. 3, N. 23
VIII. Pride, hauteur, loftiness, elevation (Stolz, Hochmuth, Herschsucht)	Organ of pride (wyniosłości), Fig. 3, N. 26 Also referred as organ of pride (wyniosłości)
IX. Vanity, ambition, love of glory (Eitelkeit, Ruhmsucht, Ehrgeitz)	Organ of cautiousness (ostrożności), Fig. 3, N. 7
X. Cautiousness, foresight (Behutsamkeit, Vorsicht, Vorsichtigkeit)	Organ of memory (pamięci): memory of things (pamięć rzeczy), Fig. 3, N. 10 Organ of attentiveness (uwagi czyli obserwacyi), Fig. 3, N. 19
XI. Memory of things, memory of facts, sense of things, educability, perfectibility (Sachgedächtniss, Erziehungs-fähigkeit)	Organ of memory (pamięci): memory of locality (pamięć mieysc) Fig. 3, N. 12
XII. Sense of locality, sense of the relations of space (Ortsinn, Raumsinn)	Organ of memory (pamięci): memory of persons (pamięć osob), Fig. 3, N. 11
XIII. The faculty of distinguishing and recollecting persons (Personen-sinn)	Organ of memory (pamięci): memory of words (pamięć słów), Fig. 3, N. 13
XIV. Faculty of attending to and distinguishing words; recollection of words, or verbal memory (Wort-gedächtniss)	Organ of memory (pamięci): memory of languages (pamięć ięzyków), Fig. 3, N. 14
XV. Faculty of spoken language; talent of philology, etc. (Sprach-Forschungs-sinn)	Organ of talent of painting and perceiving colors (malarstwa, albo raczéy kolorów), Fig. 3, N. 17
XVI. Faculty of distinguishing the relation of colors; talent for painting (Farben-sinn)	Organ of memory (pamięci): memory of tonality or organ of music (pamięć tonów czyli organ muzyczny), Fig. 3, N. 16
XVII. Faculty of perceiving the relation of tones, talent for music (Ton-sinn)	Organ of memory (pamięci):
XVIII. Faculty of the relations of numbers	Organ of memory (pamięci): memory of numbers (pamięć liczb), Fig. 3, N. 15 memory of crafts (kunsztów), Fig. 3, N. 18
XIX. Faculty of constructiveness (Kunst-sinn, Bau-sinn)	
XX. Comparative sagacity, aptitude for drawing comparisons (Vergleichender Scharf-sinn)	Organ of reasoning (rozumu), Fig. 3, N. 20
XXI. Metaphysical depth of thought; aptitude for drawing conclusions (Metaphysischer Tief-sinn)	Organ of wit (dowcipu), Fig. 3, N. 21
XXII. Wit (Witz)	Organ of enthusiasm and imagination (zapału i imainacyi), Fig. 3, N. 24
XXIII. Talent for poetry (Dichter Geist)	Organ of goodness (dobroci), Fig. 3, N. 22
XXIV. Goodness, benevolence, gentleness, compassion, sensibility, moral sense, conscience (Gutmuthigkeit, Mitleiden, Moralischer-sinn, Gewissen)	Organ of love for the truth (miłości prawdy), Fig. 3, N. 27
XXV. Faculty of imitation, mimicry	
XXVI. God and religion	Organ of stability of decisions (stałości w przedsięwzięciu), Fig. 3, N. 25
XXVII. Firmness, constancy, perseverance, obstinacy	Organ of stubborness (ląbwaŋ, Fig. 3, N 5

His conclusion is perfectly understandable and sensible. He maintains that Gall's system makes sense and that it can prove to be very important and useful in many ways, but that more research is needed before all parts of it can be accepted as proven:

> It seems that the brain could be constructed of certain organs…, however, the number, location and delimitation of these organs are still just speculation…. Gall's work is worthy of respect, as long as it leads closer to the truth on the path of exploration. Over time, when scientists combine their works, and after accurate anatomical studies are performed, our knowledge will improve. (Sniadecki, 1805, pp. 40–41)

To summarize, Sniadecki was intrigued and for the most part positive about Gall's theory, although he was perplexed by the cortex being anatomically and structurally a solid and integral substance. This left him questioning whether it really could consist of so many independent organs. His cautious attitude is consistent with what he had derived from his own observations and had expressed in his own theorizing, as can be seen in his *opus magnum, Teorya Jestestw Organicznych*, where he wrote about the human brain operating like a single organ. Sniadecki recognized, however, that Gall's theory was new, promising, and already stimulating potentially important new research on the brain and behavior, and these things were important to him. These were among the reasons why he stated:

> We hope that every reader will be able to understand this theory properly and to assess it objectively, diligently selecting the ideas that are presented in this article…. The grounds on which this theory stands have the accuracy that every science should have. (Sniadecki, 1805, p. 37)

Joseph Frank's criticisms of Gall's theory and Johann Peter Frank's assessment

Joseph Frank (see Fig. 4) was born in 1771 in Rastatt, a town in the west of Baden-Württemberg, and graduated from the University of Pavia (1785–1791), where the most famous teachers at that time were Antonio Scarpa (1752–1832), Alessandro Volta (1745–1827), and his father, Johann Peter Frank (J. Frank, 1848). He was appointed Extraordinary Professor of Special Therapy at his Alma Mater in 1795. A year later, he became Chief Physician at the General Hospital of Vienna, where he remained until 1802. This appointment provided him with the opportunity to become more acquainted with Gall and his organology.

In 1802–1803, Joseph took a scientific tour of France, Great Britain, and northern Germany, in which he visited Jean-Étienne Dominique Esquirol (1772–1840), Antoine Portal (1742–1832), Jean Corvisart (1755–1821), Edward Jenner (1749–1823), and other notable men of science and medicine (J. Frank, 1848; Kondratas, 1977). In 1803, both he and his more famous father were invited to Vilnius University by the Rector.

Prior to this time, Frank's father, Johann Peter, had been a lecturer at the University of Göttingen, where in 1784 he taught special therapies, physiology and pathology, medical policing, and forensic medicine. From 1785 until 1795, the elder Frank was Professor of Clinical Medicine at the University of Pavia and director of the hospital in Pavia. Johann Peter's next positions were as Director of the General Hospital of Vienna, and Full Professor of the Practice of Medicine at the University of Vienna, both from 1795 to 1803 (J. Frank, 1848; J. P. Frank 1948). In 1804, Johann Peter arrived in Vilnius, but he spent only 10 months there. Before leaving in 1805, he served as Professor of Special

Figure 4. Joseph Frank (1771–1842). Painted by Jonas Rustemas (1762–1835). © Lithuanian Art Museum. Reproduced by permission of Lithuanian Art Museum. Permission to reuse must be obtained from the rightsholder.

Therapy, founded the Vilnius University Clinic and reorganized the Faculty of Medicine. He left when Czar Alexander I (1777–1825) invited him to St. Petersburg, where he spent 3 years as a royal physician and Professor at the Medical and Surgical Academy of St. Petersburg.

Between 1791 and 1803, his son Joseph had been greatly influenced by Scottish physician John Brown's (1735–1788) ideas about excitability and exhaustion.[10] His earliest publications were, in fact, explanations and commentaries on Brown's works and theories (see Kondratas, 1988). After he became Extraordinary Professor at the University of Pavia, he cautiously explained his principles in his first lecture on special therapy, choosing the following words:

> I have chosen vitalism (*la doctrine des forces vitales*), based on the medical system of Stahl and Friedrich Hoffmann and developed by Cullen, Brown, and Hufeland; however, we should never forget that the human body is also dependent on mechanical and chemical laws ….

---

[10]"Brunonian medicine" was based on the notion that the human organism possesses a reasonable amount of natural energy or excitability, which, if diminished, is capable of restoration to a healthy balance with proper stimulation, and, if excessive, is capable of being lowered. Brown contended that there are no specific cures for particular diseases and that his medical treatments targeted the whole body by changing its excitability, thus correcting the signs and symptoms of sthenia or asthenia (see Brown, 1788; Risse, 1988).

Finally, these principles should be followed to the extent that they are consistent with experience. (see J. Frank, 1848, p. 535)

After his scientific tour ended in 1803, Joseph rejected the Brunonian system. In 1807, Jean Corvisart, one of Napoleon Bonaparte's (1769–1821) physicians, wrote Joseph a letter approving his rejection of the Brunonian medical system, and Joseph even quoted Corvisart's letter in his *Mémoires*:

I always thought that it was dangerous to rely on any medical system in practice. Brown's and others' systems resulted in many victims. In my mind, all theories vanish at the bedside of the sick. (J. Frank, 1848, p. 535)

Joseph Frank's new approach to diseases, clinical methods, and treatment strategies was an eclectic mix of Hippocratic medicine, vitalism, common sense, and experience-based observation.

In contrast to his father's stay of less than a year, Joseph would spend almost 20 years in Vilnius. The Vilnius Medical Society, which he established in 1805, was the first scientific medical organization in Eastern Europe. He also founded several other institutions there, including the Out-Patients' Clinic (1807), the Vaccination Institute (1808), and the Maternity Institute (1809), which was the first institution of this type in Europe (Triponienė, 2012).

Frank wrote his *Mémoires Biographiques* in French, although his native language was German. This manuscript of more than 3,500 pages is a valuable document depicting the lives of Joseph, his father, their ancestors, the city of Vilnius, and its university. Joseph's friend, Dr. Jean de Carro (1770–1857), edited this lengthy work and in 1855, 13 years after Joseph's death, it was purchased by the Vilnius Medical Society (Prašmantaitė, 2013), still unpublished. Parts of the manuscript were later translated into Polish by Władysław Zahorski (1913), into Lithuanian by Genovaitė Dručkutė (2001), and into Italian by Giovanni Galli (2006) (see J. Frank, 1913; G. Frank, 2006; Frankas, 2001).

Remembering his practice at the General Hospital of Vienna, Joseph described his first impressions of Gall and his doctrine in this work. In 1801, he tells us:

I asked doctor Gall to introduce me some principles of cranioscopy, but he just wanted to make me one of his followers. He failed: although some of the facts of his doctrine seemed to be important, it was neither strong nor comprehensive enough to form a system. (J. Frank, 1848, p. 177)

Joseph also mentioned his fellow students, who seemed more interested in having a good time than in learning medicine at the hospital. Still, they conscientiously attended Gall's lectures on organology:

Many [students] could not resist the amusements of the capital and its surroundings. In the summer, some of them … appeared in the Clinic if they did not have more pleasant activities…. However, students very diligently went to Doctor Gall's lectures on cranioscopy. (Frankas, 2015, p. 289)

Joseph also described his father's journey to Munich in 1809 and the meeting he had with his "old friend Soemmering" (Samuel Thomas Sömmerring; 1755–1830). Notably, the two men talked about Gall (who had met Soemmering but was now in France), his methods, and his doctrine:

The main topic of their discussion was doctor Gall's brain dissection method. Soemmerring criticized this method and added, by the way, that the binding of the sutures formed various shapes of the skull, stretching the skull either forward or backwards. However, he admitted that there was no sample in his vast collection of skulls opposing Gall's cranioscopic doctrine. Soemmerring thought that it was quite plausible that protuberance in the upper middle part of the forehead was a sign of kindness. He said that the proof of this rule was his son! (J. Frank, 1848, pp. 450–451)

Clearly, Johann Peter was also interested in Gall's theory. Still, there we could find nothing to suggest that Johann Peter actively promoted (or criticized) Gall's ideas to his students in Vilnius or that he wished to be counted as one of Gall's followers.

In his memoirs, Joseph also brought up a person from New York named Castel, who helped popularize phrenology in Milan:

> He [Castel] was surrounded by women seeking to find out the meanings of their own and their husbands' protuberances. They brought him children inquiring about their talents and inclinations, and asked him to create a personal educational system for them. (J. Frank, 1848, pp. 310–311)

Joseph was critical of Castel, who refused to describe the features of the skulls in the anatomy cabinet in Pavia. "However, he very easily characterized the heads of the living humans; therefore, he has paid more attention to facial expressions than to protuberances in the skulls" (p. 312). Frank seemed worried that the popularization of phrenology could lead to charlatanism, quackery, and chicanery. This fear might have made him even more cautious about accepting Gall's theory and passing it on to his students.

Lastly, he mentioned an article he had written on phrenology, which he finished in 1839 while at Como, then a part of Austrian Empire (G. Frank, 1839). Here he was forthright in rejecting Gall's doctrine, perhaps because it and its offspring had largely fallen out of scientific favor by this time:

> I wrote an article on phrenology. . . . I did not foresee how much time and trouble this article would cost me. It is published in the XCIVth volume of "The Italian Library" (*Bibliotheca Italiana*). I am satisfied that, based on the arguments of anatomy, physiology and pathology, I proved *the uselessness of cranioscopy*. (J. Frank, 1848, pp. 315–316, italics added)

Joseph Frank's "Della frenologia" [On Phrenology] was actually a letter to the editors. Because he was a physician who knew Gall personally, the editors had asked him to express his opinions about Gall's theory. Frank responded that phrenology is not scientific theory, that it could be compared to astrology, that the phrenologists do not adequately explain what happens to a patient when a certain brain region is damaged, and that some members of the Catholic Church actually seemed supportive of the doctrine, since some of Gall's ideas could be regarded as a proof of the spirituality of the soul (G. Frank, 1839). His conclusion was damning:

> Phrenology and craniology, already rejected by psychologists. . . cannot be rescued, and indeed are contradicted by the anatomical knowledge and pathological facts. . . . I will conclude by encouraging the youth to keep on guard against the value of such a theory and especially urge doctors not to waste time on such trifles, and to remember the saying: *vita brevis, ars longa*.[11] (G. Frank, 1839, p. 376)

---

[11]This is a Latin translation of the Hippocratic aphorism, which translates as "Art is long, life is short."

Thus, Joseph made it clear that he was not one of Gall's accolades but quite the contrary. He feared that the popularization of Gall's theory might lead to all sorts of mischief, and he even wrote a paper exposing the absurdity and uselessness of cranioscopy. With his decidedly negative bias, and unlike Andrew Sniadecki, he neither taught Gall's system to students nor presented it to the public in Vilnius.[12]

Epilogue

Gall's theory of organology with its primary method of cranioscopy became well known and was generally accepted in the beginning of the nineteenth century in Vilnius and at its university, much as was true in the German cities where Gall lectured and gave demonstrations during his 1805–1807 sojourn. For example, Johannes Andreas Loebenwein (1758–1820), a Professor of Anatomy at Vilnius University, had a rich collection of skulls of suicide victims and dead criminals in his anatomical cabinet and used them for teaching Gall's cranioscopic doctrine to his students (Pavilonis, 1997). Similarly, physician Stanisław Morawski (1802–1853), a graduate of Vilnius university and one of Joseph Frank's most distinguished students, described his medical education and his doctoral studies in his memoirs. During what was probably 1823, he mentioned living in two cells in Vilnius' Piarists College, where his books were scattered everywhere, there were some human skeletons, and, also, some mummies stolen from the Franciscan Church basements lying near the wall. Of greater importance for us, Morawski also drew attention to "dozens of human skulls, white as ivory, with all the organs marked using the system of Gall" (Moravskis, 1994, p. 173).

Because many professors and university students supported the ideals of the national uprising of 1831, Czar Nicholas I (1796–1855) issued a decree on May 1, 1832, closing Vilnius University (Bumblauskas et al., 2004). This was a terrible blow to Lithuanian science, and its culture more generally. Moreover, Lithuania experienced another terrible blow after the uprising of 1863, when its press was prohibited from using the Latin alphabet.[13]

Vilnius had possessed one of the oldest universities in Central and Eastern Europe and the largest in Russian Empire. It was a place where the natural sciences were supported and where new theories and explorations from the centers of Europe reached not only scientists and medical professionals but also the general public. But now Lithuania was deprived not only of its name, its university, and its press but also of primary national schools and the Lithuanian language for publications. Ironically, the Vilnius Medical Society, established by Joseph Frank in 1805, remained the only scientific medical institution still open for doctors, pharmacists, and veterinarians following the closure of university (see Triponienė, 2012).

Hence, Vilnius was no longer the progressive city open to new ideas that it had been when Andrew Sniadecki and Joseph Frank, two well-trained and experienced physicians,

---

[12]Whether his negative attitude might have had something to do with the fact that he had been a follower of John Brown but then lost faith in Brunonian system, making him more reluctant to follow another faddish idea into uncharted waters, takes us beyond the written material at hand. What is not speculative is that Frank chose to play it safe, reverting back to "time-tested" Hippocratic medicine, common sense, and experience-based observations for his teaching and clinical practice.

[13]The Lithuanian press ban was a ban on all Lithuanian language publications printed in Latin letters from 1865 until 1904; however, Lithuanian language publications that used Cyrillic were allowed (Šapoka, 1936).

who had spent time with Gall in Vienna, made their way to the city to accept faculty appointments at its famed university. Then again, the excitement surrounding Gall's skull-based doctrine during the opening decades of the nineteenth century had also dissipated with new attention being focused on what could be learned about the brain and behavior by paying considerably more attention to brain-damaged patients — a better and more easily accepted method than Gall's cranioscopy had been for physicians interested in probing some of nature's deepest secrets.

Acknowledgments

We would like to thank Arleta Bublevič for her assistance in translating the article by Andrew Sniadecki from Polish and Lorenzo Lorusso for his help in translating Joseph Frank's Italian essays.

References

Ackerknecht EH (1958): Contributions of Gall and the Phrenologists to knowledge of brain function. In: Poynter FNL, ed., *The Brain and Its Functions*. Oxford, Blackwell, pp. 149–153.

Akert K, Hammond MP (1962): Emanuel Swedenborg (1688–1772) and his contributions to neurology. *Medical History* 16: 255–265.

Arnemann J (1805): A concise account of Dr. Gall's new doctrine of the brain, and the faculties of the mind. *The Medical and Physical Journal* 14: 327–336.

Bairašauskaitė T, Medišauskienė Z, Miknys R (2011): *Lietuvos istorija. Devynioliktas amžius: visuomenė ir valdžia. VIII tomas I dalis* [History of Lithuania. Nineteenth century: The Society and the Authority. First Part of the Eighth Volume]. Vilnius, Baltos lankos.

Bendžius A, Grigonis J, Kubilius J, Merkys V, Piročkinas A, Šidlauskas A, Tornau J, Vladimirovas L (1977): *Vilniaus Universiteto istorija 1803-1940* [History of Vilnius University from 1803 till 1940]. Vilnius, Mokslas.

Bischoff CHE (1805): *Darstellung der Gall'schen Gehirn- und Schädel-Lehre; nebst Bemerkungen über diese Lehre von Christoph Wilh. Hufeland* [Presentation of Gall's Brain and Skull Theory; with Comments on this Theory by Christoph Wilh. Hufeland]. Berlin, L. W. Wittich.

Bisius S (1772): *Responsum Stephani Bisii Philosophiae et Medicinae Doctoris ad Amicum Philosophum De melancholia, mania et plica-polonica sciscitantem* [A Reply of the Doctor of Philosophy and Medicine Stephanus Bisius to the Question from the Philosophical Society about Melancholia, Mania and Plica Polonica]. Vilnae.

Biziulevičius S (1997): Medicinos mokslai senajame Vilniaus universitete 1781–1842 m. [Medicine in Old Vilnius University in 1781–1842]. In: Andriušis A, ed., *Vilniaus medicinos istorijos almanachas*. Vilnius, Medicina Vilnensis, pp. 31–98.

Bogušis V (1997): Medicina Vilniaus universitete iki XVIII a. vidurio [Medicine in Vilnius University till the Middle of the 18th Century]. In: Andriušis A, ed., *Vilniaus medicinos istorijos almanachas*. Vilnius, Medicina Vilnensis, pp. 15–29.

Bojanus L (1801): Encephalo-Cranioscopie. *Magasin Encyclopédique*, Année VIII, Tome I: 445–472.

Bojanus L (1802): A short view of the craniognomic system of Dr. Gall of Vienna. *Philosophical Magazine* 14: 77–84, 131–138.

Brown J (1788): *The Elements of Medicine or, A Translation of the Elementa Medicinae Brunonis. With Large Notes, Illustrations, and Comments. By the Author of the Original Work. In Two Volumes*. London, Printed for J. Johnson, No 72, St. Paul's church-yard.

Bumblauskas A, Butkevičienė B, Jegelevičius S, Manusadžianas P, Pšibilskis V, Raila E, Vitkauskaitė D (2004): *Universitas Vilnensis 1579 – 2004* [Vilnius University in 1579 – 2004]. Vilnius, Spauda.

Cooter R (1984): *The Cultural Meaning of Popular Science: Phrenology and the Organization of Consent in Nineteenth-Century Britain*. Cambridge, Cambridge University Press.

Critchley M (1965): Neurology's debt to F. J. Gall (1758–1828). *British Medical Journal* 2: 775–781.

Davies JD (1955): *Phrenology: Fad and Science*. New Haven, Yale University Press.

Doornik JE (1804): *De Herssen-Schedelleer van Frans Joseph Gall getoetst aan de Natuurkunde en Wijsbegeerte* [The Brain-Skull Theory of Franz Joseph Gall, Tested to Physics and Philosophy]. Amsterdam, Holtrop.

Eling P, Finger S (2015): Franz Joseph Gall on greatness in the fine arts: A collaboration of multiple cortical faculties of mind. *Cortex* 71: 102–115.

Fedorowicz Z (1958): Ludwik Henryk Bojanus. *Memorabilia zoologica* 1: 1–45.

Finger S (1994): *Origins of Neuroscience: A History of Explorations into Brain Functions*. Oxford, Oxford University Press.

Forster TIM (1815): Sketch of a new anatomy and physiology of the brain and nervous system of Drs. Gall and Spurzheim considered as comprehending a complete system of phrenology. *Pamphleteer* 5: 219–244.

Frank G (1839): Della frenologia [On Phrenology]. *Bibliotheca Italiana* XCIV: 349–376.

Frank G (2006): *Memorie I* [Memoirs, the First Volume], G. Galli, trans. Milano, Cisalpino-Istituto editoriale universitario.

Frank J (1848): *Mémoires Biographiques de Jean-Pierre Frank et de Joseph Frank son fils* [Biographical Memoirs of Johann Peter Frank and his son Joseph Frank]. Leipzic.

Frank J (1913): *Pamiętniki d-ra Józefa Franka Profesora Uniwersytetu Wileńskiego* [Memoirs of Doctor Joseph Frank, Professor at Vilnius University], W. Zahorski, trans. Wilno, Druk Józefa Zawdzkiego.

Frank JP (1948): Biography of Dr. Johann Peter Frank, Imperial and Royal Court Councillor, Hospital Director and Professor of Practical Medicine at the University in Vienna, member of various learned societies, written by himself. *Journal of the History of Medicine and Allied Sciences* 3(1): 11–46. Trans. from the German by G. Rosen.

Frankas J (2001): *Atsiminimai apie Vilnių* [Memoirs about Vilnius], G. Dručkutė, trans. Vilnius, Mintis.

Frankas J (2013): *Vilnius XIX amžiuje. Atsiminimai. Pirma knyga* [Vilnius in the Nineteenth Century. Memoirs. The First Book], G. Dručkutė, trans. Vilnius, Mintis.

Frankas J (2015): *Atsiminimai. Antra knyga* [Memoirs. The Second Book], G. Dručkutė, trans. Vilnius, Mintis.

Froriep LF (1802): *Darstellung der ganzen auf Untersuchungen der Verrichtungen des Gehirnes gegründeten Theorie der Physiognomik des Dr. Gall in Wien* [Account of Dr. Gall's Physiognomical Theory, Entirely Based on Investigations of the Functions of the Brain]. Wien, 3rd edition, Weimar, Verlage des Industrie-Comptoirs.

Gall FJ (1798): Schreiben über seinen bereits geendigten Prodromus über die Verrichtungen des Gehirns der Menschen und der Thiere, an Herrn Jos. Fr. von Retzer [Letter from Dr. F. J. Gall, to Joseph Fr[eiherr] von Retzer, upon the Functions of the Brain, in Man and Animals]. *Der neue Teutsche Merkur* 3: 311–332.

Gall FJ (1822–1825): *Sur les Fonctions du Cerveau et sur celles de chacune de ses Parties* [On the Functions of the Brain and on Each of its Parts]. Paris, J.-B. Baillière.

Gall FJ (1835): *On the Functions of the Brain and Each of its Parts: With Observations on the Possibility of Determining the Instincts, Propensities, and Talents, or the Moral and Intellectual Dispositions of Men and Animals, by the Configuration of the Brain and Head, 6 Volumes*, W. Lewis, Jr. Marsh, trans. Boston, MA, Capen and Lyon.

Gall FJ, Spurzheim J (1810–1819): *Anatomie et Physiologie du Système Nerveux en Général, et du Cerveau en Particulier* [Anatomy and Physiology of the Nervous System in General and of the Brain in Particular]. Paris, F. Schoell.

Hagner M (2003): Skulls, Brains, and Memorial Culture: On Cerebral Biographies of Scientists in the Nineteenth Century. *Science in Context* 16: 195–218.

Haller A (1762): *Elementa physiologiae corporis humani. Tomus quartus. Cerebrum. Nervi. Musculi* [Elements of Physiology of the Human Body. Forth Volume. Brain, Nerves, Muscles]. Lausannae, sumptibus Francisci Grasset.

Kaczkowski C (1821): *Dissertatio inauguralis medico-practica de plicae Polonicae in varias, praeter pilos, corporis humani partes vi et efectu* [Medical-Practical Inaugural Thesis on the Effects of

Plica Polonica in Various Parts of the Human Body]. Vilnae, Typis Josephi Zawadzki Universit. typographi.

Karenberg A (2009): Cerebral localization in the eighteenth century—An overview. *Journal of the History of the Neurosciences* 18(3): 248–253.

Klajumaitė V (2013): The phenomenon of Plica Polonica in Lithuania: A clash of religious and scientific mentalities. *Acta Baltica Historiae et Philosophiae Scientiarum* 1(2): 53–66.

Kondratas RA (1977): *Joseph Frank (1771-1842) and the Development of Clinical Medicine: A Study of the Transformation of Medical Thought and Practice at the End of the Eighteenth and the Beginning of the Nineteenth Centuries.* PhD thesis. Harvard University, Massachusetts.

Kondratas R (1988): The Brunonian influence on the medical thought and practice of Joseph Frank. *Medical History* 8: 75–88.

Moravskis S (1994): *Keleri mano jaunystės metai Vilniuje. Atsiskyrėlio atsiminimai (1818-1825)* [Several Years of my Youth in Vilnius. The Memoirs by Hermit (1818-1825)], R. Griškaitė, trans. Vilnius, Mintis.

Neuburger M (1897): *Die historische Entwicklung der experimentellen Gehirn- und Rückenmarksphysiologie vor Flourens* [The Historical Development of the Experimental Brain and Spinal Cord Physiology]. Stuttgart, Enke.

Noel PS, Carlson ET (1970): Origins of the word "phrenology." *American Journal of Psychiatry* 127: 154–157.

Norrsell U (2007): Swedenborg and localization theory. In: Whitaker H, Smith CUM, Finger S, eds., *Brain, Mind and Medicine: Essays in Eighteenth-Century Neuroscience.* Boston, Springer, pp. 201–208.

Otto AW (1831): Ludwig Heinrich von Bojanus. *Nova Acta Halle* 15(I): XXXIX–XLV.

Pavilonis S (1997): Anatomijos, histologijos ir antropologijos katedra [Department of Anatomy, Histology and Anthropology]. In: Andriušis A, ed., *Vilniaus medicinos istorijos almanachas.* Vilnius, Medicina Vilnensis, pp. 121–125.

Prašmantaitė A (2013): Jozefas Frankas ir jo "Atsiminimai" [Joseph Frank and his "Memoirs"]. In: Puluikienė I, ed., *Jozefas Frankas. Vilnius XIX amžiuje. Atsiminimai. Pirma knyga.* Vilnius, Mintis, pp. 5–19.

Railienė B (2004): Letter to editor. *Acta Biochimica Polonica* 51: 1087–1090.

Railienė B (2005): *Andrius Sniadeckis.* Vilnius, Vilniaus universiteto leidykla.

Risse GB (1988): Brunonian therapeutics: New wine in old bottles? *Medical History* 8: 46–62.

Rush B (1811a): Lecture XI. On the utility of a knowledge of the faculties and operations of the human mind, to a physician. In: *Sixteen Introductory Lectures to Courses of Lectures upon the Institutes and Practice of Medicine.* Philadelphia, Bradford and Innskeep, pp. 256–273.

Rush B (1811b): Lecture XII. On the opinions and modes of practice of Hippocrates. In: *Sixteen Introductory Lectures to Courses of Lectures upon the Institutes and Practice of Medicine.* Philadelphia, Bradford and Innskeep, pp. 274–294.

Šapoka A (1936): *Lietuvos istorija* [History of Lithuania]. Kaunas, Švietimo ministerijos knygų leidimo komisijos leidinys.

Sniadecki J (1804): *Teorya Jestestw Organicznych. Tom I* [Theory of the Organic Entities. The First Volume]. W Warszawie, w Drukarni No 646 przy Nowolipiu.

Sniadecki J (1805): Krótki Wykład Systematu Galla z przyłączeniem niektórych uwag nad iego Nauką [Short Lecture on the System of Gall with some Comments about his Science]. *Dziennik Wileński* 1: 16–42.

Spurzheim JG (1818): *Observations sur la Phraenologie, ou la Connaissance de l'Homme Moral et Intellectuel, Fondée sue les Fonctions du Système Nerveux.* Paris, Treittel at Wurtz.

Strojnowski J (1965): Teoria lokalizacji mózgowej Jędrzeja Śniadeckiego [The Theory of Cerebral Localization by Andrew Sniadecki]. *Roczniki Filozoficzne* 13(4): 75–92.

Swedenborg E (1740-1741): *Oeconomia regni animalis in transactiones divisa: quarum … prima, de sanguine, ejus arteriis, venis, et corde … (secunda, de cerebri motu et cortice et de anima humana, agit); anatomice, physice et philosophice perlustrata … Accedit, Introductio ad Psychologiam Rationalem, 2 Volumes.* Amsteldami, Franciscum Changuion.

Swedenborg E (1882–1887): *The Brain, Considered Anatomically, Physiologically, and Philosophically, 2 vols.*, R. L. Tafel, ed., trans., and ann. London, James Speirs.

Swedenborg E (1938–1940): *Three Transactions on the Cerebrum, 2 vols. and a third of anatomical plates.* Philadelphia, Swedenborg Scientific Association/New Church Book Center.

Temkin O (1947): Gall and the phrenological movement. *Bulletin of the History of Medicine* 21: 275–321.

Triponienė D (2012): *Prie Vilniaus medicinos draugijos versmės* [At the Source of Vilnius Medical Society]. Vilnius, Vilniaus universiteto leidykla.

Van Wyhe J (2002): The authority of human nature: The *Schädellehre* of Franz Joseph Gall. *British Journal of the History of Science* 35: 17–42.

Van Wyhe J (2004): *Phrenology and the Origins of Victorian Scientific Naturalism.* Aldershot, Ashgate Publishing Ltd.

Van Wyhe J (2007): The diffusion of phrenology through public lecturing. In: Fyfe A, Lightman B, eds., *Science in the Marketplace: Nineteenth-Century Sites and Experiences.* Chicago, University of Chicago Press, pp. 60–96.

Viliūnas D (2014): *Filosofija Vilniuje XIX amžiaus pirmoje pusėje* [Philosophy in Vilnius in the first half of the ninteenth century]. Vilnius, Lietuvos kultūros tyrimų institutas.

Vrolik G (1804): *Het leerstelsel van Franz Joseph Gall* [The doctrine of Franz Joseph Gall]. Amsterdam, Holtrop.

Walther P (1802): *Critische Darstellung der Gallschon anatomisch-physiologischen Untersuchungen des Gehirn- und Schädel-baues* [Critical account of Gall's anatomical-physiological investigations of the brain and skull]. Zürich, Ziegler.

Wien Geschichte Wiki (n.d.): *Altes Allgemeines Krankenhaus.* Retrieved from https://www.wien.gv.at/wiki/index.php?title=Altes_Allgemeines_Krankenhaus

Young R (1970): *Mind, Brain and Adaptation in the Nineteenth Century.* New York, Oxford University Press.

Zinn JG (1749): *Experimenta.* Inaugural Dissertation. Göttingen.

Ludwig Heinrich Bojanus (1776–1827) on Gall's craniognomic system, zoology, and comparative anatomy

Eglė Sakalauskaitė-Juodeikienė, Paul Eling and Stanley Finger

ABSTRACT
Most of what was known about Franz Joseph Gall's (1758–1828) organology or *Schädellehre* prior to the 1820s came from secondary sources,
including letters from correspondents, promotional materials, brief
newspaper articles about his lecture-demonstrations, and editions and
translations of some lengthier works of varying quality in German.
Physician Ludwig Heinrich Bojanus (1776–1827) practiced in Vienna's
General Hospital in 1797–1798; attended some of Gall's public lectures;
and, in 1801–1802, became one of the first physicians to provide
detailed reports on Gall's emerging organology in French and English,
respectively. Although Bojanus considered the human mind to be indivisible and did not entirely agree with Gall's assumption that the brain
consists of a number of independent organs responsible for various
faculties, he provided valuable information and thoughtful commentary
on Gall's views. Furthermore, he defended Gall against the charge that
his sort of thinking would lead to materialism and cautiously predicted
that the new system would be fruitful for developing and stimulating
important new research about the brain and mind. Bojanus became
a professor of zoology in 1806 and a professor of comparative anatomy
in 1814 at Vilnius University, where, among other accomplishments, he
established himself as a founder of modern veterinary medicine and
a pioneer of pre-Darwinian and pre-Lamarckian evolutionism.

Franz Joseph Gall (1758–1828)—whose life's work spanned medicine, neuroanatomy, natural philosophy, anthropology, and other disciplines—is best remembered for his attempts to identify various faculties of mind and to localize them in distinct parts of the cortex using cranial bumps and other cortical markers (Finger and Eling 2019). In 1798, he introduced the principles of what would become known as his *Schädellehre* in German and *organologie* in French in a letter to Viennese censor Joseph Friedrich Freiherr von Retzer (1754–1824) that he published in German (Gall 1798). In this public letter, he suggested that there are many faculties of mind, each dependent on a specific brain territory or organ, and that these territories could be discerned by studying features of the skulls of exceptional individuals (e.g., criminals, lunatics, geniuses) and animals.

Color versions of one or more of the figures in the article can be found online at www.tandfonline.com/njhn.

What Gall was promoting dispensed with metaphysics, went against the older notion of just a few abstract faculties (usually perception, cognition, and memory), and drew attention to the cortex. Hence, his research program signified a major break with the past. In retrospect, it can be regarded as a sustained, early attempt to account for individual differences in behavior, and as the starting point for the tradition of localizing higher functions in the cerebral cortex. Indeed, it was Gall's theory that led Jean-Baptiste Bouillaud (1796–1881) and Paul Broca (1824–1880) to their breakthrough clinical–pathological studies on language disorders, which demonstrated more definitively that the cerebral cortex is not functionally uniform (see Finger 1994; Finger and Eling 2019; Young 1970).

Starting in 1796, Gall began presenting his new ideas about the mind, brain, and behavior to physicians, students, officials, clergy, and other interested parties in lectures, demonstrations, and private meetings at his home in a fashionable part of Vienna. There he housed a growing collection of human and animal skulls and casts, and other supporting materials. He would eventually settle on 27 faculties of mind, but the number, names, and locations of his faculties varied while he was practicing medicine and pursuing his new science in Vienna. The same held true while he was making his way through the German states and other parts of Western Europe during his 1805–1807 lecture tour, prior to his settling in France.

It was in Paris that he finalized his system and published his four-volume *Anatomie et Physiologie du Système Nerveux en Général, et du Cerveau en Particulier* (*Anatomy and Physiology of the Nervous System in General and of the Brain in Particular*) with its accompanying atlas between 1810 and 1819 (Gall and Spurzheim 1810–1819). In 1825, he completed a less expensive edition of the set of books without an atlas and his extensive neuroanatomical research, while leaving his 27-faculty doctrine and his evidence for these faculties basically unchanged (Gall 1822–1825). A less expensive edition of his *Sur les Fonctions du Cerveau et sur Celles de Chacune de ses Parties* was translated into English seven years after his death in 1828 as *On the Functions of the Brain and on Each of Its Parts* (Gall 1835).

Most of what was known about Gall's system in the English-speaking world prior to 1835 did not come from his first set of books in French, which were prohibitively expensive, or from his more affordable *Sur les Fonctions du Cerveau*, but from secondary sources. His doctrine was discussed and transmitted orally, summarized in letters and notes from foreign correspondents, conveyed in short newspaper and magazine articles or translations of pieces first published in German (and later French), and revealed in a few books written in English or translated into English (e.g., Blöde 1807; see also Cooter 1984; Finger and Eling 2019).

Ludwig Heinrich Bojanus (Ludwik Henryk Bojanus, Louis Henri Bojanus, Ludovicus Henricus Bojanus; 1776–1827) was responsible for one of the most important documents reaching English and French audiences prior to Gall's arrival in Paris. The primary purpose of this article is to examine Bojanus's early contributions to the literature on Gall's new doctrine. Another purpose is to present Bojanus's life and professional career more generally, to provide a better picture of who he was as a physician, anatomist, veterinarian, and academic, as he was influential in the creation of comparative anatomy and modern veterinary medicine, especially in Vilnius, an important center of learning at the time. We will begin with Bojanus's early life and education.

Figure 1. Ludwig Heinrich Bojanus (1776–1827). Engraving by Vilnius University professor and artist Motiejus Pšibilskis (Mateusz Przybilski; 1795–1867), 1835 (from the National Museum of Lithuania, with permission).

Bojanus's early life and medical studies

Ludwig Heinrich Bojanus (Figure 1) was born on July 16, 1776, in Bousville (Buchsweiler), Alsace, then part of Hanau-Lichtenberg (Edel 2011). The majority of Alsace's inhabitants at that time were Catholics or Jews. Only about a third of the population was Protestant, and Bojanus was born Lutheran. Alsace was also a bilingual region with both French- and German-speaking people, and Bojanus was fluent in both languages (Edel 2011).

From 1786 to 1792, Bojanus studied at the Bousville Gymnasium. After the French Revolution of 1789, when the French army reconquered Alsace, the Bojanus family left Alsace and sought refuge in what is now Germany. In 1792, they settled in Darmstadt, where

Ludwig completed his secondary education. He then went to the University of Jena to pursue his medical career. Two of his teachers would play especially significant roles in Gall's career and in disseminating his ideas.

Justus Christian Loder (1753–1832)—then professor of anatomy, surgery, and obstetrics at Jena—was one of Bojanus's influential mentors. Loder would witness some of Gall's skull readings, most notably with convicted criminals, and Gall would stay at his house during his tour of the German states in 1805 (Finger and Eling 2019). Loder also assisted Gall with dissections. A strong supporter of Gall, he maintained at the time, "Gall's discoveries about the anatomy of the brain are of the utmost importance," and, "I am convinced that it is possible to recognize prominent mental faculties and such emotional characteristics from external features on the skull" (Ebstein 1920).

Christoph Wilhelm Hufeland (1762–1836) was another of Bojanus's important teachers. Hufeland appended a thoughtful commentary to Berlin surgeon Christian Heinrich Ernst Bischoff's (1781–1861) 1805 book on Gall and his "skull theory," which was titled *Gall'schen Gehirn-und Schädel-Lehre* (Bischoff 1805). This book was translated into French in 1806 and, a year later, into English with added commentary by Henry Crabb Robinson (1775–1867), now titled as *Some Account of Dr. Gall's New Theory of Physiognomy Founded upon the Anatomy and Physiology of the Brain and the Form of the Skull with the Critical Strictures of C. W. Hufeland* (Bischoff 1806, 1807). Hufeland, who was initially opposed to Gall, wrote that he accepted Gall's concept of distinct cortical organs but not his contention that "these individual organs are always intimated by elevations of the skull" (Bischoff 1807, 162).

In 1797, a year after Gall began lecturing in Vienna, Bojanus became a doctor of medicine and surgery (Edel and Daszkiewicz 2015, 2016). He then traveled to Vienna and, during 1797–1798, practiced in the General Hospital under the supervision of Director Johann Peter Frank (1745–1821), also professor of the practice of medicine at the University of Vienna from 1795 to 1803 (Frank 1848; Frankas 2015). Johann Peter Frank also got to know Gall and his theory, but unlike his son, Joseph Frank (1771–1842)—who was chief physician at Vienna's General Hospital from 1796 until 1802, hence overlapping Bojanus—we could find nothing to suggest that the elder Frank wished to be counted as one of Gall's followers or detractors at this time. As for Joseph Frank, he would go on to criticize phrenology, albeit not during his time in Vienna or shortly thereafter, but in 1839 (Frank 1839; Sakalauskaitė-Juodeikienė, Eling, and Finger 2017).

During his stay in Vienna, Bojanus attended Gall's public lectures, and he probably discussed some of Gall's ideas directly with Gall (Biziulevičius 1997). Clearly, he was intrigued enough to take extensive notes for subsequent articles about Gall and his doctrine. Additionally, and in the social sphere, it was in the Austrian capital that he met his future wife, Vilhelmine Roose (1777–1826) (Edel and Daszkiewicz 2016). In 1798, Bojanus returned to Darmstadt, where he maintained a private practice until 1800.

The Landgrave of Hesse helped Bojanus financially. This support allowed him to partake in study trips to universities and veterinary schools throughout Western Europe. His travels between 1801 and 1803 took him to Germany, Austria, England, Denmark, and France. While in Paris, he met many leading scientists. One such person was Georges Cuvier (Georges Léopold Chrétien Frédéric Dagobert, Baron Cuvier, 1769–1832), zoologist, naturalist, and "father of paleontology." Bojanus and Cuvier were both interested in principles of comparative anatomy, among other things (Fedorowicz 1958). It was during this visit to Paris, where he was stimulated by what he was seeing and hearing, that he wrote his article in French summarizing Gall's system.

Early treatises by physicians on Gall's system

Bojanus was one of the first foreign physicians or advanced medical students to report publicly on Gall's lectures. Others, however, attended Gall's lecture-demonstrations in Vienna, and two, in particular, also aspired to introduce his new science of man to a wider audience.

Ludwig Friedrich von Froriep (1779–1847) was a physician who, like Bojanus, had studied medicine at Jena before heading to Vienna in 1799 (for six months) to continue his learning. In 1800, Froriep published a *Kurze Darstellung* (*Short Account*) of Gall's material and, in 1802, a longer treatise titled *Darstellung der neuen, auf Untersuchungen der Verrichtungen des Gehirns gegründeten, Theorien der Physiognomik des Dr. Gall in Wien* (*Account of the New Theories of the Physiognomy of Dr. Gall in Vienna Based on Investigations of the Brain*), which circulated widely and had a French translation (Froriep 1800, 1802a, 1802b). Froriep did not, however, present Gall's various higher faculties of mind in detail, and he provided only vague indications of the parts of the brain associated with each. He mentioned 22 organs of mind in his longer *Darstellung*.

Philipp Franz von Walther (1782–1849), who came to Vienna from a small German town near Karlsruhe to study medicine and specialize in ophthalmology, also desired to provide more information about Gall's new theory than could be found in Gall's 1798 published letter to Retzer. In 1800, he attended Gall's lectures and let him assess his skull, and he left Gall's house a firm believer in the new doctrine. He maintained that what Froriep had written was little more than loose notes. Intent on doing better, he published his *Critische Darstellung der Gall'schen anatomisch-physiologischen Untersuchungen des Gehirn- und Schädel-baues* (*Critical Account of Gall's Anatomical-Physiological Investigations of the Brain and the Form of the Skull*) in 1802, the same year as Froriep's most widely read publication (Walther 1802).

Walther's report was, in fact, far more detailed. He discussed how the organs develop and are affected by aging; introduced some of Gall's thoughts about gender, race, and nationality differences; provided numerous examples of people with dominant personality traits; included a short biography of Gall; and even covered the Emperor's letter from December 1801, which was aimed at limiting Gall's public activities, along with Gall's responses to it. Walther did not identify himself as the author, however, providing just the initials "W-R." He would complete his medical doctorate from the University of Landshut a year later and come forth with a second treatise in 1804, in which he defended Gall, whose public lecturing, he explained, had been restricted for being materialistic and, supposedly, because he was lacking a license for these activities (Walther 1804).

Bojanus, it should be noted, published his treatise in French a year before the seminal Froriep and Walther publications. Coming out in 1801 and 1802, the Bojanus, Froriep, and Walther publications are reflective of the excitement Gall had created in conservative Vienna. Nonetheless, whether Bojanus, Froriep, and Walther were communicating with one another at the time is uncertain. What is clear is only that they would have had ample opportunity get to know one another, given the nature of Vienna's medical community, how the General Hospital attracted physicians and functioned, and how Gall strove to bring visiting physicians into his home and circle.

Bojanus on Gall's 'craniognomic system'

Bojanus did not publish his essay in Vienna. This accomplishment occurred in Paris, where he was invited to lecture on Gall's system to the members of the *Société de Médecine de Paris*. His presentation was published in the *Magasin Encyclopédique, an. VIII*, the year from the *Calendrier Républicain* (or *Révolutionnaire*) *Français* that would correspond to our 1801 (Bojanus 1801). It was called *Encephalo-Cranioscopie. Aperçu du Système Craniognomique de Gall, Médecin à Vienne*. An English translation of his article appeared a year later in the *Philosophical Magazine*. Titled "A Short View of the Craniognomic System of Dr. Gall of Vienna," the author was identified as "L. Bojanus, M.D., Member of the Medical Society of Jena and Paris, and of the Society of the Observers of Man" (Bojanus 1802). Thus, unlike Froriep and Walther, Bojanus was able to convey what he knew about Gall and his knowledge of his new system to both French and English readers.

The French version of Bojanus's article included three illustrations showing Gall's organs on posterior, lateral, and anterior views of the human skull. In contrast, there were no figures in the English version. In other respects, these two articles are virtually identical, allowing us to use the English version for the quotations and commentary that follow.

Bojanus's "Short View of the Craniognomic System of Dr. Gall" opens on a philosophical note. He maintains that Gall's work fits into the broader historical context of trying to understand the mind of man on the basis of physical traits—that is, the tradition known as physiognomy. Indeed, Gall categorized his work as physiognomy in his published letter to Retzer, explaining, "I am nothing less than a physiognomist." (Gall 1835, vol. 1, 18). "At all periods," Bojanus wrote, almost certainly paraphrasing Gall, "a desire to find in the exterior of man certain marks indicative of his interior faculties, his passions, his morals, etc., has induced the learned to establish the systems of physiognomy more or less satisfactory" (Bojanus 1802, 77).

Bojanus explained Gall's basic premises. First, the brain is the material organ of the internal faculties. Second, the brain contains different organs that are independent of one another for these various faculties. Gall cited individual differences as one type of evidence for this independence, leading Bojanus to write: "There are some men who have a great deal of genius without having a memory, who have courage without circumspection, and who possess a metaphysical spirit without being good observers" (1802, 78). Moreover: "The phaenomena of dreaming, of somnambulism, of delirium, etc. prove to us that the internal faculties do not always act together; that there is often a very great activity of one, while the rest are not sensible" (1802, 78).

Bojanus, however, expressed concerns about some of the things Gall was contending. He expressed the worry that, if the mind is no longer considered as an indivisible entity, we might not be able to save ourselves from falling into materialism. This thought was bothering many people in conservative Vienna and elsewhere. For example, the same issue had been raised by Andrew Sniadecki (Jędrzej Śniadecki, Andrzej Śniadecki, 1768–1838), who had worked in Vienna's hospitals and, like Bojanus, had attended Gall's lectures. Sniadecki became professor of natural sciences at Vilnius University in 1797 and published an article on Gall's system in the journal *Dziennik Wileński* three years after Bojanus's English article (Sniadecki 1805). He stated that the brain and its cortex—being anatomically and structurally a solid, integral substance—is designed to operate like a single entity. Like Bojanus and some of Gall's adversaries, particularly those with conservative religious beliefs, he questioned whether the mind and brain really could consist of many independent functional entities (Sakalauskaitė-Juodeikienė, Eling, and Finger 2017).

Bojanus now turned to another basic principle of Gall's new system: the expansion of the organs in the cranium being in direct proportion to the dynamics or forces of their corresponding faculties. According to Gall, the different cortical organs could be physically assessed by examining the exterior shape of the cranium, especially in extreme cases, with bumps signifying well-developed organs and, conversely, depressions signifying weak organs. Bojanus (1802, 79–80) provided an analogy, specifically that,

> [A]s we do not judge of the muscular force of a man or an animal by the volume of their members, but by the development of the muscles, we ought, in like manner, to judge of the strength of the faculties by the development of the relative organs [in the brain].

Nonetheless, he noted that there could be irregularities of the skull, such that there might not be perfect correspondences between the internal and external surfaces of the skull and the underlying brain. This discordance, he noted, could render Gall's cranioscopic methods uncertain. The same criticism would be raised by some of Gall's more vocal contemporaries. For example, John Gordon (1786–1818) and Sir William Hamilton (1788–1856), both Scots, would argue that large frontal sinuses could obscure the picture painted by Gall and render judgments about the development of some of his smaller organs particularly absurd (e.g., Anon./Gordon 1815; Hamilton 1831, 1845). The consequences of illnesses, as noted earlier by Bojanus, constituted another troubling factor.

Bojanus agreed with Gall that man is the "most perfect animal," and that the organs under the anterior and superior parts of the frontal and the parietal bones are essential for the faculties that belong exclusively to humans. He also agreed with him about people sharing an even larger number of faculties with animals lower on the ladder to perfection. In contrast to Froriep, who mentioned 22 faculties and associated organs, Bojanus listed 33 organs in his publications (see Table 1). The listings of these two auditors of Gall's lectures, combined with how Gall would go on to present 27 organs in his finalized system (Gall 1822–1825, 1835, Gall and Spurzheim 1810–1819), shows that his system was very much a work in progress while he was still in Vienna (Finger and Eling 2019).

Gall's first three faculties at this moment in time, Bojanus wrote, did not involve the cerebrum. Gall was then contending that the medulla oblongata was the seat of the organ responsible for *tenacity of life*, as "there are no speedier means of killing an animal than to cut the medulla oblongata" (Bojanus 1802, 81). Similarly, *self-preservation* is localized "a little further forward in the medulla oblongata, at the place where it leaves the brain" (Bojanus 1802, 81). But, Bojanus continued, Gall was unsure about this faculty and is currently collecting more evidence for it. Additionally, Gall localized the organ for *choice of nourishment* in the quadrigemini tubercles (colliculi): the anterior tubercles being larger in carnivores, the posterior in graminivores (i.e., "grass-eating" or herbivorous animals), and the tubercles being of equal size in omnivores.

Gall localized the *cerebral organs of the external senses*, then his fourth organ, in the middle part at the base of the brain, and the one for *instinct and copulation* at the base of the occipital bone, noting it expands during puberty and also how, "In animals, castrated before the age of puberty, the expansion of this organ does not take place" (Bojanus 1802, 82). In his finalized system, Gall would list *reproductive instinct* first (always beginning his list with the most

Table 1. Summary of the 27 faculties, as described in the 1835 translation of Franz Joseph Gall main works, and the 33 organs, mentioned by Ludwig Heinrich Bojanus in his reviews, published in *Magasin Encyclopédique* (1801) and *Philosophical Magazine* (1802).

Gall's faculties, 1835	Bojanus's list of Gall's organs,[a] 1801 and 1802
I. Instinct of generation, reproduction, instinct of propagation, etc.	*Organe de l'instinct de l'accouplement* (Instinct of copulation)
II. Love of offspring	*Organe de l'amour réciproque des parens et des enfans* (Reciprocal love of parents and children)
III. Attachment, friendship	*Organe de l'attachement, de l'amitié* (Attachment and friendship)
IV. Instinct of self-defense; disposition to quarrel; courage	*Organe du courage* (Courage)
	Organe de l'instinct de sa propre conservation (Instinct of self-preservation)
	Organe de la ténacité de la vie (Tenacity of life)
V. Carnivorous instinct; disposition to murder	*Organe pour le choix de la nourriture* (Choice of nourishment)
	Organe de l'instinct d'assassiner (Instinct to assassinate)
VI. Cunning, trick, tact	*Organe de la ruse* (Cunning)
VII. Sense of property; instinct of providing, covetousness; propensity to steal	
VIII. Pride, hauteur, loftiness, elevation	*Organe del l'instinct de s'élever* (Instinct of exalting oneself)
IX. Vanity, ambition, love of glory	*Organe de l'amour de la gloire* (Love of glory)
X. Cautiousness, foresight	*Organe de la circonspection* (Circumspection)
	Organe de l'esprit d'observation (Spirit of observation)
XI. Memory of things, memory of facts, sense of things, educability, perfectibility	*Organe du sens pour les faits (sensus rerum)* (Sense of facts)
XII. Sense of locality; sense of the relations of space	*Organe du sens de localité* (Sense of locality)
	Organes cérébraux des sens extérieurs (External senses)
XIII. The faculty of distinguishing and recollecting persons	*Organe de la mémoire pour les personnes* (Memory for persons)
XIV. Faculty of attending to and distinguishing words; recollection of words, or verbal memory	*Organe de la mémoire verbale* (Verbal memory)
XV. Faculty of spoken language; talent of philology, etc.	*Organe du sens pour les langues* (Sense for languages)
XVI. Faculty of distinguishing the relation of colors; talent for painting	*Organe de peinture, le sens pour les couleurs* (Painting and sense for colors)
XVII. Faculty of perceiving the relation of tones; talent for music	*Organe du sens musical* (Musical sense)
XVIII. Faculty of the relations of numbers	*Organe du sens pour les nombres* (Sense of numbers)
XIX. Faculty of constructiveness	*Organe du sens pour la mécanique* (Sense of mechanics)
XX. Comparative sagacity; aptitude for drawing comparisons	*Organe de l'esprit comparatif* (Spirit of comparison)
XXI. Metaphysical depth of thought; aptitude for drawing conclusions	*Organe de l'esprit métaphysique* (Metaphysical spirit)
XXII. Wit	*Organe pour l'esprit de la satyre* (Spirit of satire)
XXIII. Talent for poetry	
XXIV. Goodness, benevolence, gentleness, compassion, sensibility, moral sense, conscience	*Organe de la bonté* (Mildness)
	Organe de l'amour pour la vérité (Love of truth)
	Organe de la libéralité (Liberality)
XXV. Faculty of imitation, mimicry	*Organe de musique ou du talent théatral* (Theatrical talents)
XXVI. God and religion	*Organe de la théosophie* (Theosophia)
XXVII. Firmness, constancy, perseverance, obstinacy	*Organe de la persévérance* (Perseverance)

[a]Bojanus also mentioned two unknown organs (*Organes inconnus*) in temporal region, whose functions were yet unknown (Bojanus 1801, 1802).

primitive faculty of mind), and he localized it in the cerebellar cortex—it being his only organ for such a faculty not associated with the cerebral cortex.

Gall contended that caring for offspring is closely associated with the instinct to reproduce, and he associated the organ for *reciprocal love of parents and children* in the nearby posterior and superior part of the occipital cortex. In his lectures, he told his listeners that it is generally more striking in women than in men (whereas the urge to copulate dominates in males), and that this organ is well developed in pigeons, birds that care for their young, in contrast to cuckoos, which are almost entirely destitute of this organ and never rear their young.

Gall situated the organ for *attachment and friendship*, then numbered seven, under the posterior and middle part of the parietal bone, which was well developed in social animals, especially dogs. The next faculty and organ on his list related to *courage*, and he thought it dependent on a part of the brain between the two ears. This area, he maintained, is well developed in the hyena, wolf, and lion. His next organ, *instinct to assassinate*, was found in all carnivorous animals and in some executed criminals.

As shown by Bojanus, what Gall was conveying about his first few faculties of mind and their material substrates stemmed in part from his research on animals. Unlike many others at the time, Gall believed in a great chain of being that united humans and animals, not the prevailing system that sharply separated them, because only humans have an immortal soul. Gall's great chain was naturalistic and devoid of metaphysical entities frequently placed above humans: it lacked angels and did not have God at its top.

Bojanus further explained that Gall was still dealing with several unknown organs. He pointed out that the functions of two such organs, both situated in the temporal bones, remained mysterious to him. Gall could be arrogant and dogmatic, but even in his much later 1825 publication, he would continue to state that his system was by no means complete, and that it would require further refinements if warranted by new discoveries.

Bojanus continued with Gall's organ for *cunning*, with its material substrate under the anterior and inferior parts of the parietal bone. Gall had commented that it was especially prominent in the fox and the domestic cat. Next came the organ for *circumspection*, which has its marker in the middle of the parietal bones and is particularly striking in the goat and roebuck, animals that never travel along an unknown road without taking great precautions. *Instinct of exalting oneself*, number 13 with its marker in the middle of the interior edge of the parietal bone, he related, is observed in men distinguished by their pride. Although Gall defined pride as a desire to be superior to others, Bojanus also related that everything is directed upward in a proud man: the hair is highly frizzled, the head is elevated, the eyebrows are arched upward, and the eyelids are raised! Moreover, such a man will show a proclivity to walk on the tips of his toes and look down upon everything beneath him, whereas a modest man's body language will reflect submission. Closely related to pride, Gall then spoke about a faculty and organ for *love of glory*.

The organ for *love of truth*, which Gall had first localized on the surface of the brain below the posterior and superior angle of the parietal bones, was proving more problematic. Gall related that he had not yet collected a sufficient number of facts to be sure of its precise locus. Bojanus inserted that he (Bojanus) felt he needed more evidence to be convinced that there could even be such an organ, because he had observed individuals with crania contradicting Gall's assertion. One displayed great veracity and the other distinguished himself with extraordinary falsehoods. Both had protuberances, rather than depressions or even average cranial development, in the place where Gall argued the organ for *love of truth* seemed to be.

Gall, and Bojanus faithfully conveying his system, now shifted to the organs below the anterior and inferior parts of the frontal bone. These organs are higher on the ascending list than the ones just mentioned, although we still share them with other animals. Here Gall placed the organ responsible for the *sense of locality*, which is prominent in people who have a strong remembrance for places and a great desire to travel. This, his 16th organ, is also well developed in migrating birds. As for his 17th organ, which governs the *sense of facts* and is crucial for a good education, Gall noticed it is well developed in elephants.

Gall maintained that some cortical organs could be related to the arts while also serving more basic functions critical for a species' survival. He stated that the organ for the *sense for colors* is prominent in all painters of great talent. And with regard to the faculty for *musical sense*, which provides memory for recollecting sounds and facilitates harmonious new combinations of tones, its organ is not only evident in the heads of Gluck, Mozart, and Haydn but easily can be discerned by comparing in singing and nonsinging birds. The nearby organ for *mechanics* is prominent in animals and men who distinguish themselves in different arts (e.g., building a home or nest) requiring manual labor.

Bojanus now presented Gall's organ for *verbal memory*, his 22nd organ at that time. He situated it behind "the interior of the orbit" of the eye in those of his schoolmates, who were distinguished for this talent and also had large, cow-like eyes (Bojanus 1802, 135). Gall considered *sense for languages* a separate but related faculty and numbered it 23, noting how it is striking in great philologists.

Here it should be mentioned that Gall and his contemporaries could only guess at how different parts of the brain might work together to achieve goals in nonchaotic ways, this being well before the advent of neuron theory or the concept of synapses. Gall saw no need to go deeper into the underlying physiology, which at the time was highly speculative. He assumed, however, that the closer two organs are to each other, the easier it would be for them to function together. In other words, proximity was an important principle underlying the layout of faculties. Surprisingly, Gall would ascribe this facet of cortical anatomy and physicology to intelligent design, despite the disdain he showed elsewhere in his writings for metaphysics (Eling and Finger 2015; Eling, Finger, and Whitaker 2015).

Interestingly, Gall did not see his word and language faculties as distinctly human at this time or, for that matter, in his later writings. The same can be said for his earlier 19th faculty, *sense for numbers*. Without question, how he sequenced and numbered his organs of mind in 1797–1798 begged for refinement. To cite but one more example, it is hard to understand why he ranked *memory for persons* 24th, above his faculties for *verbal memory, sense of languages,* and *sense of numbers*, after pointing out how well developed it is in horses and dogs!

Gall had nine more organs on his list, and a number of them would transition into his final system as organs unique to humans, although not necessarily under the same numbers or even names. For comparison, the eight distinctly human faculties mentioned in the 1835 translation of his 1822–1825 volumes are *wisdom* (20), *sense of metaphysics* (21), *satire and wit* (22), *poetic talent* (23), *kindness and benevolence* (24), *mimicry* (25), *religious sentiment* (26), and *firmness of purpose* (27). Bojanus called Gall's 25th organ *liberality*, which might correspond, at least in part, with Gall's later *kindness and benevolence*. His next organ was *spirit of comparison*, a name that seems to have much in common with Gall's later *comparative sagacity*. This faculty was followed by *metaphysical spirit* (27), which Gall retained as *sense of metaphysics*, and Bojanus said Gall associated it with a round eminence in the middle of the forehead and found exemplary in the heads of Socrates and Kant. Organ 28, translated in 1802 as *spirit of observation*, was said to extend behind the whole anterior part of the frontal bone, and can be prominent in observers of all ages. Bojanus wrote that, "the celebrated physician Frank [almost certainly Johann Peter Frank, not his son Joseph] is endowed with it in an eminent degree; and Dr. Gall himself is evidently furnished with it" (Bojanus 1802, 137). This exhaulted faculty would not survive the cut.

Turning to the last few faculties and organs on the list that Bojanus provided in 1801–1802, *spirit of satire* (localized behind the frontal protuberances) would later become *wit; mildness* (in the middle of the forehead, prominent in the heads of Christ and Mary as painted by Raphael and Coreggio [sic]) was destined to be named *goodness, benevolence, gentleness, compassion, sensibility, moral sense,* and *conscience; theatrical talents* (in the summit of the frontal bone and prominent in the heads of the "great actors of the different theaters in Paris"; Bojanus 1802, 137) would become *mimicry;* and *theosophia* would be called *religious sentiment* in the 1835 English edition of Gall's 1822–1825 volumes.

Bojanus summarized Gall's thoughts about the latter faculty with the following words, which convey a sense of how succinctly and clearly he managed to summarize Gall's ideas:

> The organ of theosophia occupies the most elevated part of the frontal bone. All the representations of the old saints preserved to us afford very instructive examples; and if there be one destitute of this character, it is certain that it is also void of expression.
>
> An *excessive* expansion of it is observed in religious fanatics, and in men who become religious by superstition. It is the seat of this organ that, according to Gall, has induced all nations to consider their gods as above them in an elevated place in the heavens. When we consider, indeed, this object with a philosophic eye, there is no more reason for placing the deity above the globe than below it (Bojanus 1802, 138).

Thus, in addition to being longer and more informative than the brief periodical pieces then appearing in the British press, Bojanus presented Gall's thinking and his system in more detail, along with commentary in his pieces. Not wanting readers to confuse what Gall was stating with his own thoughts, Bojanus used markings for separating his thoughts from Gall's views. Furthermore, there is neither broad acceptance nor ridicule in his informative articles. Instead, we find him accepting much but not necessarily all that Gall was proclaiming, awaiting more research, and defending him against the charge that his sort of thinking "leads immediately to materialism" (Bojanus 1802, 138).

Bojanus's assessments were honest, brave, and perhaps even provocative. Only a few years had passed since Joseph Freiherr von Stifft (1760–1836), personal physician to Francis II (1768–1835)—supported by conservative Roman Catholic clergy—had advised the Emperor to prohibit Gall from publicly lecturing, because he was promoting dangerous, destabilizing materialism. The Emperor, in turn, stood behind a "general regulation" in 1801 that prohibited all private lectures without special permission, effectively putting an end to Gall's public lecturing and ultimately leading him to leave the Austrian capital, never to return (Finger and Eling 2019). It is also revealing that Bojanus suggested that it would now be up to the author of the new doctrine "to furnish us with further details on the subject," and to convince his audiences, in an incontrovertible manner, "of the truth of his system, a detail of which cannot be satisfactory in a treatise so incomplete" (Bojanus 1802, 138).

Life after Gall: veterinary medicine, comparative anatomy, and Vilnius university

In 1804, Bojanus participated in a competition for the position of head of the Veterinary Department at Vilnius University, for which he presented a supportive document titled

Über die Tierarzneykunst (*About Veterinary Art*). Two years later, he arrived in Vilnius and became professor of zoology (Piročkinas and Šidlauskas 1984).

Vilnius was the capital city of the Grand Duchy of Lithuania until 1795, and the center of Vilnius Governorate of the Russian Empire during the nineteenth century. In the early nineteenth century, the city's university was one of the most important institutions in Eastern Europe and the largest in the Russian Empire (Bumblauskas et al. 2004). In 1803, certain reforms were enacted by Czar Alexander I (1777–1825), a supporter of limited liberalism, who wanted to transform Vilnius with its university into a gateway from the Russian Empire to enlightened Europe (Kvietkauskas 2009). The number of departments now doubled, and foreign professors were invited to lecture. Prominent scholars and scientists from Austria, Germany, Italy, England, and France traveled in increasing numbers to Vilnius (Edel 2011). As Joseph Frank, who had become professor of special therapy and clinical medicine at Vilnius University, wrote in his *Mémoires* in 1808, "the Emperor gave the students at the Institute of Medicine everything I asked for. … The money … was to be paid … in silver" (Frankas 2013, 250).

Bojanus was popular as a lecturer during this period of academic excellence and expansion. His Latin was fluent and expressive, his presentations were richly illustrated with drawings and animal preparations, and he was even able to perform revealing autopsies before his audiences (Adamowicz 1835). Physician Stanisław Morawski (1802–1853), a graduate of the Vilnius University, classified Bojanus as "one of the most educated and prominent European people," adding that physically he was "an imposing figure" with "a beautiful face" (Moravskis 1994, 402).

Bojanus's most important works during these years were on veterinary medicine and comparative anatomy. In 1805, shortly before his arrival, he had published *Über den Zweck und die Organization der Thierarzneyschulen* (*On the Purpose and Organization of the Veterinary Schools*), then, 10 years later and in French, *Des Principales Causes de la Dégénération des Raçes des Chevaux et des Régles à Suivre pour les Relever* (*The Main Causes of the Degeneration of the Breeds of Horses and Rules to Follow to Raise Them*).

Bojanus was a friend of Lorenz Oken (1779–1851), a publisher of the periodical *Isis*. Many of his articles appeared in this acclaimed journal, some being *Die Anatomie des Blutegels* (*Anatomy of Leeches*, 1817), *Kurze Nachricht über Cerkarien und ihren Fundort* (*Message about Cercaria and Its Locality*, 1818), *Versuch einer Deutung der Knochen im Kopfe der Fische* (*Attempt to Interpret the Bones in the Head of the Fish*, 1818), and *Zweifel über das Gefässystem des Krebses* ("Doubts about the Vascular System of Crayfishes," 1822; see Fedorowicz 1958; Otto 1831).

In his 1815 *Introductio in anatomen comparatam* (*Introduction to Comparative Anatomy*; see Figure 2), one of his most important works, he described plant and animal similarities and differences. He covered appearance, habitation, generation, nutrition, respiration, temperature, irritability, sensation, voluntary movements, locomotion, body structure, and more. As Gall had been doing in his lectures, Bojanus began with lower forms of life and ended with humans, arguing that the human body is "at the highest peak of organization, the most developed and perfect" (Bojanus 1815, 51). Also as Gall was already doing in his books, he considered the great chain of being, ladder of life, or what he considered evolution, a process—one in which simple structures become more complex (*a simpliciore fabrica ad magis compositam et elaboratam ascendat*; see Bojanus 1815, 5).

With regard to the comparative anatomy of the nervous system, he noted that much lower animals do not have nervous systems resembling our own. Worms, insects, and mollusks have

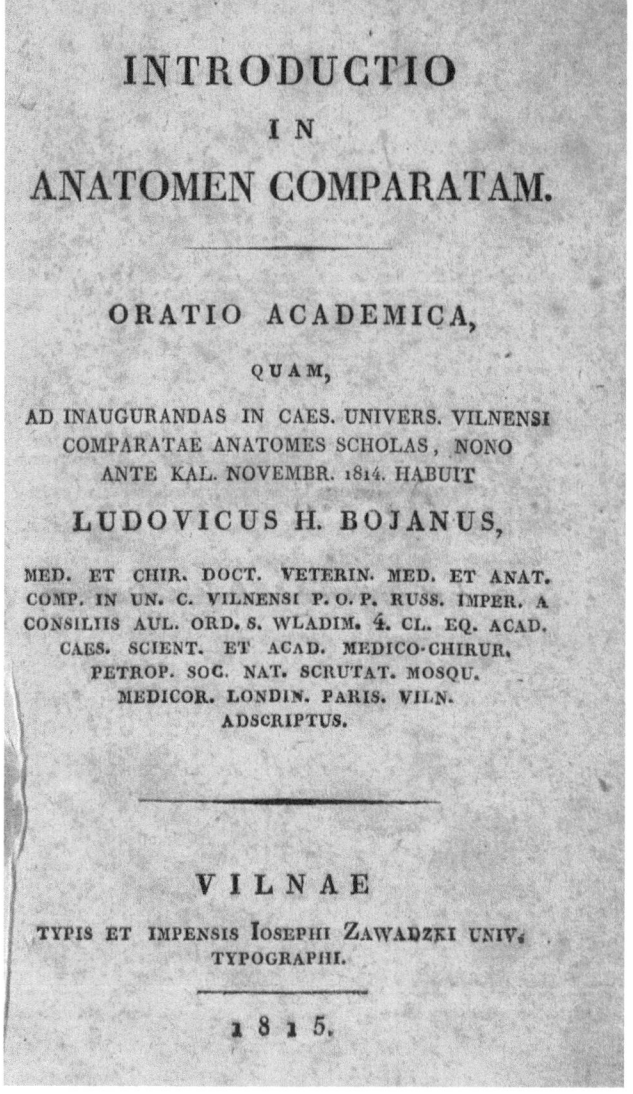

Figure 2. Title page of *Introductio in anatomen comparatam* (1815) by Ludwig Heinrich Bojanus (from the Wroblewski Library of the Lithuanian Academy of Sciences, with permission).

nodes or ganglia with filaments, whereas vertebrates have a medulla spinalis, optic nerves, and a brain. Furthermore, as Gall had begun to argue in his Vienna lectures, Bojanus stated that the noblest part of the nervous system is the cerebrum, the seat of intellect and soul, which is more highly developed in humans than in any other animal. He referred directly to Gall when comparing the structure and functions of brain in humans and lower animals, writing:

> And if we, along with *Gall* and others, would think that the brains of lower animals differ from the human brain to the extent that these animals lack the human mind's abilities and, on the contrary, that the structure of this organ is more similar to humans' in those animals with abilities closer to those of the human soul, be careful when you investigate traces of mind and perception … that you do not confuse the *divine* with simple matter! (Bojanus 1815, 26)

Bojanus referred to the works of Albrecht von Haller (1708–1777), Curt Sprengel (Kurt Polycarp Joachim Sprengel, 1766–1833), Johann Friedrich Blumenbach (1752–1840), Otto Friedrich Müller (1730–1784), and many other notables in his *Introducio*. Still, he considered Georges Cuvier—whom he had met in Paris, where he wrote about Gall and his doctrine—his most important teacher, or at least one of his most important mentors. Unlike Bojanus, however, Cuvier strongly opposed theories of evolution and was a proponent of catastrophism, the idea that there are cyclic creations and destructions of life forms associated with global events, such as massive flooding (Cohen 2017). Bojanus, in contrast, expressed a different opinion in his *Introductio*, standing out as a pioneer of pre-Darwinian and pre-Lamarckian evolutionism at Vilnius University and, more broadly, in early-nineteenth-century Europe (Piročkinas and Šidlauskas 1984; Viliūnas 2014).

Bojanus remained loyal to the Czar when Napoleon's army marched into Russian territory in 1812. He associated the outbreak of war with the 1789 French Revolution and its subsequent destruction and chaos. The war also elicited memories of his family fleeing Alsace and seeking refuge in Darmstadt. Along with some members of the academic community, he opted to leave Vilnius for St. Petersburg, whereas others remained and even joined Napoleon's Army, hoping to restore the Polish–Lithuanian Commonwealth (Edel 2011).

When Bojanus returned to Vilnius in 1814, he worked to make comparative anatomy an independent discipline, becoming the first university professor to do so in the Russian Empire. He also ran the zoological museum in Vilnius and created the first specialized collection of animal anatomy in Eastern Europe (Fedorowicz 1958). Additionally, he established a veterinary school and veterinary hospital in Vilnius in 1823 (Bumblauskas et al. 2004).

At this time, his research gaze was on a reptile (see Figures 3 and 4). His scholarly, two-volume *Anatome testudinis europaeae* (*Anatomy of the European Turtle*) was published in 1819 (vol. 1), and in 1821 (vol. 2; Bojanus 1819; 1819–1821). It contained 40 engravings and 213 figures, and was dedicated to Cuvier, who, after receiving a copy, remarked, "Je le trouve admirable, aucun animal ne sera mieux connu, qui celui-la" ("I find it admirable, no other animal will be known better"; Otto 1831). Bojanus had worked for about a decade on this project, in which he dissected some 500 turtles and used various anatomical techniques to prepare his specimens for study, including maceration, boiling in various solvents, injections of coloring agents, coloring with mercury, gelatin, and so on (Nelson 2014). He even made the drawings, although Friedrich Leonhard Lehmann (1787–1835?), who came to Vilnius from Hesse, engraved the copper plates for publication. The printing of the 80 copies of the first edition cost Bojanus 5000 rubles, then approximately two years' of salary (Edel and Daszkiewicz 2016). Although praised as a scientific success, his acknowledged masterpiece—like Gall's *Anatomie et Physiologie du Système Nerveux en Général, et du Cerveau en Particulier*, which came out at about the same time—proved too costly to become a commercial success (Nelson 2014; Piročkinas and Šidlauskas 1984).

Being a respected professor and loyal to the czarist government, Bojanus was asked to investigate the clandestine activities of certain student societies in Vilnius from 1820 to 1822. Some suspected students had considered themselves "lovers of knowledge," and in 1817, they had formed a secret organization they called *Towarzystwo Filomatów*, meaning the Philomath Society. Even though it had been established for self-education, it had become increasingly active in spreading the seeds for restoring an independent Polish–Lithuanian Commonwealth. Having learned more about it, the Russian imperial authorities charged its members with anti-

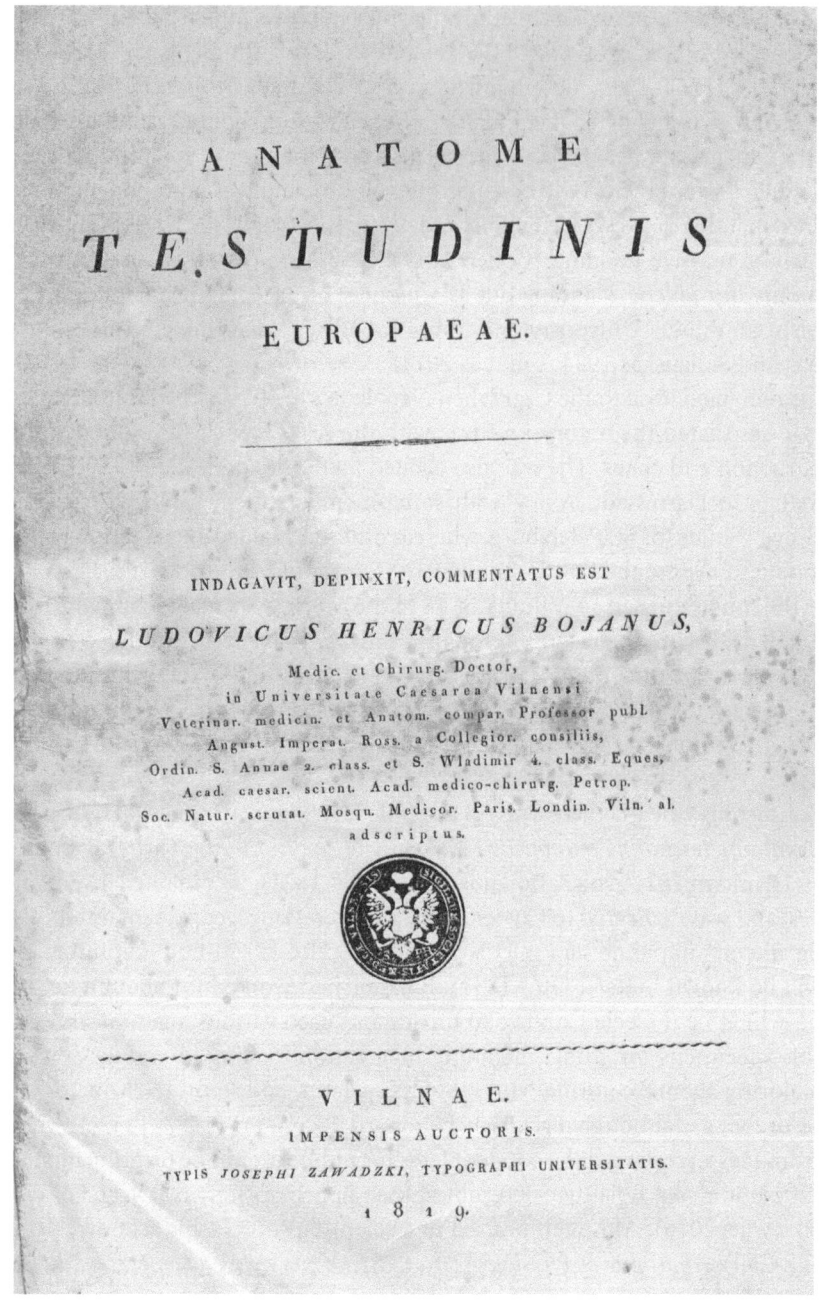

Figure 3. Title page of *Anatome testudinis europaeae* (1819) by Ludwig Heinrich Bojanus, with Vilnius Medical Society stamp (from the Vilnius University Library, Rare Book department, with permission).

czarist activities, leading to one of the largest student trials in Europe during these turbulent times (Bumblauskas et al. 2004). Bojanus reported that the results of his investigation did not reveal anti-czarist activities, and he suggested that the accused students and professors should be acquitted (Edel 2011). Nonetheless, many students were imprisoned and exiled to Siberia, and several faculty members were dismissed.

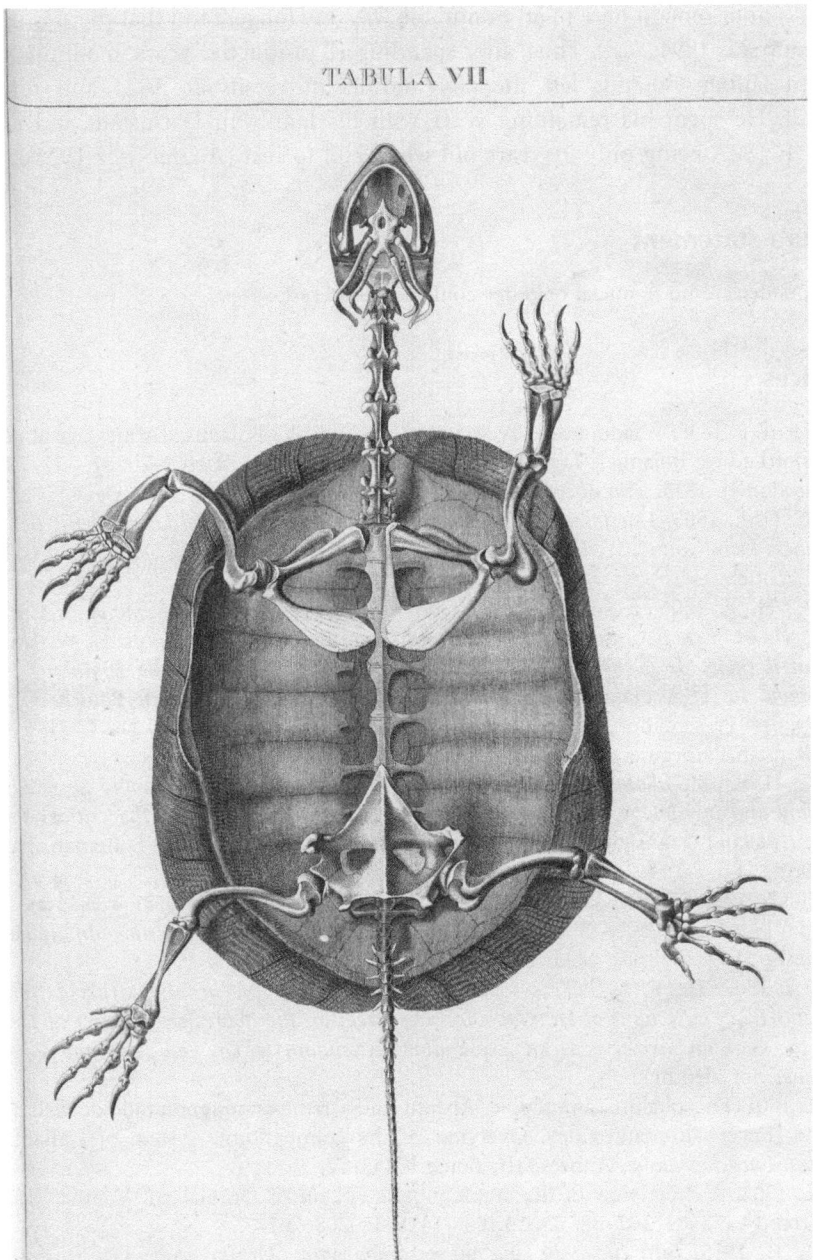

Figure 4. The seventh figure of the turtle in *Anatome testudinis europaeae* (1819) by Ludwig Heinrich Bojanus (from the Vilnius University Library, Rare Book department, with permission).

Dismayed, Bojanus turned down an opportunity to become rector of Vilnius University when the position was offered to him in 1822. Stanisław Morawski visited him at this time and found him devastated. He wrote that the exhausted professor, who had been one of his mentors, was "dressed in a robe, sitting down with his large monograph about a turtle, pale, tired, and suffering because of the decline of the University." Furthermore, "He stated he cannot remain here in an honorable way any longer, and that there is no need to stay" (Moravskis 1994, 218). Thus, after spending 18 productive years in Vilnius, and with his health failing, Bojanus left the city and its university in 1824 and returned to Darmstadt. He spent his remaining years with his family in Darmstadt and died there on April 1, 1827, being only 51 years old when laid to rest (Adamowicz 1835).

Disclosure statement

The authors declare no financial or other conflicts of interest.

References

Adamowicz A. F. 1835. Wiadomość o życiu i pismach Ludwika Bojanusa [A message about life and works of Ludwig Bojanus]. *Tygodnik Petersburski* 80: 462–64; 81: 469–70; 82: 477–78; 83: 483.

Anon. [Gordon J]. 1815. The doctrines of Gall and Spurzheim. *Edinburgh Review* 25: 227–68.

Bischoff C. H. E. 1805. *Darstellung der Gallschen Gehirn- und Schädel-Lehre, nebst Bemerkungen über diese Lehre von C. W. Hufeland* [Presentation of Brain and Skull Doctrine of Gall]. Berlin: L. W. Wittich.

Bischoff C. H. E. 1806. *Exposition de la Doctrine de Gall sur le Cerveau et le Crâne par le Dr. C. H. E. Bischoff … Suivie de Remarques sur cette Doctrine par le dr. C. W. Hufeland …, et d'un Rapport de la Visite de Gall dans les Prisons de Berlin et de Spandau. Traduit de l'Allemand sur la Seconde Édition avec des Notes, des Remarques et une Planche Représentant les Organes* [Presentation of the Doctrine of Brain and Skull of Gall by Dr. C. H. E. Bischoff]. Par Germain Barbeguière. Berlin: C. Quien.

Bischoff C. H. E. 1807. *Some account of Dr. Gall's new theory of physiognomy founded upon the anatomy and physiology of the brain and the form of the skull with the critical strictures of C. W. Hufeland.* Translation and Preface by H. C. Robinson. London: Longman, Hurst, Rees, and Orme.

Biziulevičius S. 1997. Medicinos mokslai senajame Vilniaus universitete 1781 – 1842 m. [Medicine in Old Vilnius University in 1781 – 1842]. In *Vilniaus medicinos istorijos almanachas,* ed. A. Andriušis, 31–98. Vilnius: Medicina Vilnensis.

Blöde C. A. 1807. *Dr. F. J. Gall's system of the functions of the brain: Extracted from Charles Augustus Blöde's Account of Dr. Gall's lectures, Held on the abore [sic] Subject at Dresden. Tr. from the German, to Serve as an Explanatory Attendant to Dr. Gall's Figured Plaster-Sculls.* Publisher not identified.

Bojanus L. 1801. Encephalo-Cranioscopie. Aperçu du Système craniognomíque de Gall, mèdecin à Vienne [Encephalo-cranioscopy. Overview of the craniognomic system of Gall of Vienna]. *Magasin Encyclopédique,* Année VIII, Tome I. 445–72.

Bojanus L. 1802. A short view of the craniognomic system of Dr. Gall, of Vienna. *Philosophical Magazine* 14 (77–84):131–38. doi:10.1080/14786440208676173.

Bojanus L. H. 1815. *Introductio in anatomen comparatam. Oratio academica* [Introduction to comparative anatomy]. Vilnae: typis et impensis Josephi Zawadzki.

Bojanus L. H. 1819. *Anatome testudinis europaeae* [Anatomy of European turtle]. Vilnae: impensis auctoris, typis Josephi Zawadzki, typographi Universitatis.

Bojanus L. H. 1819–1821. *Anatome testudinis europaeae* [Anatomy of European turtle]. Vilnae: impensis auctoris, typis Josephi Zawadzki, typographi Universitatis.

Bumblauskas A., B. Butkevičienė, S. Jegelevičius, P. Manusadžianas, V. Pšibilskis, E. Raila, and D. Vitkauskaitė. 2004. *Universitas Vilnensis 1579 – 2004* [Vilnius University in 1579 – 2004]. Vilnius: Spauda.

Cohen C. 2017. "How nationality influences opinion": Darwinism and palaeontology in France (1859-1914). *Studies in History and Philosophy of Biological and Biomedical Sciences* 66:8–17. doi:10.1016/j.shpsc.2017.10.003.

Cooter R. 1984. *The cultural meaning of popular science: Phrenology and the organization of consent in Nineteenth-Century Britain.* Cambridge: Cambridge University Press.

Ebstein E. 1920. *Ärzte-Briefe aus vier Jahrhunderten* [Medical reports from four centuries]. Berlin: Springer.

Edel P. 2011. Les "émigrés" français face à la tourmente napoléonienne: le cas de Louis Henri Bojanus en Lituanie [French emigrants during the Napoleonic Turmoil: The case of Louis Henri Bojanus in Lithuania]. *Darbai ir Dienos (Vytauto Didžiojo universitetas)* 55:115–26.

Edel P., and P. Daszkiewicz. 2015. *Louis Henri Bojanus.* Strasbourg: Vent d'Est.

Edel P., and P. Daszkiewicz. 2016. *Liudvigas Heinrichas Bojanus.* Vilnius: Gamtos tyrimų centras.

Eling P., and S. Finger. 2015. Franz Joseph Gall on greatness in the fine arts: A collaboration of multiple cortical faculties of mind. *Cortex* 71:102–15. doi:10.1016/j.cortex.2015.06.017.

Eling P., S. Finger, and H. Whitaker. 2015. Franz Joseph Gall and music: The faculty and the bump. In *Progress in brain research. Music, neurology, and neuroscience: Historical connections and perspectives*, ed. E. Altenmüller, S. Finger, and F. Boller, 3–32. Amsterdam: Elsevier.

Fedorowicz Z. 1958. Ludwik Henryk Bojanus. *Memorabilia Zoologica* 1:1–45.

Finger S. 1994. *Origins of neuroscience. A history of explorations into brain functions.* Oxford: Oxford University Press.

Finger S., and P. Eling. 2019. *Franz Joseph Gall: Naturalist of the mind, visionary of the brain.* New York: Oxford University Press.

Frank G. 1839. Della frenologia [On phrenology]. *Bibliotheca Italiana* XCIV. 349–76.

Frank J. 1848. *Mémoires Biographiques de Jean-Pierre Frank et de Joseph Frank son fils* [Biographical Memoirs of Johann Peter Frank and His Son Joseph Frank], *manuscript.* Leipzic.

Frankas J. 2013. *Vilnius XIX amžiuje. Atsiminimai. Pirma knyga* [Vilnius in the Nineteenth Century. Memoirs. The first book]. Trans. from the French by G. Dručkutė. Vilnius: Mintis.

Frankas J. 2015. *Atsiminimai. Antra knyga* [Memoirs. The second book]. Trans. from the French by G. Dručkutė. Vilnius: Mintis.

Froriep L. F. 1800. Kurze Darstellung der vom Herrn D. Gall in Wien auf Untersuchungen über die Verrichtungen des Gehirns gegründeten Theorie der Physiognomik [Short account of the theory of physiognomy and investigations of the functions of the brain founded by Dr. Gall from Vienna]. *Magazin für den neuesten Zustand der Naturkunde mit Rücksicht auf die dazu gehörigen Hülfswissenschaften* 2:411–68.

Froriep L. F. 1802a. *Darstellung der ganzen auf Untersuchungen der Verrichtungen des Gehirnes gegründeten Theorie der Physiognomik des Dr. Gall in Wien* [Account of Dr. Gall's physiognomical theory, entirely based on investigations of the functions of the brain]. 3rd ed. Wien.

Froriep L. F. 1802b. *Exposition de la Nouvelle Theorie de la Physionomique du Dr. Gall de Vienne, Fondée sur la Recherche des Operations du Cerveau* [Presentation of the new theory of physiognomy of Dr. Gall of Vienna, based on research of brain operations]. Leipsic: Comptoir d'Industrie.

Gall F. J. 1798. Schreiben über seinen bereits geendigten Prodromus über die Verrichtungen des Gehirns der Menschen und der Thiere, an Herrn Jos. Fr. von Retzer [Letter from Dr. F. J. Gall, to Joseph Fr[eiherr] von Retzer, upon the functions of the brain, in man and animals]. *Der neue Teutsche Merkur* 3:311–32.

Gall F. J. 1822–1825. *Sur les Fonctions du Cerveau et sur celles de chacune de ses Parties* [On the functions of the brain and on each of its parts]. Paris: J.-B. Baillière.

Gall F. J. 1835. *On the functions of the brain and each of its parts: With observations on the possibility of determining the instincts, propensities, and talents, or the moral and intellectual dispositions of men and animals, by the configuration of the brain and head.* Vol. 6. Trans. from the French by W. Lewis, Jr. Marsh.. Boston, MA: Capen and Lyon.

Gall F. J., and J. Spurzheim. 1810–1819. *Anatomie et Physiologie du Système Nerveux en Général, et du Cerveau en Particulier* [Anatomy and physiology of the nervous system in general and of the brain in particular]. Paris: F. Schoell.

Hamilton W. 1831. *An account of experiments on the weight and relative proportions of the brain, cerebellum, and tuber annulare, in man and animals, under the various circumstances of age, sex, country, etc. Prefix to Monro A, The anatomy of the brain, with some observations on its functions*, 4–8. Edinburgh: J. Carfrae.

Hamilton W. 1845. Researches on the frontal sciences, with observations on their bearings on the dogmas of phrenology. *Medical Times* 12:159–60. 177–179, 371, 379.

Kvietkauskas M. 2009. Pratarmė [Preface]. In *Imperinis Vilnius (1795 – 1918): kultūros riboženkliai ir vietinės tapatybės* [Imperial Vilnius (1795-1918): Cultural boundaries and local identities], ed. J. Konickaja and R. Kvaraciejienė, 7–10. Vilnius: Lietuvių literatūros ir tautosakos institutas.

Moravskis S. 1994. *Keleri mano jaunystės metai Vilniuje. Atsiskyrėlio atsiminimai (1818 – 1825)* [Several years of my youth in Vilnius. The Memoirs by Hermit (1818 – 1825)]. Trans. from the Polish by R. Griškaitė. Vilnius: Mintis.

Nelson E. C. 2014. The will of Ludwig Heinrich Bojanus (1776–1827), an interesting Nineteenth-Century natural history document. *Archives of Natural History* 41 (1):164–67. doi:10.3366/anh.2014.0221.

Otto A. W. 1831. Ludwig Heinrich von Bojanus. *Nova Acta Halle* 15 (I):XXXIX– XLV.

Piročkinas A., and A. Šidlauskas. 1984. *Mokslas senajame Vilniaus Universitete* [Science at the Old Vilnius University]. Vilnius: Mokslas.

Sakalauskaitė-Juodeikienė E., P. Eling, and S. Finger. 2017. The reception of Gall's organology in early-Nineteenth-Century Vilnius. *Journal of the History of the Neurosciences* 26 (4):385–405. doi:10.1080/0964704X.2017.1332561.

Sniadecki J. 1805. Krótki Wykład Systematu Galla z przyłączeniem niektórych uwag nad iego Naukę [Short lecture on the system of Gall with some comments about His science]. *Dziennik Wilenski* 1:16–42.

Viliūnas D. 2014. *Filosofija Vilniuje XIX amžiaus pirmoje pusėje* [Philosophy in Vilnius in the first half of the 19th Century]. Vilnius: Lietuvos kultūros tyrimų institutas.

Walther P. F. 1802. *Critische Darstellung der Gallschon anatomisch-physiologischen Untersuchungen des Gehirn- und Schädel-baues* [Critical account of Gall's anatomical-physiological investigations of the brain and skull]. Zürich: Ziegler.

Walther P. F. 1804. *Neue Darstellungen aus der Gall'schen Gehirn- und Schedellehre, als Erläuterungen der vorgedruckten Verteidigungsschrift des D. Gall eingegeben bei der niederösterreichischen Regierung. Mit einer Abhandlung über den Wahnsinn, die Pädagogik und die Physiologie des Gehirns nach der Gall'schen Theorie* (New Presentation on Dr. Gall's Brain and Skull Doctrine …). München: Scherer (Fleischmann).

Young R. 1970. *Mind, brain and adaptation in the Nineteenth Century.* New York: Oxford University Press.

Matter over mind? The rise and fall of phrenology in nineteenth-century France

Marc Renneville (iD)

ABSTRACT

The history of phrenology in France has a number of unique features. It was in that country that F. J. Gall sought refuge; and it was, above all, in France that phrenology would subsequently attempt to establish its credentials as a new physiological science of the mind. Up until the 1840s, phrenology expanded rapidly in the country, a growth that coincided with attempts to provide this new field with the trappings of respectable scientific endeavor—courses of lectures, learned societies, journals, and so on. This ambitious intellectual project, despite its controversial nature, made a major cultural impact in the nineteenth century, both through its influence on the written word—from learned journals to the novel—and via its striking visual imagery (sculpture, anatomical diagrams and models, engravings, caricatures, and so on). However, as the scientific impact of phrenology declined, allusions to it lost much of their cultural force. On the borderline between respectable science and mere quackery, phrenology in France represented an attempt to construct a whole new intellectual universe based on scientific principles, and as such had a profound impact on its period.

Analysis is like casting a lead line into the ocean. Throw it deep into the water and it will leave the weak-willed behind, frightened and helpless. But for the strong-hearted, those who are prepared to grasp it firmly with both hands, it offers both guidance and reassurance.
(Doctor Noir, in Alfred de Vigny, *Stello, Première consultation*, 1832)

Written off as pseudoscience for more than a century, phrenology is now undergoing something of a reappraisal. Indeed, some have argued that it should be considered as the first veritable human science, as it constitutes the first attempt to link psychological reasoning to the physiological study of the brain (Changeux 1994, 17). For other scholars, in contrast, phrenology should be seen rather as an early example of the kind of narrowly reductionist approach characteristic of modern neurobiology (Andrieu 2002, 19–40). What these otherwise contradictory readings have in common is that they see in the phrenological project a foreshadowing of concerns and questions that preoccupy contemporary science. It could be argued that this paradoxical and problematic legacy results, in part at least, from major ambiguities contained within the original theory itself, one that occupied an ambiguous position on the frontier between mainstream science and mere quackery.

A new science

Phrenology originated with Franz Joseph Gall (1758–1828; Figure 1), a doctor who had trained in Strasbourg with professor Jean Hermann (1738–1800) and others, and in Austria's imperial capital, Vienna, at the end of the eighteenth century. Gall always maintained that his theories had been a development of his own observations and experiences. Having noticed during his youth training that students with protruding eyes were notable for their excellent memory, he later made the leap from physical trait to mental faculty. Starting from the assumption—still widely contested at the time—that the brain was the seat of cognition, he decided to see if other mental skills could be linked to the external contours of the subject's head (Finger and Eling 2019). Gall reasoned that both humans and animals are born with particular mental faculties and instincts. It followed from this that these faculties are not the product of an individual's education, of acquired needs, or the result of the impact of other forces such as climate. However, those extraneous factors do have the capacity to prevent the manifestation of inherited faculties, or modify the form they take when they do appear.

Starting from such assumptions, Gall embarked on a vast empirical project, involving the close scrutiny of both animals and humans in an effort to isolate their distinctive traits. In the latter case, poets, mathematicians, musicians, actors, and cooks were all subjected to the phrenological gaze, as were the insane, thieves, and murderers. His method for pinning down the cranial site of particular faculties was to identify a subject with a "prominent" mental trait, and then to attempt to fix it to a matching physical location on the head. Gall also worked from head to traits. In this way, Gall was ultimately able to

Figure 1. Bust of Franz Joseph Gall (1828). Molding made on nature. Collection of the Musée de l'Homme (Paris). Photo taken by Marc Renneville.

draw up a list of 27 brain functions, running from a strong sexual drive and comparative perspicuity to a talent for music or a tendency to steal (Gall and Spurzheim 1810–1819).

The knowledge of humans obtained by this technique had not just scientific but also major social and political ramifications (McLaren 1981). So although the starting point of "phrenology" (a term that Gall never accepted) was a particular conception of the way of the brain functions, it also involved a broader theory of human behavior as well, as a distinctive diagnostic tool, the palpation of the head or skull.

The close examination of the head, it was claimed, allowed practitioners to reveal the deep-seated aptitudes and inclinations of any individual, offering the prospect of a society organized along rational, scientific lines, one that would make room for the "immense variety in the development, excitability, and relative proportions, of our cerebral organs" (Gall 1835, vol. 6, p. 293). Within this immense human variety, Gall devoted particular attention to the geniuses, the insane, and the criminals he met, respectively, in the Paris salons, mental asylums, and prisons of his day. In the latter case, his views on what drove certain individuals to commit crime caused an outcry, for he argued that there existed an inherited tendency to commit murder. This could be seen in the natural world in the behavior of carnivorous animals, but controversially, Gall extended the principle to a number of rulers from European history, among them the Roman emperors Caligula and Nero, Philip II of Spain, Charles IX of France, Richard the Lionheart, Mary Tudor, and Catherine de Medici (Gall and Spurzheim 1810–1819, vol. 3, p. 258).

Gall's theories provided a physiological anchor for the metaphysical notion of "moral liberty," limiting in practice its exercise to humans of sound body and mind. In this way, he laid the groundwork for a whole raft of specialisms in the field that would become pathological psychology. Gall provided, in effect, a biological answer to a metaphysical question, arguing that a person capable of exercising free will is a person in good health, whereas a person whose actions are determined for him or her is a person who is deficient, or at least in an abnormal physiological state, one showing more animal-like propensities.

In Gall's view, the overdevelopment or underdevelopment of a particular organ could lead to an irresistible impulse to engage in certain forms of behavior. Such a conception was in contradiction with the French Penal Code of 1810. This code excluded the punishment of "madmen" (art. 64) and established a graduated scale of punishments according to the gravity of the offense; what would now be termed the "dangerousness" of the offender did not enter into the equation (Renneville 2006).

By taking sides in the philosophical and theological debate that pitted determinists against free-will supporters, phrenologists were accused by some of having effectively undermined the fundamental legal principle of the right to punish. For Louis-Emmanuel Fodéré (1764–1835), professor of forensic medicine in Strasbourg, the new theory was only an "obscure *galimathias*" (Fodéré 1832, VII). Others, by contrast, considered that phrenology offered the key to a rational response to crime, and perhaps to deviant behavior more generally: for example, one of Jean-Etienne Dominique Esquirol's (1772–1840) students, Étienne Georget (1795–1828), who defended the medical approach to famous trials (Georget 1825); Félix Voisin (1794–1872), who sought to improve assistance to idiotic or delinquent children (Voisin 1832); the writer Benjamin Appert (1797–1873), who visited prisons (Appert 1832); and, last but not least, Guillaume Ferrus (1784–1861), Inspector General of Asylums and Prisons (Ferrus 1850). In either case, it constituted one of the first scientific doctrines claiming to offer a physiological answer to questions of social norms.

Let us consider the case of property. Drawing on the work of researchers who had observed how animals would tenaciously defend their own territory in the natural world, Gall reasoned that property was an "institution of nature" that found its reflection in a specific faculty of the brain. When the organ corresponding to that faculty functions within the boundaries of normality, and thus of morality, Gall argued, it regulates "the sentiment of property or the propensity to make provisions," what he called a "fundamental quality." However, when overdeveloped, that same faculty is also linked to the "propensity to theft," an impulse that in an extreme form can lead to the pathological condition of kleptomania (Gall and Spurzheim 1810–1819, Vol. 1, 142–149).

A controversial science

Prevented from teaching in his home city, Gall left Austria in 1805 and commenced a peripatetic existence, criss-crossing Europe as he presented his new theories in a series of well-attended lectures. He arrived in France in 1807, at a time when the country was in the grip of a trade blockade with England, and the press was subject to strict censorship at the hands of the imperial authorities.

Gall's doctrines provoked strong reactions in the Parisian press. On one hand, there was the view, as reported in the *Journal des Débats*, that said the doctor "walked a fine line between the odious and the ridiculous," and could "rarely be found in the middle ground between the two." On the other hand, there was the *Gazette de santé*, which rejected vigorously any suggestion that Gall's theories implied fatalism or materialism. The *Courrier de l'Europe* came down on the side of the proregime organ, the *Journal de l'Empire*, whereas the *Archives littéraires*, the *Journal de médecine pratique*, and the *Bulletin des sciences médicales* defended Gall (Renneville 2000, 84). Raised in that fertile Viennese soil, which had already given birth to mesmerism, phrenology appeared to the public of early-nineteenth-century France as a cross between animal magnetism and physiognomy, combining the therapeutic optimism of the former with the diagnostic simplicity of the latter.

Discussions of phrenology were by no means limited to the political or medical press, however: They resonated widely in the popular culture of the period. The new doctrine was the subject of a short, satirical play, the *Système bossu ou doctrine biscornue de maître Gallimat* (*The System of Bumps or the Quirky Doctrines of Master Gallimat*), and other popular treatments of the subject were quick to follow (Wegner 1988). The bookseller Martinet sold colored satirical prints entitled *Docteur Gall … imatias*, and a certain Martin, on Rue du Petit-Lion-Saint-Sulpice, churned out bawdy doggerel on the subject of "Docteur Gall à Cythère" (Renneville 2000, 85–87). The new "craniology" was even given the festive treatment in the 1808 Paris Carnival.

As we leave Parisian popular culture behind, and journey to the capital's fashionable society soirées, it must be acknowledged that Gall's theories would never command unanimous approval during his lifetime. Phrenology *did* receive the backing of Klemens Wenzel von Metternich (1773–1859), Élie Louis Decazes (1780–1860), François Broussais (1772–1838), and Jean-Nicolas Corvisart (1755–1821), but Napoleon considered its teachings no better than the quackery pedaled by Alessandro Cagliostro (1743–1795), Franz Anton Mesmer (1734–1815), and Johann Caspar Lavater (1741–1801). François-René de Chateaubriand (1768–1848) too poured scorn on Gall in his memoirs, and Gall's works were also condemned by the Catholic Church. Many argued that Gall's materialism placed him beyond the pale of respectable science.

Phrenology's fortunes improved somewhat during the 15-year period that followed the Restoration of the Bourbon Monarchy in 1815, its anticlerical associations finding favor with the country's liberal opposition. And during the reign of Louis-Philippe (1830–1848), its supporters enjoyed privileged access to the movers and shakers in the government. One of the most active of these conduits was Benjamin Appert (1797–1873), appointed after the July revolution to the post of secretary to Queen Marie-Amélie (Maria-Amalia; 1782–1866), and who also served in the household of Princess Marie of Orleans (1813–1839), one of the royal princesses, until her early death in 1839 (Appert 1846).

These royal duties did not stop Appert from pursuing his phrenological researches. He consulted regularly with his friend Charles-Henri Sanson (1739–1806), executioner to the French court, and also with "Papa Jules," otherwise known as Eugène-François Vidocq (1775–1857), the former criminal turned Paris *Sûreté* chief, who now ran a private detective agency. Sanson provided his friend with detailed accounts of the last hours of condemned prisoners, and even regularly handed over their frock coats for Appert's personal collection. As for Vidocq, he would take the royal courtier on tours through some of the seedier districts of the capital. A precursor of the social explorers of the second half of the nineteenth century, Appert wished to gain first-hand experience of the lives of his contemporaries, studying their behavior *in situ*. To that end, he would often don disguises for his explorations of the Paris underworld, pacing the streets incognito, like a phrenologically minded version of Rudolph, the character from Eugène Sue's *Les mystères de Paris* (*The Mysteries of Paris*; 1842–1843).

When he was not occupied with his royal duties or ethnological forays, Appert liked to organize society dinners on Saturday evenings. Here there was no need for disguises—or, rather, the disguises were of a different kind, for this well-connected phrenologist's soirées attracted a glittering array of personalities from the artistic and intellectual elite of the day. Among their number were the novelists Honoré de Balzac (1799–1850) and Alexandre Dumas (1802–1870; the latter was, for a time, Appert's personal secretary); the philosopher Charles Fourier (1722–1837) and his associates Victor Considerant (1808–1893) and Charles Harel (1890–1846); Dr Pierre Jean Chapelain (1788–1867), a supporter of magnetism; and Dr Casimir Broussais (1803–1847), son of the famous surgeon François Broussais (1872–1838). Both Sanson and Vidocq were frequent visitors. Appert's guests might have been lucky enough to be entertained by the piano playing of Franz Liszt, (1811–1886), who made a brief appearance at the courtier's house during his first trip to Paris. His host seized on the occasion to make a plaster cast not of the hands of the young prodigy but of his head.

The hunt is on

In his lectures and published works, Franz Joseph Gall lost no opportunity to impress on his listeners and readers the importance of building a theory on solid empirical foundations. This determination always to link analysis to literally "palpable" facts explains why phrenologists devoted so much time and energy to hunting down skulls, together with plaster casts taken—like Liszt's—from life or after death (as with Schiller). Even the alienist Esquirol, who had major reservations about phrenology, took the decision early on in his career to conserve the skulls of his former patients in order to test the validity of Gall's theories. Gall's second substantial collection was acquired by the Paris Natural History Museum in 1831 (Ackerknecht and Valllois 1955) and is now kept at the Musée de l'homme. The human parts of his first collection that were left behind in Vienna went to Rollet Museum in Baden, near Vienna, and the animal parts went to a veterinary facility there. In France, other significant "phrenological" collections remain in Lyon, Rouen, Rochefort, and Aix-en-Provence, but not all of them are accessible to the public (Renneville 1998).

As early as 1815, Paris's Faculty of Medicine began building up a representative collection of skulls and plaster casts. The job of building up the faculty's collection was given to Alexandre Pierre Marie Dumoutier (1797–1871), who worked on the project between 1815 and 1831, before being appointed to the position of official molder to the Paris Phrenological Society (Ackerknecht 1956). In 1833, Dumoutier did a cranioscopy in a criminal case. He analyzed the skull of the widow Houet, and this examination stirred controversies on the evidentiary value of phrenology and medicine in criminal investigations (Bertomeu-Sánchez 2015).

After Gall's death, Dumoutier went on to take molds from the heads of the murderous poet Pierre-François Lacenaire (1803–1836), General Jean-Maximilien Lamarque (1770–1832), Casimir Perier (1777–1832), the chef Antonin Carême (1784–1833), the regicide Giuseppe Fieschi (1790–1836), and many other celebrities. The resulting collection of anatomical specimens formed the basis for a phrenological museum on the Rue de Seine in Paris's Latin Quarter, which opened its doors on January 14, 1836.

The inauguration took place during an extraordinary public meeting of the Paris Phrenological Society, founded five years previously. The high point of the meeting, announced in advance by the press and via advertising bills, was the phrenological analysis of the skull of the celebrated criminal Lacenaire, barely a week after his execution. Who could hope for a more topical application of phrenology? The palpation of this celebrated artifact, as was often was the case, did not offer anything particularly new, but the event served its purpose: The museum became *the* destination for anyone wishing to admire the skulls and head casts of the great, the good, and the not-so-good.

Among the phrenological booty on display were busts of balloonist Jean-Pierre Blanchard (1753–1809); those of conspirators Georges Cadoudal (1771–1804) and twins César and Constantin Faucher (1760–1815); composer Joseph Haydn (1832–1809); celebrated priest and politician Abbé Henri Grégoire (1750–1831); artist César Ducornet (1806–1856), who painted with his feet; Théroigne de Méricourt (1762–1817); and many others whose biographies are unknown today, as the killer Bouhours, the thief Jacques Valotte, a crooked lawyer called Arnaud de Fabre, and a sculptor who worked in his sleep by the name of Pora. In addition to the stars of the collection, a partial death mask of Napoleon and that inconclusive skull of Lacenaire, the museum also held the skull of the Marquis de Sade (1740–1814), a fragment of the skull of composer and poet Bartolomeo Sestini (1792–1822), the death mask of King Charles XII of Sweden (1682–1718), and the top of the skull of another murderer named Reveillon. One of the most prestigious pieces was obviously Gall's bust, which made it possible to present the organ of causality (Figure 2). By 1837, the museum's collection totaled some 600 head casts, 300 skulls, 200 brain molds, some mummified heads, and skulls of various "racial types," along with representatives of four classes of vertebrates (Renneville 1998). From 1837 to 1840, Dumoutier developed his technique to a fine art during a scientific expedition to Oceania led by Jules Dumont d'Urville (Renneville 1996).

The museum aroused the admiration of foreigners, and there were frequent written requests from abroad for reproductions of its holdings. For those wishing to be initiated

Figure 2. Drawing representing Gall's profile indicating the location of the organ of causality (Fossati 1845, 505).

into the science of the phrenological organs, the curator was on hand to offer his advice. Perhaps you wished to know more about an organ responsible for cunning, shrewdness, and know-how—which, if overdeveloped, could lead to dissimulation and intrigue. You would be told that the faculty in question was called "secretiveness," and that this was a feature of the heads of the leading actresses of the day, such as Mademoiselle Rachel and Marie Dorval, as well of those of Fouché and Talleyrand. Or, then again, perhaps you were interested in knowing the cause of a strong development of the organ responsible for obtaining and choosing particular foods. The example of Henrion de Pansey would be cited, famous for his gastronomic feasts, and the busts of well-known epicureans Mirabeau and Brillat-Savarin would be indicated. An interest in poetry? Your attention would be drawn to the organ located on both sides of the head, "above the temples, close to the lower edge of the temporal fissure of the frontal bone, in between the organs for the fabulous, for music and property, just above those for construction and touch" (Fossati 1845, 379). Your guide might add that Tasso, Shakespeare, Milton, Schiller, Sestini, Michelangelo, and Victor Hugo all possessed this organ in abundance, whereas it was seriously lacking in Locke, Hume, Flourens, and the indigenous population of "New Holland" (Australia).

A science with a myriad of applications

Phrenology was at the peak of its popularity in the 1830s. At that time, it bore all the trappings of a legitimate science. Its study was the object of a learned Parisian society, which counted not only upward of 200 doctors among its members, but also men of letters

(Pierre Dubosc, Michel-Auguste Dupoty, Charles Brugnot, and Théophile Dinocourt), together with artists and sculptors (François Gérard, David d'Angers, Denis Foyatier, and Henri Lemaire). The presence of the latter group would have pleased Gall, who had always maintained that his discoveries could render a valuable service to art. A work of art that respected the principles of phrenology, he had taught, benefited from increased realism. Ignore those principles and there was a risk of depicting Venus as a scatterbrain, as Botticelli had allegedly done. As Gall wrote:

> We may judge now, what a want of fidelity to nature, has been shown by artists, who, according to imaginary proportions of beauty, give to their statues, and especially to their Venus, so small a head. These artists, as well as their admirers, are utterly ignorant of the laws of the cerebral influence over the exercise of the intellectual faculties. (Gall 1835, 2, 219)

The French artist of this period whose work shows the greatest influence of phrenology was undoubtedly David D'Angers (1788–1856; see Baridon and Guédron 1999).

The tenets of the new doctrine were disseminated in Paris and elsewhere in France via numerous programs of talks and public lectures. Other learned societies were set up in Toulon, Epinal, and Saint Brieuc; Gall's teachings also attracted followers in the penal colonies of Brest, Toulon, and Rochefort, as well as in the cities of Bordeaux, Montpellier, Nancy, Metz, and Rouen (Renneville 2000).

Some 30 phrenological societies in all sprang up in Britain during this period—a considerably higher number than in France—but the terms of the debate between supporters and opponents of the doctrine were similar on both sides of the Channel. Gall's disciples were accused of materialism in France. They had to respond both to attacks from philosophers, such as Victor Cousin (1792–1867), and from churchmen (Forichon 1840). An interesting aspect of the British response to phrenology is the favor it found with Owenite socialists, which was in some ways similar to that of Charles Fourier and his school in France (Rignol 2002).

There was also support for Gall's ideas in Scandinavia, Spain, Germany, and the United States, where Johann Spurzheim (1776–1832) taught phrenology during his last trip in 1832. Even the Canadian province of Quebec, a center of Catholic orthodoxy, proved amenable to the doctrine of phrenology, hosting a series of lectures by surgeon Jonathan Barber (1767–1864), a member of the Boston Phrenological Society in 1836–1837. The city fathers of Quebec were particularly welcoming, loaning Barber municipal premises for the purpose (Renneville 2000, 186).

Decline and marginalization

The first signs of the decline of the phrenological movement initiated after Gall's death appeared in France in the 1840s. One of the reasons for this was a question of demographics, with the deaths in the space of just a few years of a whole raft of leading practitioners, among them Spurzheim, who died in Boston in 1832, and also Bailly de Blois (1837), François Broussais (1839), and Jules Dumont d'Urville (1842). The period was also marked by a new offensive from the opponents of phrenology.

In a piece written in 1842, Pierre Flourens, the permanent secretary to the Académie des sciences, provides us with a useful summary of the principal bones of contention. According to Flourens, Gall's doctrines failed to convince on three separate levels. He condemned the phrenologists' rejection of the unity of the mind as a "psychological" error, the division of the brain into autonomous functional regions as a "physiological" error, and the notion that reason and willpower are merely the consequences of those functions as a "moral" error.

Academic phrenology proved unable or unwilling to respond to this barrage of criticism, and the field gradually came to be dominated by minor works, only marginally related to Gall's original doctrines, building their theories out of various combinations of physiognomy and craniology. By the reign of Napoleon III (1852–1870), this corpus of popular works, originally peripheral to the phrenological mainstream, had taken center stage, with the result being that the palpation methods used by phrenologists gradually came to be assimilated with the various divinatory "sciences" practiced by those trading in hopes and promises in the country's traveling fairs and shows.

There is, however, another factor that needs to be taken into account. Phrenology claimed to provide a holistic approach to the human condition at a time when research in the natural and human sciences was becoming increasingly specialized, a process accompanied by the creation of a well-policed disciplinary division of labor. In the field of anthropology in particular, phrenological doctrines faced competition from the taxonomy of races established by the Société d'ethnologie de Paris, founded in 1839 (Staum 2003). At the same time, the phrenologists' case was damaged by the influence of eclectic philosopher Victor Cousin. Not only did some phrenologists spend a great deal of time and energy in a failed attempt to counter his arguments, they also undermined the coherence and originality of their own materialist case with their support for his sort of spiritualism. The result was that the doctrines of Gall were effectively marginalized, considered of little relevance by the up-and-coming generation of philosophers—Ernest Renan (1823–1892), Hippolyte Taine (1828–1893), Théodule Ribot (1839–1916), and the others—who would lay the intellectual groundwork in France for the new discipline of psychology.

A normative science on the margins

Challenged and then rejected by mainstream science, did phrenology nevertheless contribute to the dissemination of a new conception and representation of the mind? For followers of Gall, the body offered vital clues to mental processes. This notion of a deterministic causal relationship between organic structure and mental functions would form the basis of Bouillaud's clinical cases. Jean-Baptiste Bouillaud (1796–1881) was close to Gall's view, and he defended the principles of his brain physiology at the Académie nationale de médecine (Bouillaud 1865). Traces of these principles can be found in Bénédict-Augustin Morel's (1809–1873) theory of degeneration in the 1850s, in Paul Broca's (1824–1880) famous observations of aphasia during the 1860s, in Cesare Lombroso (1835–1909) criminal anthropology in the 1870s, and in the taxonomy of "professional types" developed by French medicine at the end of the century (Renneville 2013). Phrenology thus shared in a more general obsession with identification in the scientific community of the period—an obsession that would, of course, continue to be influential in the persistence of morphological and racial typologies in twentieth-century social and medical science.

Phrenology also played its part in undermining the legal notion of the autonomous rational subject, despite the fact that the latter had been strengthened by the new legal codes put in place in France following the Revolution. Throughout the nineteenth century, the concept of free will was under attack from all sides, with an assortment of physically determined pathological states seen as undermining an individual's ability to exercise control over his or her own actions. Phrenology was the first theory to ask the question that would give rise to the birth of criminology at the end of the century: Was crime a form of sickness, the perpetrators of which required treatment rather than punishment? In this way, phrenology contributed to a convergence of medical and judicial norms that have continued up to the present—observable, for example, in the contemporary debate in France surrounding the ethics of imposing compulsory medical treatment on offenders and ex-offenders.

The phrenological method gave priority to the visible over the invisible, to supposedly detached observation over introspection. Gall firmly rejected the philosophical introspection as a method of analyzing the human mind. He participated, in this respect, in the dissemination of a new scientific culture based on the figure of the neutral, objective "expert," far from the French philosophical tradition. Discussion of the essence of the human condition thus centered increasingly during the nineteenth century on the concrete and the measurable (the latter preferably by means of a vast array of precision instruments) rather than on the metaphysical and the divine. In other words, understanding humans, exploring their limits and their potential, was now seen to require a close familiarity with the contours of the skull.

This approach had two important consequences for the future. First, the experimentation would become the epistemological cornerstone of the new, "scientific" psychology. Second, the refusal to tie scientific practice to any external system of values would lead to a conception of scientific ethics that would veer toward scientism.

Gall's theories also have to be seen against a background of the growth of an individualizing, narcissistic, middle-class culture in the nineteenth century. For several decades, the palpation of the head would be considered a potent tool for the psychological analysis of the self. Novelists provide many examples of such a usage for phrenology, first treating it—like Balzac—as a vehicle for concrete self-knowledge and later transforming it into an object of ridicule. Alfred de Vigny's (1797–1863) disillusioned character Stello, for example, describes his symptoms to Dr. Noir in considerable detail using the language of phrenology (Vigny 1832/1993, 498–501). Later in the century, Gustave Flaubert's (1821–1880) Bouvard and Pécuchet would get into the act as well, trying their hand at the palpation of heads (Flaubert 1881/1999, 373–378). The apparent insights offered by the phrenological method also provided a rich seam of material for satirists and caricaturists, but as the golden age of the "bumpologists" faded into the past, the references to phrenology became increasingly superficial, cut loose as they were of the original intellectual and cultural context in which they were elaborated.

Ironically, it was probably the very strength and pervasiveness of that hostility and mockery that enabled phrenology, from its position on the margins of respectable science, to contribute to reinforcing a biological conception of the mind. Although it is important not to neglect the place of phrenology in the history of science, it was above all the vitality of the opposition to its doctrines that guaranteed its large-scale dissemination and lasting influence. Thus, despite the spirited refutation from the likes of Hegel (1777–1831), who mockingly condemned phrenology for having reduced the human mind to the status of a mere "bone," the imagery of the phrenologists survived and, indeed, is still all around us.

Disclosure statement

No potential conflict of interest was reported by the author.

ORCID

Marc Renneville ⓘ http://orcid.org/0000-0001-5231-1167

References

Ackerknecht, E. H. 1956. P. M. A. Dumoutier et la collection phrénologique du Musée de l'Homme. *Bulletins et Mémoires de la Société d'anthropologie de Paris* 7 (5–6):289–308. doi:10.3406/bmsap.1956.9731.

Ackerknecht, E. H., and H. Valllois. 1955. François-Joseph Gall et sa collection. *Mémoires du Muséum national d'Histoire naturelle (zoologie)* 10:1–92.

Andrieu, B. 2002. *L'invention du cerveau*. Paris: Press Pocket.

Appert, B. 1832. De la phrénologie appliquée à l'amélioration des criminels. *Journal de la Société phrénologique* 1 (2):144–51.

Appert, B. 1846. *Dix ans à la cour du roi Louis-Philippe et souvenirs du temps de l'Empire et de la Restauration*, Vol. 3. Berlin, Voss, Paris: Renouard.

Baridon, L., and M. Guédron. 1999. *Corps et arts. Physionomies et physiologie dans les arts visuels.* Paris: L'Harmattan.

Bertomeu-Sánchez, J. R. 2015. El esqueleto de la viuda Houet: Frenología y Medicina legal en Francia durante la década de 1830. *Criminocorpus. Revue Hypermedia*. http://journals.openedition.org/criminocorpus/2927.

Bouillaud J. 1865. Discussion sur l'organologie phrénologique en général et sur la localisation de la faculté du langage articulé en particulier. Discours prononcé dans les séances de l'Académie de médecine les 4 et 11 avril 1865. *Bulletin de l'Académie impériale de médecine* 30:575–638.

Changeux, J.-P. 1994. *Raison et plaisir*. Paris: O. Jacob.

Ferrus, G. 1850. *Des prisonniers, de l'emprisonnement et des prisons*. Paris: Germer-Baillière.

Finger, S., and P. Eling. 2019. *Franz Joseph Gall: Naturalist of the Mind, Visionary of the Brain*. New York: Oxford University Press.

Flaubert, G. 1881/1999. *Bouvard et Pécuchet*. Paris: Gallimard.

Fodéré, L.-E. 1832. *Essai médico-légal sur les différentes espèces de folie….* Strasbourg: Le Roux.

Forichon, L. 1840. *Le matérialisme et la phrénologie combattus dans leurs fondements, et l'intelligence étudiée dans son état normal et ses aberrations, dans le délire, les hallucinations, la folie, les songes et chez les animaux*. Paris: P.-J. Loss.

Fossati, G. 1845. *Manuel pratique de phrénologie*. Paris: Germer Baillière.

Gall, F. J. 1835. *On the Function of the Brain and of Each of Its Parts*, 6 vols. Translated by Winslow Lewis. Boston: Marsh, Capen & Lyon.

Gall, F. J., and J. C. Spurzheim. 1810–1819. *Anatomie et physiologie du système nerveux en général, et du cerveau en particulier*, 5 vols. Paris: Imprimerie Haussmann et d'Hautel, Librairie Schoell.

Georget, E.-J. 1825. *Examen médical des procès criminels des nommés Léger, Feldtmann, Lecouffe, Jean-Pierre et Papavoine, dans lesquels l'aliénation mentale a été alléguée comme moyen de défense; suivi de quelques considérations médico-légales sur la liberté morale*. Paris: Migneret.

McLaren A. 1981. A prehistory of the social sciences: Phrenology in France. *Comparative Studies in Society and History* 23 (1):3–22. doi:10.1017/S001041750000966X.

Renneville, M. 1996. Un terrain phrénologique dans le Grand Océan (Autour du voyage de Dumoutier à bord de l'Astrolabe en 1837–1840). In *Le terrain des sciences humaines (Instructions et enquêtes. XVIIe-XXe siècles)*, ed. C. Blanckaert, 89–138. Paris: L'Harmattan.

Renneville, M. 1998. Un musée d'anthropologie oublié: le cabinet phrénologique de Dumoutier. *Bulletins et Mémoires de la Société d'Anthropologie de Paris* 3–4:477–84. doi:10.3406/bmsap.1998.2533.

Renneville, M. 2000. *Le langage des crânes. Une histoire de la phrénologie.* Paris: Empêcheurs de penser en rond.

Renneville, M. 2006. The French revolution and the origins of French criminology. In *The Criminal and Their Scientists: The History of Criminology in International Perspective*, ed. P. Becker and R. Wetzell, 281–92. Cambridge: Cambridge University Press.

Renneville, M. 2013. Lombroso in France. A paradoxical reception. In *The Cesare Lombroso Handbook*, ed. P. Knepper and P.-J. Ystehede, 281–92. London: Routledge.

Rignol, L. 2002. La phrénologie et l'école sociétaire. Science de l'homme et socialisme dans le premier XIXe siècle. *Cahiers Charles Fourier*, 13. http://www.charlesfourier.fr/spip.php?article54.

Staum, M. 2003. *Labeling People: French Scholars on Society, Race, and Empire 1815–1848.* Montreal and Kingston: McGill-Queen's Press.

Vigny, A. D. 1832/1993. *Les Consultations du docteur-Noir. Première consultation. Stello., Œuvres complètes, Vol. 2: Prose.* Paris: Gallimard.

Voisin, F. 1832. Application de la physiologie du cerveau à l'étude des enfants qui nécessitent une éducation spéciale. *Journal de la Société phrénologique de Paris* 1 (1):112–31.

Wegner P.-C. 1988. Franz Joseph Gall in der Zeigenössischen französischen Karikatur. *Medizinhistorisches Journal* 23:106–22.

Franz Joseph Gall came to Copenhagen, and for a brief moment the brain was the talk of the town

Jacob Lauge Thomassen and Simon Beierholm

ABSTRACT

When the inventor of phrenology, Franz Joseph Gall, came to Denmark in the fall of 1805, he was met with great enthusiasm and fascination among the general public, as well as within the scientific community. His visit was an event that was covered by the newspapers unlike any other scientific lecture. However, as soon as Gall left, public interest in phrenology almost instantaneously vanished. Different theories have been put forth in the attempt to answer the question as to why phrenology never found a audience in Denmark. The Danish phrenologist Carl Otto explained it by referring to the poor quality of the Danish phrenological publications. Danish historians have argued that phrenology was too incompatible with the dominant scientific paradigm, *Natürphilosophie*. This article argues that the newspaper coverage of phrenology was more about sensational news stories than about science, and ultimately phrenology was a fad that wore off when the newspapers shifted their focus to other news.

Research into the history of phrenology in Denmark is sparse. This is probably due to the fact that phrenology as a phenomenon never seemed to capture the imagination of the Danish public in the same way it did in many other places. The limited research that exists on Danish phrenology largely centers on physician Carl Otto (1795–1879)—the only "true" Danish phrenologist—who published the phrenological journal *Tidsskrift for Phrenologien* (*Journal of Phrenology*) in 1827–1828.

Scientific historian Anja Skaar Jacobsen has made the most important contribution to the history of phrenology in Denmark with her two articles on Carl Otto (Jacobsen 2004, 2007). Historian Sidsel Eriksen and historian of medicine Egill Snorrasson have also published articles about Otto (Eriksen 2002; Snorrason 1974). Additionally, historian Ole Sonne has contributed to the history of Danish phrenology with an introductory chapter in an anthology on cultural and scientific changes and innovations in the Romantic era, *Romantikkens Verden* (*The World of Romanticism*), and historian Jes Fabricius Møller briefly touched up phrenologists in the second half of the nineteenth century in his Ph.D. thesis (Møller 2002; Sonne 2008). The general consensus on the question as to why phrenology never became an established science within the Danish scientific community is that it was incompatible with the intellectual school of German *Natürphilosophie*, which was prevalent among the scientific and cultural elite at the time (Jacobsen 2004, 154–56).

What would later become known as "phrenology" was launched by German physician and anatomist Franz Joseph Gall (1758–1828) in Vienna in the 1790s and rested on the idea that the brain is composed of different centers, each mediating a specific personality trait (Jacobsen 2004, 135). Combined, these centers make up the entirety of a single personality. The size of each individual center was believed to shape the scalp in ways that could be recognized and associated with behavioral features.

In 1805, Gall introduced his theories in Denmark as part of his great European tour. It was a sensational event that was covered extensively by the newspapers of the day (Christensen 1922; Jacobsen 2004; Sonne 2008). Gall stayed in Denmark for approximately two months—from September to the beginning of November—during which he delivered a series of public lectures. He received a warm welcome from both the public and the scientific community. However, public interest in Gall and his theories vanished almost as quickly as it had emerged. For the two months of lecturing, the Danish public seemed spellbound by phrenology; yet after Gall left the country, public interest in the phenomenon never reemerged. Hence, phrenology did not develop its own scientific field of research as it did in United States and the United Kingdom (Modern 2011; van Wyhe 2004). Instead became all but forgotten in Denmark (Jacobsen 2004, 135). This curious paradox will be further examined in this article.

Franz Joseph Gall's tour of Denmark

Franz Joseph Gall arrived in Copenhagen on September 2, 1805. On September 30, an announcement was published in the newspaper *Kjøbenhavns Adressecomptoirs Efterretninger* (usually referred to as *Adresseavisen*), a local, daily newspaper with a large print run. The announcement read:

> Næste Onsdag Eftermiddag den 2den October kl. 5 til 7 agter Dr Gall at begynde sine Forelæsninger over Læren om Hjernen og Hjerneskallen, dernæst de paafølgende 10 Dage at fortsætte samme i samme Timer, undtagen Tirsdagen og Løverdage, da han vil læse fra Kl. 6 til 8. Man behager at lade Indgangsbilletter afhente i Stormgaed No. 283 paa 1ste Sal. Billetten koster 10 Rd.
> [Next Wednesday afternoon October 2nd from 5 to 7 pm, Dr Gall will begin his lectures on the study of the brain and the skull, and he will continue this, in the same hours, the following 10 days, except Tuesday and Saturday, when he will lecture from 6 to 8 pm. Tickets are for sale in Stormgade 283, 1st floor. Tickets cost 10 Rd.]

Gall's lecture tour was a sensational event throughout most of Europe, and the same was true in Denmark. There was no lack of interest from high-ranking members of cultural, political, and scientific societies, or from the general public, and approximately 150 people attended his first lecture. During his time in Copenhagen, he was invited to dine with the secretary of finance, Heinrich Ernst Schimmelmann (1747–1831), and his wife Charlotte in their home, which was a frequent meeting place for the Danish cultural elite (Sonne 2008).

Accompanied by several high-ranking physicians, Gall visited Copenhagen's prisons and poorhouses on October 10, when he got to showcase his theories by examining the heads of several prisoners. He found the "center for destruction" to be prominent among prisoners convicted of murder, and the "center for the love" of children to be missing among child murderers (Christensen 1922). The press also reported how Gall examined the prisoner Peder Michelsen (1762–?), who was well known to the public, and how Gall verified information about this prisoner that correlated with public perceptions of him (Christensen 1922).

On another occasion during his visit, Gall was shown a bust of the twelfth-century bishop Absalon (1128–1201) and, without prior knowledge of Danish history or of Absalon's life as both a bishop and an army commander, Gall determined that the "center of bravery" was one of Absalon's dominant traits. As with his examination of Peder Michelsen, this was seen as a proof that he was not making up his conclusions as he went along. Who would have expected to find evidence of bravery, a characteristic associated with an army commander, in the brain of a bishop, without prior knowledge that he was indeed both? Impressed with Gall's conclusions, a journalist noted in a newspaper how Gall's doctrine was not only a window into the mind of the individual, it was history coming alive (*Nyeste Skilderie af Kjøbenhavn* 1805, 124–25).

During the following weeks, talk about Gall and his doctrine spread like wildfire, and when he held his second lecture on October 14, around 200 people attended. Among them were several high-ranking men from the Danish government, along with the Prince of Denmark. A group of Swedish scientists had also heard about Gall's visit and journeyed to Copenhagen to see the lectures for themselves (Sonne 2008, 162).

On October 31, Gall's lecture at the School of Veterinary Science illustrated the difference between a human brain and an animal brain. A week later, on November 6, he left Copenhagen and headed for Kiel, where he gave his final lectures in Denmark before traveling to Hamburg, Germany.

Public reception of Gall

As briefly noted, Gall's visit was closely followed by the local press, especially by the periodical *Adresseavisen*. The same day Gall announced his first lecture, *Adresseavisen* ran an extensive introductory piece about the German physician and the theory behind his organology. However, according to the newspaper reporter, Gall did not need an introduction at all:

> I den nyere Tid har næppe nogen Opfinder opvakt almindeligere Sensation blandt Lærde og Ulærde, Høje og Lave, end Dr. Galls Lærebygning om Hjernen og dens Dannelse.
>
> [In recent memory, there is hardly any other inventor who has caught the interest among scholars and commoners alike in such a sensational fashion as Dr. Gall's theory about the brain and its formation.]

According to the piece, Gall's doctrine had been the talk of town, even before he arrived in Copenhagen, and everyone soon had an opinion on him or his teachings. Educated and uneducated people alike were gossiping about Gall's public lectures, and nothing could rival him as the hot topic of the daily conversations.

An article in *Nyeste Skilderie af Kjøbenhavn* described how it seemed that all one could hear anyone talk about was brains and scalps. The author of the article noted that he was even dreaming about Gall's lectures. (*Nyeste Skilderie af Kjøbenhavn* 1805, 1616).

On October 11, *Adresseavisen* ran another article about Gall, concluding now that the citizens of Copenhagen had had time to familiarize themselves with Gall's theory, even his most vocal opponents could not look away from the undeniable proofs of his work. Gall's thinking was "saa grundede paa sund Fornuft, at enhver, selv hans ivrigste Modstandere, indtages for ham" [so

grounded in common sense that everyone, even his biggest opponents, are captivated by him]
(Christensen 1922, 218).

The newspapers further described an enthusiasm surrounding Gall that almost resembled
a mania. Stores were selling plaster cast skulls illustrating Gall's system, and people even used
the numbers of the different centres of the brain to play the lottery (Christensen 1922, 219).

Gall's theories—which would later be known as phrenology—seemed to have come to
Copenhagen to stay, and on December 7, a month after Gall's departure, the periodical
Adresseavisen ran a piece in which they declared, "Den Enthusiasme, hvormed Gall er bleven
modtagen i København, er grænseløs" [The enthusiasm Gall has been met with here in
Copenhagen is without limits] (Christensen 1922, 219). However, this enthusiasm proved to
be limited. The periodical only mentioned Gall's tour one more time after this, on December 14,
in a small notice about his arrival in Hamburg (Christensen 1922, 220). Seemingly, this was the
last trace of the great enthusiasm the Danes had had for Gall while he was in Denmark.

Gall's legacy in Copenhagen

On January 4, 1806, only two months after Gall's departure, an account of Gall's imprint on
Copenhagen titled "Et brev fra København til en ven på landet" ("A Letter to a Friend in the
Country") was published in *Adresseavisen*. Despite the widespread sensation surrounding his
visit, the author of the piece was not impressed with the legacy Gall had left in Copenhagen:

> For nogle Maaneder siden havde Gall og hans Hjerneskaller leveret Æmne nok til Samtaler;
> men siden hans Afrejse synes hans Lære at have forladt København tillige med ham. Om der
> end aldrig havde eksisteret nogen Hjerne eller Hjerneskal, om end Gall aldrig havde holdt
> offentlige Forelæsninger, kunde der ikke herske større Ligegyldighed for hans Lære end netop
> nu.
> [A few months ago, Gall and his skulls were subject of conversation all over the city; but ever
> since his departure this theory seems to have left the country along with him. As if a skull or
> a brain has never existed, or as if Gall never gave never public lectures, so is the complete
> indifference to his theory now.]

The author seemed to confirm the enthusiastic public mood portrayed in the newspapers.
Yet he also highlighted one of the more curious aspects of the history of phrenology in
Denmark—namely, how it only seemed to capture the public imagination for a short time
before disappearing from the public sphere. The plaster cast skulls many people had
bought during Gall's visit to Copenhagen had now been either discarded or become toys
for children.

Not long after Gall's departure, the daily conversations and the newspapers moved on to
other subjects. Gall's stay in Copenhagen was like a meteor, the author concluded—visible
and shiny, but vanishing as quickly as it had appeared. During his visit, people who had
never shown any previous interest in the functions of the brain suddenly became experts on
the workings of the mind. But after Gall's departure, no one seemed to care much about
skulls or brains anymore (Christensen 1922, 222).

On October 18, 1806, an article chronicling "the characteristics of Copenhagen" specifically mentioned Gall's visit and Copenhagen's brief love affair with him as a prime example
of the city's tendency to admire anything new and exotic without a sense of critique, as long
as it was foreign:

Da Gall f. Eks. sidst kom her til Staden, vare alle, Indvidede og Uindvidede, Lærde og Ulærde, Læger og Ikke-Læger saa indtagne for hans Lære, at den forfægtedes med en Slags Fanatisme, og at enhver, der vovede at drage dens Ufejlbarhed i Tvivl, betragedes som en Dosmer. Hans Forelæsninger var som en Slags Komedie, som enhver, der ikke vilde anses for en Dumrian eller en Gniepind, absolut maatte see. - Nu kaldes Galls System offentlig i "Allgemeine Litteraturzeitung" en hjerneløs Hjernellære, uden at nogen hos os bryder sig derom. Ja, fra det Øjeblik af, da Gall var uden for Vesterport, bekymrede sig intet Menneske mere om at undersøge hans Hjernelære.

[For example when Gall came to the city, everyone, scholars and commoners, physicians and non-physicians, were so receptive to his theory that it turned into a kind of fanaticism, and anyone who dared question its infallibility was seen as a dunce. His lectures were a kind of comedy that anyone who didn't want to be considered stupid or cheap just had to see.—Now Gall's system is being called a brainless brain theory in "Allgemeine Litteraturzeitung" that none of us care about. Yes, from the moment Gall was outside Vesterport, no one cared about his brain theory any further.]

In fact, some of the earlier enthusiasm had now turned into a kind of resentment. There was a feeling of having been spellbound by a con artist who came, performed some cheap tricks, then left with his pockets full:

Saalænge man hos Dannermand skal møde Ældgamle Godmod, Gjæstfrihed og Føde, Brav bulet Pung og bulet Hjerneskal, Saalænge fattes ham vist ej en Doctor Gall, Der føler Bulerne, - seer flux hvad hver bar inde, Og naar hver Bule har betalt sin Dom, Hver Hjerne er justeret, hver Pung er tom, Saaer Iystig did igen, hvorfra han kom.

[As long as you meet old fashioned good heartedness, hospitality and food from the Danes, a wallet with bumps and a scalp with bumps, so will a doctor like Gall come along and feel the bumps—see what each has inside, and when every bump is judged, every brain is adjusted, every wallet is empty, than will he joyfully return to where he came from.]

In these descriptions, which show the changing public interest in Gall, his lectures were branded more like an act of comedy than real science—a farce, in which people feigned interest in order to not seem ignorant or out of touch with the currents of the time. In brief, Gall's doctrine seemed to have been nothing more than another short-lived fad. It was something exciting and exotic for the people of Copenhagen to talk about for a brief moment in time, and when it was no longer news, casual conversations drifted on to other topics.

Gall and contemporary science

A handful of publications on Gall's teachings emerged in the wake of his visit. Most were translations from German publications or short introductions to Gall's teachings, but none proved efficient in spreading and maintaining his ideas.

Within the scientific establishment, only a few individuals would comment further on Gall, now that his teachings were out of the public eye. Among the professors at the University of Copenhagen, physician Johan Daniel Herholdt (1764–1836) introduced the theory in his lectures, but only to denounce it, whereas philosopher and psychologist F. C. Sibbern (1785–1872) had a more favorable view on the issue. The headmaster at Odense Katedralskole (Odense Cathedral School), Hans Outsen Björn (1777–1843), also had a positive view of Gall, and in January 1810, he opened the school with a speech in which he praised Gall's theories for being a science based on empirical proof and not on *a priori* philosophizing (Björn 1810). Apart from these instances, Danish phrenology did not seem to amount to anything more than a two-month craze surrounding a visit from a German physician who was all the rage at the time.

It was not until physician Carl Otto (see Figure 1) tried to reintroduce phrenology into the public consciousness in the 1820s, that phrenology—the term itself now widely circulated and accepted (Finger and Eling 2019)—became a topic of conversation again. Although not entirely successful in his mission, Otto was definitely the most successful phrenologist in of Denmark— so much so, that he is considered to be the only real Danish phrenologist. To critics and colleagues alike, Otto was simply called "the phrenologist," which he took as a compliment (Jacobsen 2007). Otto had been introduced to phrenology as a young student of medicine

GEORG E. HANSEN & COMP.   KGL HOF PHOTOGRAPH

Figure 1. Carl Otto.

through both Herholdt's and Sibbern's lectures, and although they did not share the same views about phrenology, Herholdt nonetheless became Otto's mentor (Sonne 2008, 161–62)

Otto published two journals during his time as a practicing phrenologist, and *Tidsskrift for Phrenologien* is still considered to be the most extensive contribution to Danish phrenology. The journal covered all aspects of the science, from its practical uses to philosophical questions and general news about phrenologists all around Europe.

In an article published in *Tidsskrift for Phrenologien*, Otto tried to answer the question as to why earlier attempts to introduce phrenology into the Danish scientific establishment had not been successful. According to Otto, this was either because every other phrenological publication in Denmark was too short and barely scratched the surface of phrenology or because the publications were too flawed and inaccurate in their descriptions. In Danish, they were *upaalidelige*, meaning unreliable. Furthermore, many of these publications had a tendency to mix phrenology with other seemingly related sciences, such as pathognomy (i.e., the study of facial expressions), which, according to Otto, was not as strongly founded in facts as phrenology, and thus damaged its reputation (Otto 1827, 353).

According to Otto, another reason for phrenology's lack of general acceptance in Denmark was the simple fact that phrenology was too complicated to fully grasp after only one or two lectures. No wonder so many people misunderstood the specifics, he explained, as Gall had to keep his lectures short and could only highlighted the main ideas of his theories:

> At danske Videnskabsmænd endnu derfor kun kunde erholde meget svage og undertiden aldeles falske Begreber, er naturligt; selv [Gall's] Tilhørere kunde umuligen af hans Forelæsninger erhverve sig en sand og Klar Anskuelse af Videnskabens Væsen, da Gall paa Rejsen altid var nødt til at fatte sig kort og maatte indskrænke sig til blot at meddele Hovedideerne af hans Lære (Otto 1827, 353)

> [The fact that Danish scientists could only get a weak or even false idea [of phrenology] is only natural; even for [Gall's] own followers it was impossible to get a true and clear grasp on the science from his lectures alone, since Gall always had to keep his lectures short and only communicate the main ideas of his teaching.]

Otto acknowledged that the practical side of phrenology was the weakest, because it required extensive knowledge of the science to do it properly, and he warned readers to be cautious if they wanted a precise reading of a skull.

Björn put forth a similar argument in his speech to Odense Katedralskole. He believed there had been too much focus on the practical side of phrenology and not enough on its foundations. However, it was not hard to understand why the practical side of phrenology overshadowed the theoretical, Björn argued, as this was the easiest part to try for oneself. This had led many people to dismiss phrenology entirely if their own experiences failed to support the doctrine:

> At Gall imidlertid ved denne Beføling har vildledt Mange, er ikke hans men hans Læres Skyld, der som reen empirisk fordrede denne Fremgangsmaade […] just Bestemmelsen af de forskjellige Organers Sted og Antaler den svageste Side ved Galls Lære, men tillige den der meest opvækker Nysgerrigheden. Mange gav sig derfor ei Tid til at studere Grundprinciperne for hans Lære(Björn 1810, 20)

> [The fact that Gall has misled many, it is not his teaching's fault, which has a clean empirical method … it is the determining of the different faculties that is the difficult part of Gall's teaching, but also the part that wakes the curiosity. That is why many people took the time to study all the principles of his teaching.]

If one were to think it possible to simply touch another person's head and thereby be able to know everything about this particular person, one would surely be mistaken, Björn stressed. There was no point in arguing with an incompetent *dommer* [judge], who prematurely has decided to ridicule phrenology without seeking the whole truth (Björn 1810, 20).

Other people had different and more simplistic ideas as to why phrenology never had a breakthrough in Denmark. The playwrite Johan Ludvig Heiberg (1791–1860) made some amusing remarks in his satirical newspaper *Kjøbenhavns Flyvende Post* (*Copenhagen's Flying Post*) at the expense of phrenology and its naïve scientific methods:

> Gall, som havde bemærket denne samme Forhøining paa alle Saltvands-Fiskes Hoveder, og paa en Islænders, der medens hans Ophold i Kjøbenhavn solgte det bedste saltede islandske Lammekjød, dømte nu strax, at der aldrig har været og at der aldrig vil gives nogen Saltmadsspiser, som jo ikke har denne Forhøining i Panden. (Heiberg 1827, 22)

[Gall, who had noticed the same bump in the forehead of a saltwater fish, and on the forehead of a Icelandic man who sold lamb the time of Galls stay in Copenhagen, instantly proclaimed that there had never been or never will be anyone who enjoys salty food who does not have that exact bump in the forehead.] Carl Otto found Heiberg's irony not only unfunny but also personally offensive. In a letter to his friend Carl Johan Ekströmer (1793–1860), royal physician to the King of Sweden, Otto would later refer to Heiberg as his enemy (Jacobsen 2004, 147).

Otto found some support in the literary community, most notably with the Danish poet and fairytale writer Hans Christian Andersen (1805–1875). Throughout his lifetime, Andersen showed considerable interest in phrenological teachings and was examined at least twice (Mylius 1998). He made frequent references to phrenology in his early novel, *Fodreise fra Holmens Canal til Østpynten af Amager i Aarene 1828 og 1829* (*Journey on Foot from Holmen's Canal to the Eastern Headland of Amager in the Years 1828 and 1829*; Andersen 1829, 9, 31, 95). Otto and Andersen communicated through letters, and Otto invited Andersen to dinner at his house several on several occasions (Otto, date unknown).

However, Otto's argument about the quality of the phrenological publications falls short, as he himself never had any luck spreading phrenology to the general public, despite publishing extensively on the subject. He was not an outsider to the scientific and cultural community, either. He was a respected physician with a professorship at the University of Copenhagen, a high-ranking member of the Danish Order of Freemasons, and was friendly with the cultural elite.

Heiberg's satirical comment indicated a much more plausible explanation—that phrenology was, in its essence, too fantastical to really believe. It was this feature that earned phrenology its spotlight, but also the reason it disappeared so quickly.

Phrenology and Natürphilosophie

Anja Skaar Jacobsen—the historian who has analyzed the debates surrounding Danish phrenology and Carl Otto, in particular—has argued that phrenology did not get mainstream acceptance because the prevailing Danish cultural and scientific environment of the time was heavily influenced by *Natürphilosophie*, the romantic thinking of German philosopher Frederich Wilhelm Joseph Schelling (1775–1854) and his followers. *Natürphilosophie* was based on a holistic view of a unity between spirit and nature, with the human body, mind, and soul closely connected. Phrenology, with its very specific ideas of the relationship between body and soul, was simply deemed too materialistic for the *Natürphilosophie* mainstream (Jacobsen 2004).

Natürphilosophie was introduced in Copenhagen by Henrich Steffens (1773–1845) in a series of lectures in 1802 and 1803 (Steffens 1996 [1803]). The idea of *Natürphilosophie* inspired the Romantic period in Danish culture, and influenced not only science but also literature, art, and broader culture (Koch 2004, 304). During the 1820s and 1830s, physicist H. C. Ørsted (1777–1851)—famous for the discovery of electromagnetism—and his brother, A. S. Ørsted (1778–1860), were at the very center of the Danish scientific and cultural elite. H. C. Ørsted was skeptical but nonetheless curious about phrenology. On a trip to Europe, where he saw a bust of the philosopher Immanuel Kant, he wrote in his diary that the "faculty for acquisitiveness" was very much present here. Ørsted seemed to entertain the idea that some parts of phrenological theory might be accurate, as he discussed with the philosopher Johann Gottlieb Fichte (1762–1814). However, both agreed that the mental faculties could simply not develop and function independently, but could only be construed as parts of the whole (Christensen 2009).

Otto attributed the curiosity among some of the elite to philosopher and psychologist F. C. Sibbern (Otto 1836, 59). Sibbern was close to the Ørsted brothers, a supporter of both Schelling and Steffens, and a proponent of phrenology. In his magnum opus, *Menneskets Aandelige Natur og Væsen* (*On the Human Mind*; 1819–1828), on the relation between body and soul, Sibbern embraced the underlying philosophy of phrenology, emphasizing how the mind was clearly affected negatively by trauma to the brain, and consequently there had to be a connection between the brain and the soul (Sibbern 1819). However, only 300 copies of this book were printed, and the combination of Sibbern's complex style of writing and the book's 500 page span further limited its impact (Koch 2004).

Grave robbery

On one occasion Otto was successful in his attempt to get phrenology into the public conversation, but not for the reasons he had hoped. As part of his position as physician at the local prison, Otto had become fixated on the idea of examining the head of the criminal Petri Worm (1814–1838), who had been convicted of murder. Worm was from a good family and possessed a number of qualities that Otto found contradictory to Worm's criminal tendencies. Worm was, for example, described as both goodhearted and a poet by nature, hardly the characteristics of a murderer. Nonetheless, Worm denied Otto permission to examine his head after his beheading in 1838. Otto, believing his work was a sacred scientific mission, decided to resort to more radical measures. He had his assistants dig up Worm's decapitated head from the graveyard under cover of night. The assistants were disturbed in the act and got away with the head, but they did not have time to cover up the grave (Eriksen 2002).

The story of the grave robbery spread like wildfire through Copenhagen, and it was not long before Otto became the primary suspect. His work with phrenology and with criminals was well known, and once again phrenology became the talk of the town. Luckily for Otto and his assistants, the chief of police knew Otto and his work and closed the case by arguing that the entire body had not been taken (Eriksen 2002).

Phrenology always lurked on the margins of the public debate in Denmark. It was present, but at the same time just out of sight. Only when it generated sensational news for the press did people start to pay attention to phrenology again. Consequently, phrenology in Denmark had more to do with sensational news coverage than genuine interest in the new "science" of mind.

Phrenology after Carl Otto

On April 25, 1855, an announcement in *Adresseavisen* invited the public to a lecture on phrenology. The lecture would be held at the Hotel d'Angleterre in Room 63, and the lecturer was the German phrenologist and physiognomist Heinrich Bossard (1811–1877).

Fifty years after Gall's visit to Copenhagen, and 20 years after Carl Otto had published his journals, lectures on phrenology had gone from the largest lecture halls in the city, which held 200 people, to a small hotel room. Bossard was met with laughter by the Danish audience. One by one, Bossard invited volunteers to get their heads examined as he worked to expose their personalities (Goldschmidt 1859, 1–3). To the attending audience, however, this seemed more like a magic show than a scientific lecture.

However, Bossard's visit to Denmark was not all in vain. Two people in the audience saw the possibilities of phrenology and would go on to write books on the subject. One was author and member of the Danish cultural societies Meïr Aron Goldschmidt (1819–1887), and the other was Danish-German Colonel Otto von Schädtler (von Schädtler 1858).

Goldschmidt described Bossard's lecture in his book *Om Physiognomiken* (*On Physiognomy*; 1859). On one hand, Bossard's lecture summed up the entire history of Danish phrenology as entertainment rather than science. On the other, it also revealed that Danish phrenology did not end entirely with Carl Otto, but that Danish phrenologists continued to emerge sporadically all the way into the beginning of the twentieth century. Bossard's lecture had inspired both Goldschmidt and Schädtler. The painter Sophus Schack (1811–1864) published a book about physiognomy with an introduction by Sibbern in 1858. From 1888 to 1889 a phrenological consultation opened on Holger Danskes Vej in Copenhagen, managed by one C. H. Bernard, who also published a small book called *Phrenologiske Ledetraade* (*Phrenological Clues*; 1889); a rather incoherent work that combined the philosophy of phrenology with ideas about God, the Bible, the lost city of Atlantis, and theories of evolution.

Even as the scientific paradigm shifted from *Natürphilosophie* to the materialistic naturalism in the middle of the nineteenth century, Danish phrenology still did not find an audience. How many people visited Bernard's consultation is unknown, as the address was only listed as a consultation in the address books for two years. After 1889, all traces of Bernard are gone. One can therefore conclude with some certainty that Bernard's consultation never found a sufficient clientele, and that Bernard's forays similarly reflect the Danish indifference to phrenology.

Conclusions

When Franz Joseph Gall came to Copenhagen as part of his lecture tour during the fall of 1805, his reputation had preceded him. His new science was a hot topic throughout

Europe, and his lectures had captured scientific, medical, and public imaginations alike. The Danish newspapers reported what seemed like unlimited enthusiasm about Gall and his ideas, and for two months, all anyone could talk about were brains and scalps. However, by November, when Gall ended his series of lectures and left Denmark, this public fascination with phrenology had vanished. Phrenology in Denmark was like a meteor—visible and shiny, yet vanishing as quickly as it had appeared.

The only diehard phrenologist to emerge in Denmark, Carl Otto, explained the phenomenon by calling the Danish phrenological publications unreliable and flawed. Furthermore, he claimed, phrenology was a science too complicated for most ordinary people to grasp. Others would allude to sheer inconsistency and vague scientific methods of the science.There is consensus among historians that Danish phrenology never received mainstream acceptance because the scientific community was dominated by the German *Natürphilosophie*. However, this does not answer all questions. Why, when naturalism replaced *Natürphilosophie* as the dominant scientific paradigm, did Danish phrenologist still did not find an audience?

Phrenology could only capture the public interest after Gall's visit when it was associated with sensational news stories. It required a visit from an internationally famous scientist or a notable grave robbery. Ultimately, phrenology was likely nothing more than a fad in Denmark. It was something exciting and exotic that filled up newspaper pages and gave people a subject to talk about, but only for a brief moment. When phrenology was no longer news, casual conversations drifted on to other topics.

Disclosure statement

No potential conflict of interest was reported by the authors.

References

Andersen H. C. 1829. *Fodreise fra Holmens Canal til Østpynten af Amager i Aarene 1828 og 1829*. Copenhagen: Forfatterens Forlag.

Bernard C. H. 1889. *Phrenologiske Ledetraade efter Francis Joseph Gall, John Gasper Spurzheim, George Combe*. Copenhegen: Henriksens Bogtrykkeri.

Björn H. O. 1810. Historiske Efterretninger om Dr. Gall og hans Organlære. *Et Indbydelses-Skrift til Odense Carthedralskoles Höitidelighed i Anledning af Hs. Majestæt Kongens Födselsdag*, 1–34. Odense.

Christensen D. C. 2009. *Naturens tankelæser. En biografi om Hans Christian Ørsted*, Vol. 2. Aarhus: Museum Tusculanums Forlag.

Christensen V. 1922. Dr. Galls Ophold i København 1805. *Historiske Meddelelser Om København, 1921–1922* 8:217–25.

Eriksen S. 2002. Mellem medicin og moral. Eller hvorfor frenologen Carl Otto og rationalisten Carl Holger Visby begge var interesserede i den dødsdømte Petri Worms hoved. In *Historie og Historiografi. Festskrift til Inga Floto*, ed. C. Due-Nielsen, 58–78. Copenhagen: Den danske historiske Forening.

Finger S., and P. Eling. 2019. *Franz Joseph Gall: Naturalist of the mind, visionary of the brain*. New York: Oxford University Press.

Goldschmidt M. 1859. *Om Physiognomiken*. Copenhagen: Boghandler Otto B. Wroblewsky.

Heiberg J. L. 1827. *Organet for saltsands*. Copenhagen: Flyvende Post.

Jacobsen A. S. 2004. Carl Ottos forbryderhoveder. *Bibliotek for Læger* 196 (2):132–61.

Jacobsen A. S. 2007. Phrenology and Danish Romanticism. In *Hans Christian Ørsted and the Romantic Legacy in Science*, ed. R. M. Brain, R. S. Cohen, and O. Knudsen, 55–74. Dordrecht: Springer.

Koch C. H. 2004. *Den danske idealisme 1800–1880*. Copenhagen: Gyldendal.

Modern J. L. 2011. *Secularism in antibellum America: With refrence to ghosts, protestant subculture, machines and their metaphors: Featuring discussions of mass media, moby dick, spirituality, phrenology, anthropology, sing sing state penitentiary and sex with the new motive power*. Chicago: University of Chicago Press.

Møller J. F. 2002. *Biologismer, naturvidenskab og politik ca. 1850–1930*. PhD Thesis, University of Copenhagen. Nyeste Skilderier i Kjøbenhavn Copenhagen. 1804–31.

Mylius J. D. 1998. *H. C. Andersen liv. Dag for dag*. Copenhagen: Ringhard & Lindhof.

Otto C. 1827. Phrenologiens Historie i Danmark. *Tidsskrift for Phrenologien*.

Otto C. 1836. [Untitled.]. In *Testimonials on behalf of George Combe, as a candidate for the chair of logic in the University of Edinburgh*. Edinburgh: Nefel and Co.

Otto C. [Date unknown] Letter to H. C. Andersen. http://andersen.sdu.dk/brevbase/brev.html?bid= 23331

Sibbern F. C. 1819. *Menneskets aandelige Natur og Væsen. Et udkast til en Psychologie, bd. I.* Copenhagen: Gyldendal.

Snorrason E. 1974. The Danish physician Carl Otto (1795–1870) and phrenology. In *Wien und die Weltmedizin*, 146–58. Vienna: Hermann Böhlaus Nachf.

Sonne O. 2008. Frenologi - hjerneskallens forudsigelser. In *Romantikkens Verden. Natur, menneske, samfund, kunst og kultur*, ed. O. Høiris and T. Ledet, 155–172. Aarhus: Aarhus Universitetsforlag.

Steffens H. 1996 [1803]. Indledning til philosophiske Forelæsninger. Copenhagen: Det Danske Sprog- og Litteraturselskab C.A. Reitzel.

van Wyhe J. 2004. *Phrenology and the origins of victorian scientific naturalism*. Aldershot: Ashgate.

von Schädtler O. H. 1858. *Allgemein Verständliche Psychologie Auf Die AnerkanntestenUnd Thatsächlichsten Offenbarungen Der*. Hamburg: Hoffmann und Campe.

Gall's Visit to The Netherlands

Paul Eling, Douwe Draaisma, and Matthijs Conradi

In March 1805, Franz Joseph Gall left Vienna to start what has become known as his cranioscopic tour. He traveled through Germany, Denmark, and The Netherlands. In this article, we will describe his visit to The Netherlands in greater detail, as it has not yet received due attention. Gall was eager to go to Amsterdam because he was interested in the large collection of skulls of Petrus Camper. Gall presented a series of lectures, reports of which can be found in a local newspaper and in a few books, published at that time. We will summarize this material. We will first outline developments in the area of physiognomy, in particular in The Netherlands, and what the Dutch knew about Gall's doctrine prior to his arrival. We will then present a reconstruction of the contents of the lectures. Finally, we will discuss the reception of his ideas in the scientific community.

The Physiognomical Background in The Netherlands and Gall's Tour to The Netherlands

The word "physiognomy" refers to the "science" of reading someone's character from his outlook, in particular the face. Francis Bacon described it as "discovery of the disposition of the mind by the lineaments of the body" (Bacon, 1994, p. 81). Physiognomy can be traced back to Aristotle. He wrote extensively about comparisons between animal and human characteristics. The book *Physiognomonica* was believed to be written by Aristotle for a long time, but we now think it was one of his followers (Jahnke, 1997). Within this text, one can find descriptions of the basic methodology of physiognomy. For instance, physiognomic analysis should begin with the delineation of distinct, external features of the body and then relate these characteristics to specific personality types.

In the middle ages, physiognomy became linked to mystical and astrological theories. An encyclopedic work, referred to frequently well into nineteenth-century texts, is Giovanni Battista della Porta's (1535–1615) *De Humana Physiognomonia* from 1586. In the middle of the eighteenth century it became somewhat obscure and George II made it illegal in 1743. However, in 1793 Johann Kaspar Lavater (1741–1801), a minister in Zürich revitalized the tradition with his *Physiognomische Fragmente* (Lavater, 1775–1778), which was translated in English as *Essays on Physiognomy*.

Petrus Camper (1722–1789)

The Dutch were also familiar with the physiognomic tradition. Petrus Camper was well known for his anthropological studies and in particular for his studies on the "facial angle" (see Figure 1). Camper was born in 1722 in Leiden (Schuller tot Peursum-Meijer & Koops, 1989). In 1734, he was admitted to the university. Boerhaave and Gravesande's, friends of the family, provided him with a thorough training. He finished his studies in 1746 with a double dissertation and became a physician in Leiden. In 1748, he coordinated an academic tour, visiting

London and Paris. He returned to Holland in 1749 and accepted a chair at the University of Franeker. In 1754, he moved to the "Illustre School" of Amsterdam. Anatomy was his greatest passion and Amsterdam offered him much more facilities to work in this discipline. In 1761, he returned to Franeker. From 1763 to 1773 he lectured at the University of Groningen: theoretical medicine, anatomy, surgery, and botany, and he was also involved in more applied medical projects. In 1773, he returned again to Franeker and, after several years as an unemployed civilian, he directed his attention towards politics. He died in 1789. Camper has been characterized as follows: He has worked in many different areas of science, but he did not advance one of them.

A central notion in Camper's work is *analogy*; Camper believed in some underlying uniformity in nature. He particularly disliked the approach of Linnaeus. His interest in the underlying structural similarities resulted in studies in which he compared the human body with that of animals. He also analyzed differences between humans from different continents. Because the Dutch often sailed across the oceans, it was an optimal opportunity to "collect" evidence. In 1767, he began to work on a manuscript entitled *Natuurkundige Verhandeling over het verschil in de wezenstrekken in Menschen van onderscheidene lan-daart en ouderdom* (Natural essay on the difference in character in men from different countries and age). The manuscript was published in 1795 by his son after Camper's death. In this paper, Camper reported on *craniometric* studies. He developed a method to measure differences in size of different aspects of the skull. He argued that the *facial angle* was especially important: The angle is formed by drawing two lines: one horizontally from the nostril to the ear and the other perpendicularly from the advancing part of the upper jawbone to the most prominent part of the forehead. He asserted that the larger the angle, the higher the degree of development. The method Camper used to make drawings was not his original technique. Albrecht Dürer (1471–1528) had used it, as well as the physician Bernard S. Albinus (1697–1770), working in Leiden. Camper specifically applied this method for measurement of the skull to comprehend the differences in intellectual capabilities.

Figure 1. Petrus Camper.

Of particular interest in these days was the comparison between humans and orangu-tans. The well-known Dutch physician Nicolaas Tulp had introduced the anatomical study of this animal, referred to him also as Satyrus Indicus in the literature (Kruger, 2005). Be-cause of the good opportunities to get materials from remote areas, Camper had been able to get skulls from orangutans. He became internationally famous with this study, and when Gall took the trouble to travel from Amsterdam to Franeker, as we will describe later, he did not go to visit Camper, since he had died, but he was interested in the skull collection, in particular of those of the apes.

Lavater's Physiognomy in The Netherlands

In the 1780s, Lavater's work became popular in The Netherlands (Lavater, 1780) and sev-eral books were published describing Lavater's ideas (see also Noordhoek, 1925). In 1780, the *Physiognomische Catechismus, bevattende de gronden en beginselen der Gelaadkennis* (Physiognomic catechism, containing the reasons and principles of physiognomy) was published by an anonymous author in Amsterdam (Anonymous, 1780). Pieter den Hengst published a translation of Daniel Chodowiecki's (1781) book (see Figure 2). Chodowiecki (1726–1801) was a Polish-German painter and was most famous as an etcher. He pro-duced many illustrations for scientific books. Originally, Chodowiecki had published his critical essay in 1777, and an expanded version was printed the following year; den Hengst's book is a translation of the latter. Chodowiecki argues that there is no consist-ent relation between the character of the soul and features of the body. An individual may behave normally and suddenly may become aggressive. Moreover, the wide variety of appearances in peoples around the world cannot be associated with their characters. According to Chodowiecki, Lavater's work is on pathognomic features rather than phys-iognomic features. And finally, he complains that there are so few examples of women in Lavater's book.

Although Chodowiecki had been critical, Lavater had been positive about the book and apparently the Dutch readers were encouraged to become acquainted with it. In 1784, J. W. Haar published the second edition of his translation of Lavater's major work. This edition contains a list of over 500 subscribers. It is obvious that only a small minority of the sub-scribers were medical men. From this, one may infer that physiognomics was not a specific medical issue, rather a subject that attracted the attention of a large general audience. A. Fokke Simonsz, a member of the learned society Felix Meritis in Amsterdam, prepared an essay *Verhandeling over de algemeene Gelaatkunde* (Treatise on the general physiognomy; 1801). A very interesting book was published by Antonie Martini van Geffen in 1825 in 's-Hertogenbosch, entitled *De karakterkunde volgens de gelaatsleer, in verband beschouwd met de schedelleer en de menschelijke temperamenten* (Characterology according to physiog-nomy, considered in relation to craniology and the human temperaments). Martini van Gef-fen (1791–1869), stemming from an affluent family, was a minister in a village, Vught, near 's-Hertogenbosch. In this book, Martini van Geffen explicitly attempts to combine Lavater's physiognomy with Gall's skull theory.

Naturally, there are some differences, Martini van Geffen argues, but in principle the two frameworks attempt to explain human faculties and temperaments on the basis of particular physical features of the head. Martini van Geffen appreciated the possibility of integrating the two perspectives.

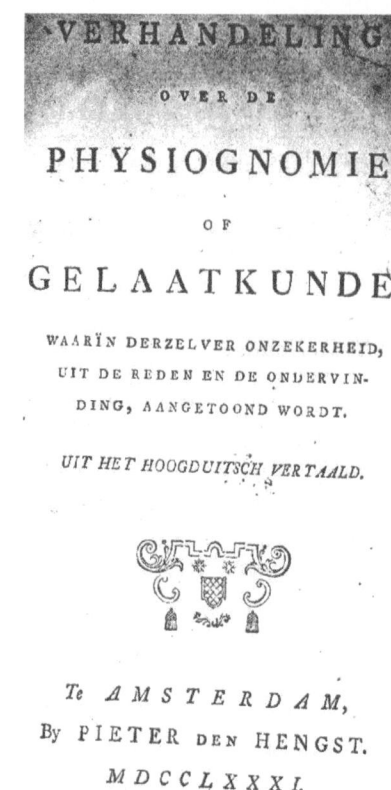

Figure 2. Title page of the translation of Chodowiecki's treatise.

In summary, the general idea that one can deduce someone's character from one's appearance, in particular from the head or the face, was familiar to a large number of people in The Netherlands and apparently was a topic that attracted some attention in the form of public lectures and books. Gall's ideas about inferring faculties from the skull can be regarded as a specific alternative theory in this area of "anthropology," very much like Camper formulated a specific model based on the facial angle.

Gall's Biography (Until 1805)

Franz Joseph Gall was born on March 9, 1758 in Tiefenbrunn, close to Pforzheim, South Western part of Germany (Lesky, 1979). His father was a tradesman (who spelled his name as Gallo). From his early childhood on, Gall had a great interest in nature, both flora and fauna. Not only does his later work attest to this interest but also the beautiful gardens at his houses in Vienna and Paris.

Originally, Gall was destined for a religious position. Commencing in 1777, he studied medicine in Strassburg where Johann Hermann impassioned him to learn comparative anatomy. From 1781 onwards, he continued his studies in Vienna. His principle teacher there was Maximillian Stoll. He completed his dissertation in 1785 and started a private practice, attracting the attention of high society. His first work, *Philosophisch-medizinische Untersuchungen über Natur und Kunst im kranken und gesunden Zustande der Menschen*

(Philosophical-medical studies on nature and art in disordered and healthy conditions), was published in 1791 and quickly displayed Gall's main interest.

From autobiographical remarks, we know that Gall already in his home in Vienna performed group studies about psychological faculties. He visited "the tower," the asylum for mental patients, to examine their physical appearance with specific attention to their heads. He was also interested in collecting skulls of important individuals. In 1796, Gall began his private lectures on *Skull Theory or Organology*, and these lectures attracted large audiences, including women of higher status and privileged backgrounds. Consequent to these lectures, the habit of feeling the lumps on the head and the production of demonstration heads developed. *Kranioskopie* became extremely fashionable.

Gall as Anthropologist

Throughout his work, Gall has indicated that his main objective was to investigate the true nature of man and to develop a "true theory of man." He hoped that such a theory would enable the creation of sociocultural changes for the benefit of everybody.

It is significant that the first real outline of his theory appeared in the form of a letter to his friend von Retzer in 1798 in a German newspaper, *Teutschem Merkur*. Baron von Retzer was an administrator working for the Vienna censorship and probably advised Gall to publish his ideas outside Austria. In his *Program*, the letter to von Retzer, Gall discussed his relationship with physiognomics (Lesky, 1979). He was very clear: "I am nothing less than a physiognomist" (Lesky, 1979, p. 57). Others have "baptized" his child cranioscopy, but he was not happy with that name. He was not primarily interested in the skull, but in the brain. Next to the brain was his interest in psychology. This probably stemmed from his observations and appreciation of human behavior as a child and an adolescent. At that time, Lavater's physiognomy became popular and Gall attempted to relate specific behavioral features to bodily characteristics; this continued to be Gall's central message in all his lecture series. It is noteworthy that his theory appeared more valuable to people working with criminals or children, than to physicians, who were interested in diseases and practical means to treating people inflicted with these diseases.

The Tour

On December 24, 1801, a handwritten order of the Emperor of Austria stated that the Chancellor of State, Earl Lazansky, prohibited the proclamation of Gall's theory, stating that people apparently ran the risk of losing their own head and that this theory could lead to materialism and to loss of interest in religion and moral principles. Many influential people tried to defend Gall, but with little success. The clerical group at the Vienna court, probably directed by the court physician Joseph Andreas Stifft, won the battle. On February 3, 1802, another decree prohibited the private lectures of Gall.

In 1805, Gall started his European tour that finished in November 1807 in Paris. The tour has been elaborately described by John van Whye (2002).[1] He then traveled through Italy, Germany, Denmark, Holland, and Switzerland. His arrivals were advertised in news-papers and we would now refer to this as "hype." The general pattern seems to be that in a particular

[1] The description can be retrieved from the internet at http://www.historyofphrenology.org.uk/texts.htm

place he presented a series of lectures, dealing with several aspects of his organology or craniology. The lectures tailored to a general audience. By and large, they were well received and the audience was very enthusiastic. If the interest for the lectures turned out extremely large, Gall stayed for another series.

Until 1810, when the first volume of his *Anatomie et physiologie du système nerveux en général, et du cerveau en particulier* (Anatomy and physiology of the nervous system in general, and of the brain in particular) (with J. C. Spurzheim, 1810–1819) rolled from a French press, most of what was known of his theories came from reports of his lectures, written by authors who took down what Gall had to say about brain function and its relation to the shape of the cranium. Between 1802 and 1810, a veritable procession of such reports appeared (e.g., Walter, 1802; Martens, 1803; Leune, 1804; Anonymus, 1805; Bischoff, 1805; Katejan Arnold, 1805; Kupf, 1805; Adelon, 1808). Not unlike present-day bootleg recordings of concerts, these reports were of uneven quality and some of them, for instance Ackermann (1806), were written with the explicit purpose of criticizing cranioscopy. From a commercial point of view, books on Gall's theory apparently were very successful: Several of them were translated or went through two or three editions.

At the same time, critics began to counteract. Among them were Ackermann, Hufeland, Walther, and Hegel (Wyhe, 2002). In Paris too, Gall had some important adversaries. Their influence was strong enough to keep Gall out of the French Academy of Sciences. A scientific committee with Tenon, Sabatier, Portal, Pinel, and Cuvier, (he, in particular, played a prominent role), to some extent influenced by Napoleon, had difficulties recognizing a clear relationship between Gall's anatomical views and his brain physiology (Lesky, 1979). Gall, on the other hand, did not want to assume an inferior role and preferred to remain independent and let the arguments in the paper and lectures do their work.

Gall's Visit to The Netherlands

Among historians of medicine, Gall's cranioscopic tour has received deserving attention, but only a passing reference is made to his sojourn in The Netherlands. Ebstein (1924) presented a minute report of Gall's dealings in promoting his theories, reconstructed from correspondence and diaries of his contemporaries, but he omitted the visit to Holland. We feel that this historiographical blank deserves to be filled. Several Dutch articles discussed Gall's visit to The Netherlands (Conradi, 1995; Steendijk-Kuypers, 1996; Heiningen, 1997). Drawing on journals and newspapers of that period, these authors have demonstrated that Gall's lectures stirred controversy among anatomists and public debates on the tenets of cranioscopy.

In the early spring of 1806, Gall visited The Netherlands. He delivered 10 lectures in Amsterdam and then five more in Leiden. Gall also proposed to give lectures in The Hague, but these were cancelled when it became clear there would not be enough subscriptions to fill a sizeable audience. Between and after the lectures, Gall visited Dutch learned societies. He met with several anatomists and inspected local anthropological collections and cabinets. In the last days of April, Gall left Holland and proceeded to Germany.

The furor that Gall had caused in Austria and Germany did not remain unnoticed in The Netherlands. As early as 1802, the *Algemeene Konst- en Letterbode* (General Magazine for Arts and Letters; see also Figure 3) published a one-page notice on Gall's views. The anonymous author stated — without comment — that the Viennese authorities had prohibited his

Figure 3. Announcement of Gall's arrival in Amsterdam in the newspaper Algemeene Konst – en Letterbode, Friday, April 4, 1806.The text indicates that Gall, the Viennese physician is travelling through Europe to explain his notorious theory. From Hamburg, he went to Munster and Utrecht and is now in Amsterdam.

public lectures and then proceeded to explain the cranioscopic theory. He mentioned some of Gall's findings in comparative anatomy, such as the discovery of the organ of musicality in the skulls of singing birds or the organ of cunningness in the skulls of cats and foxes. He also reported that Gall claimed to have found an excessively well-developed organ of musicality in Mozart's brain. The notice was phrased in the most cautious of terms: In the author's estimation cranioscopy was a "curious" theory; its validity, however, remained to be seen.

Fuller discussions of cranioscopy appeared two years later. In January 1804, Gerardus Vrolik (1775–1859), a Professor of Physics and Medicine at the Amsterdam Athenaeum Illustre, published *Het Leerstelsel van Franz Joseph Gall* (The Doctrine of Franz Joseph Gall). Vrolik's publication was based on his two lectures, given at the Felix Merites Society in Amsterdam. Vrolik referred to the facial angle of Camper and to Lavater's physiognomic system and he pointed to Gall's system as another possibility to gain more insight into the character of an individual. He offered some support for the implications it had for education and criminology, but for the present time he rejected the cranioscopically based method for diagnosing and predicting individual characteristics.

In December 1804, Jacob Elisa Doornik (1777–1837) published a highly critical examination of cranioscopy: *De Herssen-Schedelleer van Frans Joseph Gall getoetst aan de Natuurkunde en wijsbegeerte* (Franz Joseph Gall's theory on the brain and the cranium, tested by physics and philosophy; see Figure 4). Doornik held a professorship in medicine, also at the Athenaeum Illustre. He had been present at Gall's lectures in Berlin and he

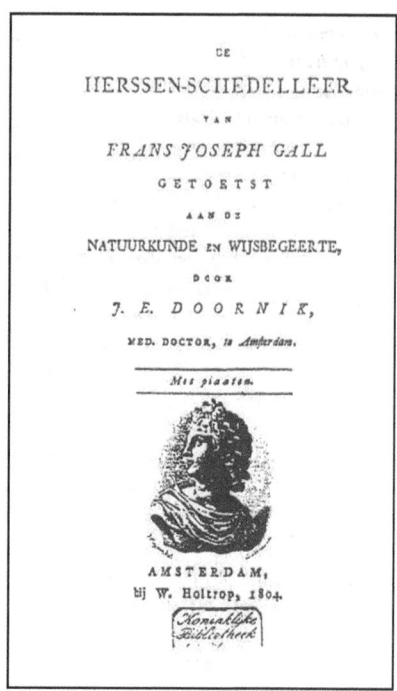

Figure 4. Title page of Doornik's book on Gall's craniology, evaluated in the light of physics and philosophy.

criticized Gall for not providing evidence for his statements, for being materialistic, and for drawing tenuous conclusions. He argued that one should not judge a man on his talents but according to his works.

Later that month, the Remonstrant theologian Martinus Stuart (1765–1804) published his Dutch translation of Leune's *Entwicklung der Gallschen Theorie über das Gehirn* (Development of Gall's Theory of the Brain) (1803), prefaced by an extended and positive introduction (see Figure 5).

Vrolik and Doornik, even if they held opposite views on the value of Gall's theories, both had their books published by the Amsterdam publisher Holtrop. Apparently, Holtrop sensed a market for books on cranioscopy that he could monopolize on, regardless of their contents; a year later (1805), he published the Dutch translation of *Doctor Gall auf die Reise,* a comedy play by Wilhelm von Freygang, satirizing the touring Viennese anatomist (Freygang, 1805). In 1806, Holtrop completed a quartet of cranioscopic books by publishing two lectures on Gall, delivered by Doornik in the Amsterdam Society of Felix Meritis during the winter of 1805–1806. Doornik's judgment was, once again, vastly critical.

Gall, in the meantime, was held up in Germany for a series of extra lectures in Münster. Appetizing the audience, the Amsterdam newspaper *De Ster* published a few extracts from Gall's letters to his friends at Amsterdam. Gall indicated that his wish was to learn as well as to teach: He announced his intention to visit the famous anthropological cabinet of Petrus Camper and asked his friends to secure him as many "apes, parrots, big sea-animals or fish, dead or alive, as they could lay their hands on," in particular "the head of this orang-outang" (De Ster, 13 maart 1806, p. 13) he had heard about. As a sea-faring nation, he

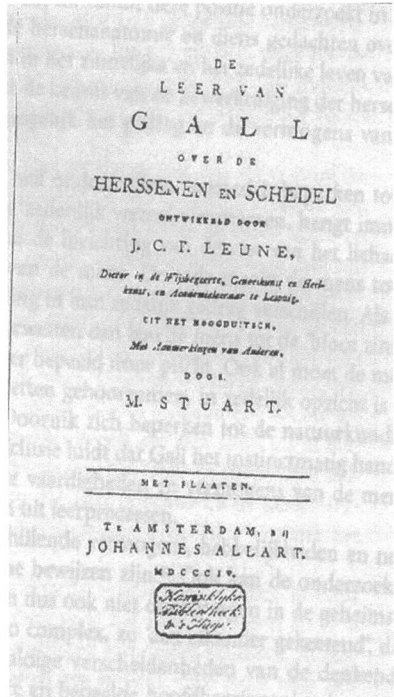

Figure 5. Title page of Stuart's book on the doctrine of Gall.

believed the Dutch were in a privileged position to gather a rich collection of rare specimens from foreign countries. Apparently, his correspondents had signaled Gall that his views met with skepticism among anatomists, for he declared that he had nothing to fear "but Nature herself." In a second letter by Gall, he wrote that his extensive collection of demonstration material made traveling an elaborate enterprise, but that his skulls and anatomical preparations were essential for his lectures. He ended by requesting "three well furnished rooms with decent people, preferably in the center of Amsterdam." Those who wished to attend the lectures were invited to sign registers in the offices of De Ster, Felix Meritis, Doctrina, and the Leesmuseum ("Reading Museum"). The opening lecture was scheduled to take place on April 8, at 6 pm in *Het Wapen van Amsterdam*. That evening, Gall welcomed approximately 100 audience members, who had paid 21 Dutch florins for a series of 10 lectures.

In conclusion, it appears that physiognomy as well as Gall's ideas were received well by theologians, while anatomically informed scientists were much more skeptical. Apparently, discussion of Gall's views took place in learned societies, opposed to physicians.

Content of the Lectures

The content of Gall's lectures in Amsterdam can be reconstructed from two sources. First, there are several short reports in the newspaper *De Ster*. In a report dated April 12, a certain G. discusses the three lectures Gall had delivered to that date. It seems G. was disappointed by Gall's style of lecturing. The long series of anecdotes, presented in a "coffee house conversation" kind of way, fell short of the standards for a "coherent, academic lecture" (De Ster, 12 April 1806, p. 118). Gall's sarcasms against anatomists and his repeated gibes at philosophers were uncalled for. He ridiculed his opponents, putting in their mouth the grotesque claim that if he, Gall, was right, midwives might very well correct nature by pushing away evil organs, or lifting good organs elsewhere. Three days later, G. wrote a second report that was much more favorable. He now praised Gall's "immortal discoveries," the fruit, no doubt, of "German tenacity and patience." Just when his knowledge of human anatomy and physiology seemed exhausted, he went on, Gall presented a "rich harvest." On April 17, *De Ster* published a report by G. on Gall's sixth lecture, dealing with the organ of procreation or amativeness (Number 1 on the cranioscopic map). G. appears to have been quite impressed: Drawing on mythology, paintings, and descriptions of nymphomanic patients, Gall had proved, to G.'s satisfaction at least, that this particular organ indeed had its seat in the back of our heads.

The second source on Gall's lectures is Stuart's book *Herinneringen uit de lessen van Frans Joseph Gall* (Memories from the lessons of Franz Joseph Gall), which appeared in 1806. Stuart had been in the audience in Amsterdam and he presented a full report of each of the 10 lectures. In the dedication (to an anonymous gentleman), he wrote that Gall spoke very fast, defying all efforts to write a verbatim report. Stuart could not convey what Gall had pointed out on the demonstration material he had with him, such as skulls, drawings, and wax models of animal and human brains. Despite these shortcomings, Stuart's report of 133 pages is fairly rich in detail.

Those in the audience, who had come to know about the exact cranial locations of the human faculties, the most notorious part of Gall's doctrine, had to sit and be patient. Before dealing with cerebral topography, Gall wished to discuss some preliminary matters. Cranioscopy, he argued, was a misnomer: His theory was intended to explain the development of the brain as an organ of our mental faculties. The brain serves our mind as its material condition, much as the eye is the corporeal condition for sight. Gall then pointed out that the various organs indicated penchants and instincts but were not the *cause* of behavior. Our acts, as he explained, are under the control of the mind and therefore guided by moral restraints. Man is by no means forced to act on his impulses. Further on in his lecture, as well as in lectures to follow, Gall referred time and again to this important point. His doctrine was partly materialistic, he granted, since it dealt with the material conditions of thought and behavior, but it was certainly not fatalistic. The material nature of our cerebral organs leaves our moral freedom intact. In other words, charges of atheism were completely unfounded. These were, of course, the words of a cautioned man. Allegations of atheism had led to the ban of his lectures in Austria and for Gall, who probably lived on this income, experienced further problems with authorities abroad with the threat of jeopardizing the tour as well as his greater cause. Therefore, Gall requested that his listeners agree that God had created a natural world and a moral world, and that a truth in the former world could not contradict a truth in the latter world.

After clearing these two misunderstandings, Gall went on to argue that the brain, like so many human organs, was a double organ. Moreover, each half consisted of specialized parts. Lesions at different locations produced a variety of impairments. These selective disturbances proved that the brain — and similarly the mind — was a divided entity. The unity of consciousness was not a valid argument against the theory of the double brain, any more than the unity of sight substantiates that we have only one eye. In this way, most of the first lecture went away with a general defense of the doctrine against the charges of philosophers, metaphysicians, fellow anatomists, and other adversaries. It also determined the blueprint for the next four lectures.

In his second lecture, Gall argued that the human soul is not the undivided entity that philosophers have made it to be. Just as our sensory perceptions originate from separate senses, the various mental faculties originate from separate organs in our brain. Each of these organs has its own neuronal connections and locations. Gall held up a wax-coated brain, pointing out the separate courses of nerve fibers. Mental faculties differ from person to person and even within the same person during the course of his life. The cerebellum for instance, organ of amativeness, is very small in childhood but shows a disproportionate growth during adolescence. The third lecture took this theory two steps further: (a) the size of an organ is a measure of its strength, and (b) size can be determined from the outside, by touching the surface of the skull, preferably with the flat hand. During the embryonic phase, the formation of the bones of the skull faithfully follows the shape of the brain. At a later stage — and even in old age — natural metabolism allows the skull to adapt its form to the size of the organs. Passing along the skull, Gall invited his audience to see for themselves that specific faculties reveal themselves in bumps.

Having demonstrated that some talents are truly outstanding in more than one sense of the word, Gall stated in his fourth lecture that there is a global relationship between the size of the brain and the level of intellect. On average, the skull of a child of superior intellect will be larger than the skull of his less-endowed peer. Skulls of people from "primitive cultures" are smaller than European skulls. Gall claimed to have performed autopsies on the brains of 300 mental patients, finding that they all had "soft brains and tough skulls" (Stuart, 1806, p. 32). Thick skulls were presumed to harm the brain. This is especially relevant in forensic matters. Gall recommended that judges should keep in mind that both infanticide and suicide may be caused by brain diseases.

The audience in the fifth lecture found the Viennese anatomist still dealing with preliminaries. First, he discussed the criticisms of Ackerman, Plattner, and others. Gall minced no words: These men were all misguided, ill-informed, or plain incompetent. If it were not for their deserved reputation in other fields, Gall confided to his audience, he would have refuted them in print long ago. His own research was methodologically impeccable. He explained that he looked for resemblances in the skulls of geniuses in one particular area and tested his hypotheses against the skulls of people, who lacked this specific talent. He also referred to his numerous autopsies, both on human and animal brains. It is clear that Gall took great pride in his understanding of comparative anatomy.

Finally, in the sixth lecture, Gall started with a discussion of separate organs and their location. The first on Gall's catalogue was the double organ for amativeness. It could be felt as two swellings at the back of the head. Why should there be such a well-protected position for such a "low" organ? Gall argued that nature may have her own scheme of values, perhaps she feels that the organ regulating reproduction is more vital to man than his organ for reflective thought. The evidence Gall put forward was essentially a heap of facts and observations, mixed with anecdotes. The growth of the cerebellum is proportional to the development of sexual maturity. A "hysteric woman" with exceptionally intense sexual urges had a "burning" back of the head. Men have broader necks than women, in accordance with their greater sexual appetite. At stud-farms, there is a marked preference for horses with broad necks. Turkish men have broad necks, which betray their polygamic nature. After castration the cerebellum shrinks. Excessive stimulation of the neck, as happens in a hanging, may cause an erection. All in all, Gall adduced no less than 13 arguments in favor of the back of the head as the seat of amativeness.

The next four lectures were devoted to the next 26 organs on Gall's list. Immediately above the two bumps of amativeness was a single bump, corresponding to the organ for the "love of children." In passing, Gall laid down the general rule that insanity will always strike at the most prominent organ. In an Amsterdam asylum, he said, he had put this hypothesis to the test in the presence of professor Vrolik. Working from the back of the head to the front, Gall presented his evidence for the various organs of memory (for instance memory for places, language, numbers, colors, etc.). An exceptional ability to memorize revealed itself in protruding eyes (the famous Amsterdam actor Cruys, Stuart added, was certainly a case in point.) Subsequently, Gall detailed all the remaining parts of the cranial map, discussing organs like acquisitiveness, reverence, mirthfulness, benevolence, and secretiveness. There could not be more organs, Gall claimed, nor less; these 26 organs just presented themselves to his inquisitive eyes; they were truly "found." If such a relatively small number of organs seemed insufficient to account for the rich variety of human behavior and personality, one should reflect that music, calculation, and all manner of writing are equally built on a limited number of elements. Ending on a pious note, Gall explained to his Amsterdam audience that we owe our human nature to the sophisticated construction of the brain, which testifies to the "perfect work" of a holy, wise, and loving God.

Reception

As indicated above, there was a clear interest in physiognomy in The Netherlands. In the 1780s, a number of books were published, suggesting that Lavater's views were well received. Lavater died in 1800 and interest in his work was resurrected. However, it is important to note that the audience was more one of educated men in general than specifically of physicians. As described above, in the second edition of the Dutch translation of Lavater's book (Lavater, 1780), a list of subscribers is included, and physicians apparently are a very small minority. Brain anatomy is clearly missing in the many books that were published during Gall's tour to promote his theory. The audience was more interested in the implications of the theory than in the structure of the brain. Physiognomy was regarded to reveal something about an individual's character or personality. It may be relevant for interpersonal communication, rather than for understanding and treating diseases. Another area, in which it played a significant role, was criminology with the central question being whether there was a biological basis for criminal behavior.

This latter aspect can also be found in the works of Gall. The basic motive for Gall's undertaking was his wish to develop a psychological framework. Moreover, he was not just interested in "classical" psychological faculties such as perception, imagination, and memory; he certainly devoted much attention to "moral" qualities. And these moral qualities could be linked to actual behavior, including misconduct. In his quest to explore morality, Gall ventured to prisons to examine the skulls of criminals. Gall's views were also tested by presenting him the skulls of criminals.

When the initial reports of a new theory on how to deduce a man's character from the skull reached The Netherlands in the early years of the nineteenth century, people were more or less familiar with the general purpose of this theory. An important distinction is that Lavater looked at the "weak parts" of the facial part of the head, whereas Gall concentrated on the skull (Martini van Geffen, 1825).

Gall's theory undoubtedly suffered from many weaknesses. His psychological organs did not follow from a well-developed philosophical system but were found more or less by serendipity. The evidence was weak: anecdotes of individuals with a bump, detected only by Gall himself. Contrary evidence of others with similar behavior, but lacking the bump at the required site, was not accepted as counterproof by Gall. Anatomically, the theory also did not rest on a convincing basis: How can one detect bumps at places where there is no cortical tissue lying beneath? Or, why are there only bumps at places that we can touch, but not at the bottom of the skull? Scientists, familiar with the human brain, not only detested Gall's materialistic and fatalistic views but they also were aware of the lack of philosophical and empirical support. The following anecdote may illustrate this. When Gall himself came to The Netherlands, he was invited by members of a learned society in Amsterdam. On his way to Amsterdam, he paid a short visit to Utrecht, where he apparently met Mathias van Geuns, a Professor of Medicine. Van Geuns apparently did not think highly of Gall and it was made very clear in his interactions with Gall. According to Steendijk-Kuypers (1996, p. 2563), Van Geuns kept Gall waiting for over an hour and then welcomed him as follows: "You are Doctor Gall, who formulated the craniology, but I must confess to you directly that I do not have a great respect for this theory" (our translation from the German citation). He then presented Gall to the head of the botanical garden and left.

The next day, Gall continued his journey to Amsterdam. From the description in the newspaper articles, it is clear that the audience was not very enthusiastic about Gall's presentations. They did not meet the expected standard of a scientific lecture. Moreover, Gall could not deal with critical questions and thus gave others the impression that he thought too highly of himself and was reluctant to accept constructive criticism. However, there exist reports that seem somewhat more favorable. History repeated itself in Leiden. His lectures were not well appreciated. Gall quickly had enough of the Dutch, and the feeling was mutual. Very little interest in his work is discernible after 1806.

References to Gall's work remain scarce indeed and are generally not found in medical contexts. In 1809, a review was published of a German play *Dr. Gall and Dr. Faust* (Oemler, 1808). The play highlights that interest in Gall's skull theory was minimal. As mentioned above, Martini van Geffen published his *Schedelleer en het Menselijke temperament* (Skull theory and the human temperament) in an attempt to integrate Lavater's and Gall's theories (see Figure 6).

In 1819, three young medical students (Broers, de Fremery, and Tilanus) went on a study tour to France and Germany (van Mesdag, 1927). The diary of that trip was never published, but short excerpts were available in an article published in a criminological journal in 1927. It appears that the three had a specific interest in criminology and also visited some jails. They attended four lectures Gall gave at the *Athenée* in Paris. The contents of these lectures resemble the lectures Gall gave on his tour between 1805 and 1807. There are no indications that Gall's lectures had a great impact on these young men and had any further influence when they pursued their medical profession upon their return home.

Schroeder van der Kolk, who is perhaps considered the founding father of psychiatry and neurology in The Netherlands (Eling, 1998) was aware of Gall's work. We only found a few places where he refers to Gall's "craniological views." It is interesting to note that Schroeder appears to have accepted the notion of localization of function — in some sense — before the decisive discussion was started by Broca in 1861. In his *Handboek van de Pathologie en Therapie der Krankzinnigheid* (Handbook of Pathology and Therapy of Madness; Schroeder, 1863), edited and published after Schroeder's death by his colleague and friend Hartsen, he wrote about the different roles or functions of various parts of the cerebral hemispheres. He

Figure 6. Title page of A. Martini van Geffen's book on physiognomy in relation to Gall's craniology.

argued that it was already suspected for a long time that the frontal part of the brain was associated with intellectual functions. Next to Camper, he refers to Gall, "who, even though he placed his organs rather arbitrarily, also localized the most important organs for higher intellectual functioning behind the forehead" (Schroeder van der Kolk, 1863, p. 37). Apparently, Schroeder too was critical of details of Gall's views, but he still regarded Gall's work as a valid reference for the localization of functions in the brain.

Capelle (1854) published an extended essay in a literary journal entitled *De Gids* (The Guide). The purpose of the essay was to introduce the views of the German physician Carl Gustav Carus (1789–1869) to a wide audience. One wonders whether Van Capelle's motivation was to get his paper published in a medical journal. In this essay, Van Capelle also mentioned Gall's "theory of organs" and claimed that it had already been rejected for a long time, as it had been proven to be built on "false" premises.

Nevertheless, Thijssen (1858) seriously attempted to raise some support for Gall's theory. He must have been familiar with it as he competently discusses many weaknesses and inconsistencies. Although it had been rejected a long time ago, Thijssen assumed that it was generally known among his readers, a nonspecializing audience. Very much like current authors have resurrected the statue for Gall as the man, who introduced the principle of localization of (more or less independent) faculties on the cortex of the brain, so Thijssen pointed out that no one could seriously refute the general principle. Of course, one may have problems with the specific mental faculties that Gall distinguishes and one may even point out the anatomical absurdities put forth by Gall; however, it is the general principle that may have served as the starting point for the study of the biological foundation of psychological functions. After having discussed various aspects, Thijssen raised the following question:

> So what remains, after all, of this entire system of Phrenology, of Gall himself? . . . I remember this sea of foolishness I had to go through, before I arrived at Gall himself, and still I recollect clearly how gratefully I admired this great man, who was the first to clearly explain how the soul reveals itself out of the cerebral and spinal nervous system. (1858, p. 198)

Thijssen's plea enjoyed little success. Interest in measuring the skull only revived when Lombroso became popular towards the end of the nineteenth century. In The Netherlands, Winkler became highly interested in this phenomenon. Winkler considered Lombroso to be his friend. He also published studies in this area (Draaisma, 1995). However, Gall did not play a role in these discussions.

In retrospect, Gall appears to have "sown" in a rather infertile Dutch soil. His audience was a bit skeptical from the beginning. Gall's personal attitude seems to have played an important role, too. It should be noted that Dutch scientists were rarely examining the relationship between brain and behavior in an empirical way. While heated discussions on the issue of localization of function took place in France, Germany, and England, hardly any serious scientific papers on this topic was written in The Netherlands (Eling, 2008).

References

Ackermann JF (1806): *Der Galllsche Hirn-Schedel- und Organenlehre vom Gesichtspunkte der Erfahrung aus beurtheilt und widerlegt*. Heidelberg, Mohr und Zimmer.

Adelon NP (1808): *Analyse d'un cours du docteur Gall*. Paris, Giguet et Michaud.

Anonymous (1780): *Physiognomische Catechismus, bevattende de gronden en beginselen der Gelaadkennis* [Physiognomic Catechism, containing the reasons and principles of Physiognomy]. Amsterdam, Gerrit Bom en Zoonen.

Anonymus (1805): *Ausführliche Darstellung des Galllschen Systems der Schädellehre*. Magdeburg, S.N.

Bacon F (1994): *The Advancement of Learning*. Whitefish, Montana, Kessinger Publishing.

Bischoff CHE (1805): *Darstellung der Gall'schen Gehirn- und Schädellehre*. Berlin, S.N. (unknown).

Broca P (1863): Localisation des fonctions cérébrales. Siège du langage articulé. *Bulletins de la Société d'Anthropologie*. Séance du 2 Avril 1863: 200–204.

Capelle H van (1854): De symboliek van den schedel [The Symbolism of the Skull]. *De Gids 18*: 433–461.

Chodowiecki D (1781): *Verhandeling over de physiognomie of gelaatkunde* [Treatise on the physiognomy of the face]. Amsterdam, P. de Hengst.

Conradi M (1995): Franz Joseph Gall in Nederland. *De Psycholoog 7*: 320–323.

Doornik JE (1804): *De Herssen-Schedelleer van Frans Joseph Gall getoetst aan de Natuurkunde en Wijsbegeerte*. Amsterdam, Holtrop.

Doornik JE (1806): *Voorlezingen over F. J. Gallls herssen-schedelleer*. Amsterdam, Holtrop.

Draaisma D (1995): De Hollandse Schedelmeesters. Lombroso in Nederland. *Feit en Fictie 2*: 50–73.

Ebstein E (1924): Franz Joseph Gall im Kampf um seine Lehre. In: Singer C, Sigerist HE, eds., *Essays on the History of Medicine*. Zürich, Seldwyla.

Eling P (1998): Jacobus Schroeder van der Kolk (17—1862): His resistance against materialism. *Brain and Cognition 37*: 308–337.

Eling P (2008): Cerebral localization in The Netherlands in the 19th century, emphasizing the work of Aletta Jacobs. *Journal for the History of the Neurosciences 17*: 175–194.

Fokke Simonsz A (1801): *Verhandeling over de algemeene Gelaatkunde* [Treatise on the general Physiognomy]. Amsterdam, S.N.

Freygang W von (1805): *Doctor Gall op reis: blijspel in één bedrijf.* van Esveldt, JS, trans. Amsterdam, Holtrop.

Gall FJ, Spurzheim JC (1810–1819): *Anatomie et physiologie du système nerveux en général et du cerveau en particulier; avec des observations sur la possibilité de reconnaître plusieurs dispositions intellectuelles et morales de l'homme et des animaux par la configuration de leur têtes,* volumes 1–4 and atlas. Paris, F. Schoell.

Heiningen T van (1997): De receptie van de hersen-schedelleer van Franz Joseph Gall in Holland kort na 1800. *Gewina 20*: 113–128.

Jahnke J (1997): Physiognomy, Phrenology, and non-verbal communication. In: Bringmann, WG, Lück H, Miller R, Early Ch, eds., *A Pictorial History of Psychology.* Chicago, IL, Quintessence Publishing Co.

Kajetan Arnold JTF (1805): *Dr. Joseph Gallls System des Gehirn und Schädelbaues.* Erfurt, Henning.

Kruger L (2005): The scientific impact of Dr. N. Tulp, portrayed in Rembrandt's "Anatomy Lesson." *J Hist Neurosci 14*(2): 85–92.

Kupf M (1805): *Beleuchtung der Gall'schen Gehirn- und SchäcZellehre durch Vernunft: und Erfahrung geleitet.* Berlin, S.N.

Lavater JK (1775–1778): *Physiognomische Fragmente zur Beförderung der Menschenkenntnis und Menschenliebe.* Leipzig – Winterthur: Bey Weidmanns Erben und Reich, und Heinrich Steiner und Compagnie.

Lavater JK (1780): *Over de Pysiognomie door J.C. Lavater.* Amsterdam, Johanes Allart.

Lesky E (1979): *Franz Joseph Gall (1758–1829): Naturforscher und Anthropologe.* Bern, Huber.

Leune JCF (1804): *De leer van Gall, over de hersenen en schedel.* Stuart M, trans. Amsterdam, Allart.

Martens FH (1803): *Leichtfassliche Darstellung der Theorie des Gehirn- und Schädelbaues des Dr. Gall.* Leipzig, S.N.

Martini van Geffen A (1825): *De karakterkunde volgens de gelaatsleer, in verband beschouwd met de schedelleer en de menschelijke temperamenten* [Characterology according to physiognomy, considered in relation to craniology and the human temperaments]. 's-Hertogenbosch, Palier en Zoon.

Noordhoek WJ (1925): Lavater und Holland. *Neophilologus 10*: 10–19.

Oemler CW (1808): *Doctor Gall en doctor Faust, of, De groote omwenteling op aarde.* Fokke Sz A, trans. Amsterdam, Timmer.

Porta G B Della (1586): *De humana physiognomonia.* Aequensis, J. Cacchius.

Schroeder van der Kolk JLC (1863): *Handboek van de Pathologie en Therapie der Krankzinnigheid* [Handbook of Pathology and Therapy of Madness]. Utrecht, van der Post.

Schuller tot Peursum-Meijer J, Koops WRH, eds. *Petrus Camper 1722–1789, onderzoeker van nature* [Petrus Camper 1722–1789, an inveastigator by nature]. Groningen, S.N.

Steendijk-Kuypers J (1996): Het succes van een dwaling. De hersen-schedelleer van Franz Joseph Gall (1758–1828) en de echo van de frenologie in Nederland. *Nederlandsch Tijdschrift voor Geneeskunde 140* (51): 2560–2564.

Stuart M (1806): *Herinneringen uit de lessen van Franz Joseph Gall.* Amsterdam, J. W. IJntema.

Thijssen HF (1858): Over de leer van Gall. *De Gids 10*: 145–214.

Van Mesdag S (1927): Een criminologische bijdrage van het jaar 1819 [A criminological contribution from the year 1819]. *Tijdschrift voor Strafrecht 37*: 189–209.

van Wyhe J (2001): [Online]. *Travels of a Craniologist: Franz Joseph Gall and his European lecture tour, 1805–1807.* Available: http://www.historyofphrenology.org.uk/texts.htm

van Wyhe J (2002): The authority of human nature: The Schäellehre of Franz Joseph Gall. *British Journal of the History of Science 37*: 17–42.

Vrolik G (1804): *Het leerstelsel van Franz Joseph Gall.* Amsterdam, Holtrop.

Walter JG (1802): *Kritische Darstellung der Gallsche Aantomisch-Physiologische Untersuchungen des Gehirn- und Schädelbaues.* Zürich, S.N.

Georg Büchner: Anatomist of the animal brain and the human mind

Michael Hagner

ABSTRACT
The writer Georg Büchner (1813–1837) is considered one of the giants of German literature. Comparatively less well known, however, is the fact that Büchner was also a gifted neuroanatomist who completed his medical studies with a dissertation on the nervous system of the barbel (a freshwater fish with a high incidence in the River Rhine) and gave a lecture on cranial nerves shortly afterward, hoping to secure a position at the University of Zurich. In the copious secondary literature on Büchner, it has often been discussed whether and how his poetic and scientific writings were interrelated. In this article, I compare Büchner's anatomical and literary views of the brain and argue that two distinct perspectives on the organ were developed here. In the literary works, human behavior was linked to the brain in a manner that betrays the influence of Franz Joseph Gall's organology. In the anatomical writing, the brain appeared as an exemplar of natural harmony and beauty. In the one case, the brain appeared as an aristocrat, in the other as a pariah. I take this stark contrast to mean that Büchner understood the brain as an epistemically slippery, contradictory object that could only be approached from different angles.

Introduction

Anyone seeking to come to terms with Georg Büchner's thoughts on the brain can approach the topic in two ways: either by reading his two texts on neuroanatomy—the *Mémoire sur le systeme nerveux du barbeau* and the *Probationary Lecture* on brain nerves he gave in Zurich—or by casting a glance at the few remarks on the brain to be found scattered throughout his literary works. To be sure, the latter contain statements he would never have permitted himself in his anatomical texts. And yet, Büchner's poetry and neuroanatomy seem to be connected by fine threads, giving rise to an understanding of the brain as a densely interwoven web of meaning. More precisely, in his neuroanatomy Büchner traced a path from below to above. He started out from the simple brains of fish in the hope of arriving, sooner or later, at an understanding of the human brain. This was the same strategy pursued by comparative anatomy and embryology from the early-nineteenth century onward.

In his literary work, by contrast, he took the opposite route, from above to below. This is not to say that the animal brain is inferred directly from the human brain. It is striking, however, that as soon as Büchner talked about the latter, his thoughts turned almost instinctively to human mental illness, brutality, and bestiality. Well before Darwin's theory of evolution, the pathological interpretation of these phenomena as signs of degeneration was already associated with humanity's animal origins. Büchner was by no means slavishly attached to this fatal logic, but it formed—along with the evolutionary ideas of Romantic *Naturphilosophie* and the materialistic aspects of Franz Joseph Gall's organology—the discursive framework for his preoccupation with the brain. *Naturphilosophie* and organology are not normally placed in such close proximity, and after 1830 they had both passed the peak of their popularity in Germany. Büchner was neither a *Naturphilosoph* nor a phrenologist, but the significance of the brain in his scientific and literary *oeuvre* can be understood only in the context of these two theories.

Georg Büchner was born in 1813 in the village of Goddelau, in the state of Hesse. His father, Ernst Büchner (1786–1861), was a physician with considerable experience in surgery, psychiatry, and forensic medicine, even publishing several of his forensic expert opinions as case studies (Büchner 2013). In Darmstadt, where Ernst Büchner held the position of municipal physician, his son assisted him in performing autopsies before commencing his own medical studies in Strasbourg in late 1831, continuing them in Giessen, and completing them in Strasbourg following his forced return there in 1835. Büchner was forced into exile because of his political activities against the reactionary monarchistic regime in the state of Hesse. The following autumn he moved to Zurich, where he died only a few months later, in February 1837. His life was too short to enter into the annals of science. Accordingly, Büchner became known not as a neuroanatomist but as a poet and social revolutionary, forced to flee Germany for French Strasbourg as a 21-year-old medical student (Figure 1).

There is no denying that the two neuroanatomical texts would hardly merit a footnote in the history of the neurosciences if Büchner did not happen to be one of the most important writers in German literature. His literary works were a "free-time activity" (Kurzke 2013, 13), but with each of the few works he wrote between 1834 and 1837, he extended the boundaries of a German literature still very much under Goethe's spell. The linguistic radicalism of a story like *Lenz* (1835), a comedy like *Leonce & Lena* (1836), a drama like *Danton's Death* (1834/1835), or the fragmentary play *Woyzeck* (1836/1837) would not be fully appreciated until the twentieth century. Would the history of literature have taken a different turn if Büchner had not died of typhus, presumably caught through infection during an autopsy? There is no way of knowing. Perhaps his considerable ambition as a researcher in the natural sciences would have led him to turn his back on literature. Although barred on political grounds from universities in German states, he had moved to Zurich in 1836 in order to submit his dissertation and be awarded a doctorate in medicine. He hoped to continue his scientific career at the newly established university there; its first chancellor, Lorenz Oken (1779–1851), was himself a famous naturalist. The probationary lecture Büchner gave in November 1836 was the precondition for him being awarded the *venia legendi* as a private lecturer. It was meant to lay the foundation for everything to come; in fact, it marked the end of his brief scientific career.

Figure 1. George Büchner. Undated pencil drawing by August Hoffmann. The original was destroyed by fire during World War II in 1944.

Büchner, the poet: The guillotine, the brain, and human vivisection

Georg Büchner had already thematized the maximal ambition of the neurosciences in his first drama, *Danton's Death*, written before he had decided to investigate the brain in his dissertation. In a formulation as well-known as it is brutal, he proposed seizing thoughts directly at their source: "We'd have to crack open the skulls and tear out the thoughts from the brain fibers" (Büchner 1992, 13). Danton, in whose mouth Büchner placed this line at the beginning of the drama, reacts with this demand to the futility of ever truly knowing another human being. Authenticity is not to be found in words and looks, gestures and deeds, only in the brain itself. It is as if thoughts were located in the hardware and, assuming they were actually there, could also be understood as such; as if the fibers of the brain—which lie beyond good and evil, falsehood and truth—could yield information about the contents of thoughts and feelings. Büchner's violent sentence on the relationship between the brain and its thoughts can be historically contextualized on three levels: a moral-vivisectionist level, which targets the anatomical and physiological practices of the time; a topographical level, interested in the localization of intellectual processes in the brain; and a utopian level, which can be understood as a first step in the neuroscientific quest to read the mind. In what follows, I will examine each of these three levels in turn.

The image of the brain being cracked open is redolent of the guillotine's reign of terror during the French Revolution, but the demand for such experiments was older. The dissection of corpses of executed criminals runs like a red thread through the history of modern anatomy (Richardson 1989), yet the Enlightenment went a step further. In the mid-eighteenth century, mathematician and philosopher Pierre-Louis Moreau de Maupertuis (1698–1759) called for experimental vivisections to be carried out on the brains of criminals condemned to death (Maupertuis 1768, 410). The practices under the guillotine after 1793 were not far removed from this. On one hand, there was interest in whether the claim made by physician Joseph-Ignace Guillotin (1738–1814) about his new method of execution was really true: that it was more merciful than the conventional method of beheading with an ax. On the other, following the *terreur* of the French Revolution and Luigi Galvani's (1737–1798) famous experiments on animal electricity, doctors conducted galvanic experiments on freshly guillotined corpses to investigate whether there was any consciousness or sensation of pain in their heads (Jordanova 1989; Hagner 1997, 185–93; Borgards 2004).

Büchner reacted with caustic irony to what executioners and doctors were up to around 1800. At the beginning of the drama, Danton wants to tear out the thoughts from skulls. By the end, staring his own death in the face, he bitterly describes the guillotine as the "best doctor," presumably because it puts an end to life's misery. And immediately afterward, Danton taunts the executioner that even he can do nothing to prevent his head from kissing that of his guillotined friend in the basket beneath the blade (Büchner 1992, 88). The kiss of the severed heads represents the poet's grotesquely exaggerated response to the galvanic electrical charges of those doctors who took a professional interest in the cerebral origins of consciousness and were prepared to accept the guillotine as one element in a macabre experimental setup. Whereas these doctors focused entirely on their research object, the brain in separation from the body, the execution apparatus, as the better doctor, had already fulfilled its therapeutic mission.

Büchner returned to the experimentalization of the human brain in *Woyzeck*. The doctor in that play tells the captain, who has sought his advice about his melancholy, that he will soon suffer a stroke: "bloated, fat, thick neck, apoplectic constitution." For the cynical experimenter, who misuses the penniless Woyzeck as a human guinea pig and throws him a few coins for his services, this offers the fairest prospects: "If it is God's will that your tongue is partly paralyzed, then we will make the most immortal experiments" (Büchner 1992, 159). The demise of the one is the triumph of the other. Just as a guillotined head can no longer say what it is like to be just a head, so a captain suffering from motoric aphasia following a stroke can no longer report the "most immortal experiments" carried out on him. Where else in the nineteenth century are the fearful perspectives linking the brain, the mind, and vivisection more clearly and pithily expressed than in these few sentences from *Woyzeck* and *Danton's Death*?

The inspiration through Gall's cerebral anthropology

The second level of observation concerns the topographical approach to the place where thoughts arise. For many brain scientists in the 1830s, and hence also for Büchner, tracing thoughts, experiences, and feelings back to physiological processes in the brain seemed natural. This was due, above all, to the controversial and influential localization theory developed around 1800 by Franz Joseph Gall (1758–1828). Gall proceeded from the assumption that different intellectual, moral, and instinctive qualities had their seats and origins in distinct areas of the brain, primarily in the cerebral cortex. In line with bourgeois codes of conduct and Late Enlightenment conventions, he attributed the various human characteristics, predispositions, and talents to separate and independent organs of the brain. Accordingly, special musical or linguistic abilities were derived from a particularly well-developed region localized in the cerebral cortex. Even today, this idea of *big is beautiful* remains one of the key conceptual foundations of the neurosciences. Gall was a superb neuroanatomist: He was the first to systematically demonstrate the significance of the cerebral cortex for higher brain functions, and he also proved that the brain had evolved from the spinal cord. His most consequential and problematic contribution, however, was his psychologically motivated project of no longer regarding humans as metaphysical beings endowed with a soul but instead seeking to understand them in their everyday behavior. There was no place in this system for the hypothesis of an indivisible and free human soul; the aim was, instead, to explain the moral and intellectual nature of the human race on the basis of cerebral functions.

Gall's popularity in the early-nineteenth century was based on the part of his doctrine that linked the dents or bumps of the skull to the development of the cerebral cortex. This made possible the notorious correlation between the shape of a person's skull and his or her intellectual qualities (or defects). The implications of Gall's theory were too serious, however, to remain confined to sensation-seeking charlatanry. Critics of Gall's materialism and determinism made their voices heard within the natural sciences as well. In particular, he was reproached for having called into question free will, and hence individual autonomy, by localizing brotherly love, murderousness, linguistic ability, or musicality in the brain. These were questions that fascinated the young Büchner. Although Gall's organology had largely fallen out of fashion in 1830s Germany (unlike in the United Kingdom), it may safely be assumed that Büchner was familiar with its key tenets, at the very least.

Büchner directly referred Gall in *Danton's Death*, in a sentence in the first scene that he subsequently deleted. Danton says there about a man whose wife is apt to "lie on her back" and shows him her "*cœur*" but other men her "*carreau*": He "wears the resulting swellings on his forehead, he takes them for humor bumps and laughs about them" (Büchner 2000, 10). The drastic sexual connotations of this passage have often been noted by Büchner commentators (e.g., Grimm 1979; Milz 2008), but the passage also offers a biting commentary on the kind of naïve palpation of the skull that was much in vogue around 1800: The husband who fails to notice that his wife is cheating on him also takes bumps on the skull for well-developed brain organs. Why did Büchner reject this passage? Probably not because he had changed his mind about Gall's cranioscopy or felt embarrassed by its sexual coarseness. The more likely reason is that Büchner was worried about historical accuracy, as in 1794, the point in time when the play is set, "humor bumps" were not yet a part of scientific or popular discourse. Gall started to give public lectures in Vienna in 1796, but he would not present his organological ideas in a programmatic essay in the famous journal *Der Teutsche Merkur* until 1798 (Gall 1979). A generation later, Büchner wanted—like almost all neuroscientists of the 1830s—to have nothing more to do with the cranioscopic part of Gall's theory, although he was far from opposed to the idea that a person's mental qualities have their seat and cause in the brain, as another scene in *Danton's Death* shows.

After Danton and his comrades have been sentenced to death, a citizen addresses the vice-president of the revolutionary tribunal and supporter of Maximilien de Robespierre

(1758–1794), René-François Dumas (1753–1794), on the sentence. There follows the dialogue: "Dumas: Indeed, it's extraordinary. But the men of the revolution have an instinct that others lack, and this instinct never lets them down. Citizen: It's the instinct of the tiger" (Büchner 1992, 77). This tiger instinct, although it makes no direct reference to organology, has plausibly been interpreted as an implicit allusion to Gall's "carnivorousness" or "murderousness" (Oehler-Klein 1990, 302–09). The bloodthirstiness the citizen in this scene finds so horrifying in the Jacobins, and which he traces back to an animalistic instinct in man, has its direct counterpart in Gall. In his chapter on the "*Instinct carnassier; penchant au meurtre*" ("Predator Instinct; Predisposition to Murder"), he cites not only murderers and tyrants such as Nero but also the bloodbath unleashed by the protagonists of the French Revolution "in spite of education, morality, religion and the law" (Gall 1823, 97–98). Gall's materialism does not claim that humans necessarily *have* to murder, but it does assert that they *can* murder regardless of social background, acculturation, or biography, simply because their natural cerebral configuration predisposes them to do so. Although this tendency may be more pronounced in some than in others, it is strong enough to be capable of overwhelming all other mental qualities.

Büchner's closeness to Gall's cerebral anthropology is made apparent in another short passage from his revolutionary drama. Danton asks his lover Julie: "What is it in us that lies, whores, steals, and murders?" (Büchner 1992, 49), to which Gall would give a clear answer just a few years later: *It is the specific constellation of your brain organs.* Of course, alternative explanatory offers were available at the time, such as sinfulness, mental illness, or brutalization through environmental conditions. In the context of the passage just cited, Büchner did not refer to any of these possibilities, including the one proposed by Gall. Yet Gall's materialistic notion that *something in us* is responsible seems closest to Büchner. What this something was could not be known with any certainty, and it could not be brought under control, as Danton goes on to remark: "We are puppets and unknown powers pull the strings; we are nothing, nothing ourselves" (Büchner 1992, 49). For Büchner, these unknown forces resided in the human brain, and for him, as for Gall, this may have been the key factor behind his special interest in this organ's anatomy.

Büchner did not just place his question in the mouth of his protagonist, he had previously addressed it to his fiancée, Wilhelmine Jaeglé, in the so-called "fatalism letter" of January 1834:

> I find in human nature a terrible uniformity, in human affairs an inexorable force, which is granted to all and to none. The individual is only foam on the wave, greatness a mere accident, the majesty of genius a puppet play, a ludicrous struggle against an iron law, which to recognize is the highest achievement, but to master, impossible. … What is it in us that lies, murders, steals? (Büchner 1999, 377)

Büchner was no fatalistic pessimist, otherwise he could hardly have written the *Hessian Land Courier* two months after this letter, an incendiary manifesto for political freedom and justice (Dedner 2016). But this did not prevent him from examining human nature without any illusions. To that extent, at least, there are grounds for assuming that he oriented himself on Gall's drives and instincts, however simplistic he may have found phrenological explanations. Büchner sensed that the puppet master, the "iron law," was more complicated than the mere dominance of certain cortical organs. Nonetheless, the "tiger instinct" and the question of the internal sources for the drives point toward the cerebralization of humankind, partly driven by Gall's organology. Put differently: Without organology, Büchner could hardly have expressed himself *in this way*.

The idea that every thought is attached to a fiber in the brain goes well beyond organology's anthropological claim to explain human behavior. Even Gall, who sharply distinguished between white and gray matter in anatomical and functional respect, never went so far as to seek to identify the exact cortical correlate for every thought and feeling by peering inside the skull of a living human being. Although this can be seen as a utopian anticipation of early-twenty-first-century neuroscientific attempts to decode thoughts by measuring cerebral activity through digital imaging, it also has historical antecedents dating back to the eighteenth century (Hagner 2006). At that time, sensualism posited the existence of a brain fiber for each individual sensory impression. This view was propounded by the Genevan naturalist Charles Bonnet (1720–1793), who saw the brain as a conglomerate of numerous individual organs or fibers (Bonnet 1769, 18–27). If Bonnet had stopped at this form of sensualist localization, he would have been forced to postulate an organizing principle for spiritual capacities and qualities that would actually have organized nothing, as all human qualities would have been constituted from monadic sensory impressions. The diversification of the psyche into countless fibers stood in irreconcilable contrast to the principle of the unity of the self. Even Bonnet did not for a moment cast doubt on the immateriality and immortality of the soul, which he derived from the uniform sensation of mental processes (Bonnet 1769, 7). Bonnet's theory was occasionally taken up and debated by his contemporaries (Hagner 1997, 46–51, 300), but it was ultimately not pursued any further, as it was unclear to an ever-more-empirically minded neuroscience how any sensual impression, still less a thought, could be connected with a single brain fiber of unknown functioning. This was to remain a utopian prospect for Büchner as well, yet at the very time he was preoccupied with the brain in both literature and anatomy, nerve fibers were making their way into brain research.

Büchner, the anatomist: The harmony and beauty of the brain

In 1836, one year after the publication of *Danton's Death*, the Breslau anatomist Gabriel Gustav Valentin (1810–1883) published a pathbreaking treatise in which he derived the structure of the entire nervous system from cells and fibers. With these two anatomical structures, he thought he had found the prototype for the nervous system of all living beings (Valentin 1836). Valentin's breakthrough owed much to the spirit of Romantic *Naturphilosophie* and especially the ideas of Johann Wolfgang Goethe (Clarke and Jacyna 1987, 69–74), inasmuch as he rediscovered the Goethean concepts of the original idea (*Uridee*) and prototype (*Urtypus*) in the microscopic anatomy of the nervous system. Although there is nothing to suggest that Büchner was able to integrate Valentin's voluminous work into his dissertation, he would no doubt have welcomed its *naturphilosophisch* ambition to understand the brain's structure "as the manifestation of a primal law, a law of beauty that produces the highest and purest forms from the simplest plans and lines." These words could have been penned by Valentin; they are, in fact, by Büchner, who continued: "For it [the philosophical method], everything, form and matter, is bound to this law" (Büchner 2008b, 155).

These lines are taken from the *Probationary Lecture*, held in the winter of 1836. Büchner expressed himself on nerve fibers only as a poet, not as an anatomist, but in this lecture he developed—like Lorenz Oken, Valentin, and other researchers animated by the spirit of *Naturphilosophie*—an esthetic perspective on the brain that stands in sharp contrast to the extreme statements in *Danton's Death* and *Woyzeck*. He spoke of the "fairest and purest forms in humankind" and of the "perfection of the noblest organs in which the spirit seems almost to break through and stir behind the most delicate veils" (Büchner 2008b, 153). This kind of rhetoric was no doubt partly intended to win over his audience, especially Oken, for whose sake Büchner had come to Zurich. Yet these words are also fully compatible with a research program that, rather than beginning with the human brain, sets out on the long march from the simplest to the most complex nervous systems (Roth 2004). In agreement with other evolutionary and comparative anatomists, Büchner explicitly ruled out the top-down option: "Starting out with the most complicated form, that is, man, must always remain a futile undertaking" (Büchner 2008b, 159).

Such a sentence is understandable when one considers that, in 1836, the numerous folds and ridges of the brain's surface still struck doctors as an impenetrable tangle. It would take another two decades before several anatomists succeeded in bringing order into the chaos of the brain's furrowed surface. The anatomy of cerebral convolutions understood itself as an antiphrenological project, the definitive overcoming of Gall's messy cortical localization theory (Hagner 2004, 122–27). Yet even once the cortex had been classified anatomically, researchers were still far from understanding its function, and this held equally true of the far smaller, countless fibers and cells Valentin had identified as its primal substance. Only in literature could the equivalence of brain fibers and thoughts be postulated. When Büchner the anatomist spoke of the beauty and perfection of the human brain, he was referring to a largely baffling object, one that did not drive naturalists to resignation only because they assumed that all life forms shared a common blueprint. The claim that such a primal law determined the brain's material composition, form, and function implies an esthetic perspective on the brain, but not in the sense that it would have dispensed with the need for the dissecting knife and the magnifying glass (which Büchner used intensively). Rather, it is esthetic to the extent that attributes such as beauty and perfection were here ascribed to the natural regularity of the cerebral structure.

Development could be studied in two ways: by comparing the brains of fish, amphibians, birds, mammals, and humans; and by investigating the different stages of development in the embryonic brain. By this means, even the "noblest organ," the fully developed brain of an adult human, would eventually be made to reveal its secrets. Oken and Carl Gustav Carus (1789–1869), frequent reference points for Büchner in his *Mémoire*, showed in exemplary fashion how such an approach could be realized. Oken had attended Gall's public lectures in Göttingen in 1805, in which Gall demonstrated the methods of comparative anatomy and claimed that the brain is a *"véritable continuation"* (real continuation) of the spinal cord (Gall and Spurzheim 1809, 129). He also stated that the skull consists of three bones. Oken was fascinated. In his 1807 manifesto, *Über die Bedeutung der Schädelknochen* (*On the Meaning of Skull Bones*), he explained the entire human being from the vertebrae. His sketch of the vertebral theory of the skull began with an analogy: The brain was "a more voluminous spinal cord, developed into more powerful organs, the cranium a more voluminous spinal column" (Oken 1807, 5–6). The idea of an evolution from below to above, for which Gall had laid the ground, could be further refined by assuming that the skull consisted of three vertebrae and the brain of three separate regions, with corresponding cranial nerves belonging to each. These nerves would have to be traced back to their origins in painstakingly detailed empirical work, an undertaking Büchner tackled with great dedication in his dissertation on the nervous system of the barbel.

Oken's analogical idea was further developed by Carus. He divided the brain into three large regions from front to back: the cerebral hemispheres, the quadrigeminal region together with the medulla oblongata, and the cerebellum (Carus 1814, 117–21, 266, 287). Carus assigned these areas corresponding functions based on the developmental stage of the species in question (including humans). Carus's ambition went so far as to encompass all mental life (Hagner 2004, 76–93), resulting in a comprehensive theory of identity. He argued:

> [O]riginally, each of the three large cranial vertebrae corresponds to a brain mass, that each brain mass likewise originally has its parallel in one of the senses (hearing, sight, smell) as the elements of our mental development, and that consequently, depending on individual human differences, one of the senses will stand out more and this prominence will be externally reflected in the more pronounced formation of the cranial vertebra pertaining to it. (Carus 1828, 176)

Carus was an avowed opponent of Gall's organology, but the explanatory claim made by his theory was hardly less immodest.

Büchner was familiar with the passages from Oken and Carus. They formed the epistemic and experimental framework for his own investigations, as emerged from the summary given in his *Mémoire*:

> I believe I have proven that there are six pairs of elemental cerebral nerves, that these correspond to six vertebrae of the skull, and that the development of brain mass proceeds in accordance with its origin, from which it follows that the head is simply the result of a metamorphosis of the medulla and the vertebrae. (Büchner 2008a, 101)

When it came to phylogenesis, then, the young anatomist stood in the tradition of Romantic *Naturphilosophie*, understood here not as a license for fanciful speculation but as the matrix for an empirical anatomy. There was one crucial difference, however: Unlike Carus, Büchner did not posit an analogy between areas of the brain, sensual perceptions, and human mental life. Nothing of the sort was mentioned in his anatomical writings, and this suggests that, despite his indebtedness to the analogical thinking of *Naturphilosophie*, he was unprepared to pursue so comprehensive an anthropology as that envisaged by Gall or Carus. Büchner stuck to what he could observe and understand in the object he knew best: the brain of the barbel. The human brain represented a faraway goal that Büchner had no wish to anticipate in his dissertation or in his lecture. In this he deviated from the path of *Naturphilosophie* trodden by Carus, in particular. If any analogies were to be drawn between form and function, then only at the level of the more easily investigated brains of fish, not at the level of the human brain.

Conclusion

The contrast could hardly be greater. If the brain is described in Büchner's anatomical writing in the discourse of *Naturphilosophie* as an island of harmony and perfection, in his literary work it appears as a battlefield on which brutality and bestiality run rampant. In one case the brain is an aristocrat, in the other a pariah, or, to put it paradoxically: The fish's brain became nobler under the dissecting knife even as the human brain became more primitive under the gaze of the psychological observer. It would be too simple to seek the explanation in the speculative freedom that the poet Büchner, in contrast to the naturalist, could allow himself. In the first place, Büchner's neuroanatomy was not free from philosophical assumptions and, second, his neuropoetry also made reference to manifest historical events, theories, and practices that, ever since the eighteenth century, had been turning the brain into a scientific, political, and cultural object.

To that extent, it is more appropriate to arrive at the underlying primal law from the two opposed perspectives: What in the fish's brain appears as orderly, harmonious, and beautiful, appears in the human brain as disorderly, perplexing, and unpredictable. Büchner died too young to become an important brain researcher, yet he was one of the first to note that, notwithstanding the assumption of a primal law, brain studies cannot proceed from a single perspective. Like society, and like human beings, the brain is a contradictory phenomenon. This irritation, which is also not entirely unknown to today's neurosciences, is the subject of Büchner's literary and scientific forays into the brain.

Acknowledgments

This text is a revised and considerably extended version of an essay originally published in the catalogue of an exhibition marking the 200th anniversary of Georg Büchner's birth in 2013 (Hagner 2013). I wish to thank Burghard Dedner for his advice and Robert Savage for his impeccable translation of this text.

Disclosure statement

No potential conflict of interest was reported by the author.

References

Bonnet C. 1769. *La palingénésie philosophique, ou idées sur l'état passé et sur l'état futur des êtres vivans*. Vol 1. Geneva: Philibert & Chirol.
Borgards R. 2004. "Kopf ab". Die Zeichen und die Zeit des Schmerzes in einer medizinischen Debatte um 1800 und Brentanos Kasperl und Annerl. In *Romantische Wissenspoetik. Die Künste und die Wissenschaften um 1800*, ed. G. Brandstetter and G. Neumann, 123–50. Würzburg: Königshausen & Neumann.
Büchner E. 2013. *Versuchter Selbstmord durch Verschlucken von Stecknadeln*, ed. H. Boehncke and H. Sarkowicz. Berlin: Insel.
Büchner G. 1992. *Sämtliche Werke. vol. 1: Dichtungen*, ed. H. Poschmann. Frankfurt a. M.: Deutscher Klassiker Verlag.
Büchner G. 1999. *Sämtliche Werke. vol. 2: Schriften, Briefe, Dokumente*, ed. H. Poschmann. Frankfurt a. M.: Deutscher Klassiker Verlag.
Büchner G. 2000. Danton's Tod. In *Sämtliche Werke und Schriften (Marburger Ausgabe), vol. 3.1: Text*, ed. B. Dedner and T. M. Mayer. Darmstadt: Wissenschaftliche Buchgesellschaft.
Büchner G. 2008a. Mémoire sur le système nerveux du barbeau. In *Sämtliche Werke und Schriften (Marburger Ausgabe), vol. 8: Naturwissenschaftliche Schriften*, ed. B. Dedner and A. Lenné, 3–117. Darmstadt: Wissenschaftliche Buchgesellschaft.
Büchner G. 2008b. Probevorlesung. In *Sämtliche Werke und Schriften (Marburger Ausgabe), vol. 8: Naturwissenschaftliche Schriften*, ed. B. Dedner and A. Lenné, 153–69. Darmstadt: Wissenschaftliche Buchgesellschaft.

Carus C. G. 1814. *Versuch einer Darstellung des Nervensystems und insbesondere des Gehirns nach ihrer Bedeutung, Entwicklung und Vollendung im thierischen Organismus*. Leipzig: Breitkopf & Härtel.

Carus C. G. 1828. *Von den Ur-Theilen des Knochen- und Schalengerüstes*. Leipzig: Fleischer.

Clarke E., and L. S. Jacyna. 1987. *Nineteenth century origins of neuroscientific concepts*. Berkeley: University of California Press.

Dedner B. 2016. Der "Fatalismusbrief" vom Januar 1834. *Georg Büchner | Portal*. http://buechner portal.de/aufsaetze/72-burghard-dedner-der-fatalismusbrief.

Gall F. J. 1823. *Sur les functions du cerveau et sur celles de chacune de ses parties*, vol. IV. Paris: Boucher et al.

Gall F. J. 1979. Des Herrn Dr. F. J. Gall Schreiben über seinen geendigten Prodromus über die Verrichtungen des Gehirns der Menschen und der Thiere, an Herrn Jos. Fr. von Retzer [1798]. In *Franz Joseph Gall. Naturforscher und Anthropologe*, ed. E. Lesky, 47–59. Bern: Huber.

Gall F. J., and J. C. Spurzheim. 1809. *Recherche sur le système nerveux en général et sur celui du cerveau en particulier*. Paris: Schoell & Nicolle.

Grimm R. 1979. Coeur und Carreau. Über die Liebe bei Georg Büchner. In *Georg Büchner I/II: Sonderband text und kritik*, 299–326. Munich: edition text + kritik.

Hagner M. 1997. *Homo cerebralis. Der Wandel vom Seelenorgan zum Gehirn*. Berlin: Berlin Verlag.

Hagner M. 2004. *Geniale Gehirne. Zur Geschichte der Elitegehirnforschung*. Göttingen: Wallstein.

Hagner M. 2006. Gedankenlesen, Gehirnspiegel, Neuroimaging. Einblick ins Gehirn oder in den Geist? In *Der Geist bei der Arbeit. Historische Untersuchungen zur Hirnforschung*, 223–45. Göttingen: Wallstein.

Hagner M. 2013. Georg Büchner – Anatom des tierischen Gehirns und des menschlichen Geistes. In *Georg Büchner – Revolutionär mit Feder und Skalpell*, ed. R. Beil and B. Dedner, 329–41. Ostfildern: Hatje Cantz.

Jordanova L. 1989. Medical mediations. Mind, body and the guillotine. *History Workshop Journal* 28:39–52. doi:10.1093/hwj/28.1.39.

Kurzke H. 2013. *Georg Büchner. Geschichte eines Genies*. Munich: Beck.

Maupertuis P. L. M. de. 1768. Lettre sur le progrès des sciences. In *Œuvres*, Vol. 2. Lyon: Bruyset.

Milz C. 2008. Eros und Gewalt in Danton's Tod. *Georg-Büchner-Jahrbuch* 11 (2005–2008):25–37.

Oehler-Klein S. 1990. *Die Schädellehre Franz Joseph Galls in Literatur und Kritik des 19. Jahrhunderts*. Stuttgart/New York: Gustav Fischer.

Oken L. 1807. *Über die Bedeutung der Schädelknochen*. Jena: Göpferdt.

Richardson R. 1989. *Death, Dissection and Destitude*. London: Penguin Books.

Roth U. 2004. *Büchners naturwissenschaftliche Schriften. Ein Beitrag zur Geschichte der Wissenschaften vom Lebendigen in der ersten Hälfte des 19. Jahrhunderts*. Tübingen: Niemeyer.

Valentin G. G. (1836). Über den Verlauf und die letzten Enden der Nerven. *Nova Acta Physicomedica Academiae Caesareae Leopoldino-Carolinae Naturae Curiosorum* 18 (1):51–240.

Phrenology as clinical neuroscience: how American academic physicians in the 1820s and 1830s used phrenological theory to understand neurological symptoms

Frank R. Freemon

In the early 19th century, doctors searched for a theoretical system to use as an explanatory framework for the baffling symptoms of brain diseases. One system briefly popular in medical science was the doctrine of phrenology, and this paper examines that doctrine in the United States. It was introduced as a fully developed theory in 1822, was popular among leading American physicians for just over a decade, and faded rapidly from medical interest by the 1840s. The American situation provides an excellent case study because the doctrine was introduced as a fully formed system at a specific date, was never considered in opposition to religion, as it was in continental Europe, and was not complicated by a rift between rising and established generations of medical scientists, as in Great Britain (Parssinen, 1974; Cooter, 1984).

The doctrine of phrenology held that human personality and behavior were resident in, or generated by, the human brain, that certain portions of the brain corresponded to different aspects of the human personality, and that the size of the specific brain region correlated with the strength of the associated mental faculty. The doctrine originated with Franz Joseph Gall (1758–1828), a medical doctor and anatomist of Vienna. Gall later reported that he first obtained a glimpse of the ideas that became the doctrine of phrenology when, as a child, he observed a playmate who had a prodigious memory and bulging eyes. When later taught in medical school that the brain was responsible for memory, Gall made a three-way correlation: (1) the portion of the brain responsible for memory was located just behind the eyes; (2) when a person's memory became highly developed, this portion of the brain enlarged, and (3) the shape of the skull was affected by the enlarged portion of the brain in such a way that the eyes were pushed forward. Gall generalized this correlation: prominent forms of behavior were produced by an enlarged portion of brain tissue which in turn changed the shape of the overlying skull (Temkin, 1947).

Gall first gained prominence as an anatomist, rather than as a theoretician correlating behavior with skull shape. Medical students and practitioners in Vienna were amazed at Gall's anatomical dissection technique, which was quite different from earlier and contemporary (as well as modern) methods. He traced functional pathways as far as possible; for example, he cut along the optic nerve and pathway to show the diffusion of the anatomical substrate for the sense of vision into the brain substance.

Gall was the first anatomist to consider that the cerebral convolutions might have functional significance. Their anatomical structure might really mean something and not just represent haphazard twists like a plate of spaghetti. When considering the full series of anatomical and psychological ideas that came to be known as phrenology, the doctrine of the brain (relating psychological faculties to the size of brain regions) must be carefully separated from the doctrine of the skull (relating the shape of the skull to these hypothetical brain enlargements). Many people who considered the latter absurd could accept a hypothesis relating mental power to the size of localized brain regions.

Because the Austrian authorities declared that Gall's ideas were contrary to revealed religion, and feared the philosophical ramifications of the idea that the brain might produce thought and voluntary behavior, Gall moved from Vienna to Paris in 1805. Gall's lectures in Paris were popular, partly because many were impressed by his new brain dissection techniques. He was joined by a young man who became his primary disciple, Johann Gaspar Spurzheim (1776–1832).[1]

Three individuals who heard Gall lecture while visiting Paris introduced phrenology into the United States, all returning from Europe in 1822. One of these physicians was John C. Warren (1778–1856) of Boston, who was undertaking a major research program comparing behavior and brain anatomy in a wide range of species. For example, the dog had a much better ability to detect odors than did the human; correspondingly, a very large protuberance occurred at the point where the olfactory nerve entered the dog brain. Warren hoped that phrenology would guide him in discovering the correspondence between brain regions and forms of behavior. The other two physicians were John Bell (1796–1872) and Charles Caldwell (1772–1853) who were both prominent Philadelphia doctors, although Caldwell had just relocated to the new medical school in Lexington, Kentucky. In fact, he had been on a European trip to buy books for the school library when he attended Gall's lectures. These two physicians founded in 1822 the Central Phrenological Society in Philadelphia to study the new science (Riegel, 1933; Davies, 1955).

John Bell gave the opening lectures at the first two meetings of the Society. A graduate of the leading medical school of the era, the University of Pennsylvania, he had continued his medical education in Europe, and had attended Gall's brain dissections and phrenological lectures. For the members of the Central Phrenological Society, Bell summarized Gall's observations, including his correlation of bulging eyes with a prodigious memory. He noted that phrenology had been criticized in continental Eurupe because it located mental activity in the biological realm of brain tissue rather than in the spiritual realm of the mind and soul. Bell stated that all those who studied the brain, not just Gall and Spurzheim, claimed that science should one day be able to discover an anatomy of the mind. As an example of the value of the new doctrine in correlating brain anatomy with abnormal behavior, Bell pointed to the report of a British physician who palpated the skull of a woman who had killed her baby, found a depression over that portion of the brain that had been identified by Gall as 'the organ of philoprogenitiveness' (love of children) and hypothesized that the unfortunate woman had committed her crime because of a defective brain (Bell, 1822).

The Central Phrenological Society was a Philadelphia organization, and most of the city's leading physicians became at least nominal members. These included Nathaniel Chapman (1780–1853), the editor of the *American Journal of the Medical Sciences*, and Philip Syng Physick (1768–1837), the continent's leading surgeon, famous for the successful removal of a bladder stone from the Chief Justice of the Supreme Court. About three-quarters of the members of the new society were physicians who hoped to learn how to understand the

---

[1] In Britain, Spurzheim usually gave his middle name as Christoph; the best biography is Nahum Capen's introduction to the Boston edition of Spurzheim's *Phrenology. . .* (1833).

brain afflictions of their patients, but the other stated aim of the new society was to aid in developing new methods of education.

The annual report of the Central Phrenological Society in 1823 complained that there were many who could not understand the new doctrine and that the initial enthusiasm showed signs of waning, although it claimed that interest had been revived by the arrival from Edinburgh of several phrenological specimens, including the busts of famous people, living and dead. Members of the society hoped to deduce from the skull shape the psychological strengths and weaknesses of these famous persons. Medical interest in phrenology was stimulated by the presentation of patients at some meetings; for example, the phrenologists tried to determine by palpations of the head which faculties were defective in a mentally retarded girl.

One of the few addresses to the Central Phrenological Society to mention the potential value of phrenology as a source of medical treatment was given by John Redman Coxe (1773–1864), a well-known Philidelphia physician on the faculty of the University of Pennsylvania. He argued that the human mind was actually the soul, eternal and perfect. What appeared to be mental retardation or mental illness was really a damaged or malformed brain. Abnormal behavior occurred because the instrument of action of the perfect soul did not operate properly. Coxe claimed that phrenology could explain the details of this operation. From the specific forms of abnormal behavior, the physician-scientist could determine which phrenological organ, which part of the brain, was poorly developed. This brain tissue could be increased in size by mental exercises, just as muscle tissue is enlarged by physical exercise. And just as the larger muscle has greater strength, so the larger phrenological organ will increase mental power. Furthermore, just as the specific muscle only increases the strength of a particular limb, so the specific brain region will improve only the particular corresponding mental faculty; the enlarged biceps strengthens the arm and the enlarged phrenological organ behind the eye strengthens the memory.[2]

Despite lectures such as those given by Bell and Coxe, despite examinations of busts and of patients, interest in phrenology in Philadelphia continued to lapse and, by 1826, many meetings were canceled because of insufficient attendance and the Central Phrenological Society was formally terminated in 1827.[3]

Although Philadelphia was the center of phrenological activities in the early 1820s, a few physicians outside that city used phrenology in their teaching or practice. John C. Warren, professor at the Massachusetts Medical College in Boston, included phrenological concepts in his student teaching and looked to the doctrine to generate hypotheses for his research in comparative brain anatomy. Gall had stated that the faculty of courage was generated by a certain portion of the brain, so Warren examined the brain of a lion, but was disappointed to find that this portion was not enlarged.

The leading American phrenologist was Charles Caldwell, on the medical faculty of Transylvania University in Lexington, Kentucky. His initial attendance at Gall's lectures in Paris had been undertaken in a spirit of derision, but he was so enthralled by the dissections and by Gall's quiet manner that he became converted to phrenology, on which he lectured, not only in Kentucky, but throughout the United States. He wrote the first American textbook

---

[2] John Redman Coxe, 'On phrenology in connexion with the soul; and as to the existence of a soul in brutes,' 1823. Vanderbilt University Medical Library.
[3] Minutes of the Central Phrenological Society, 1822–1827. The Historical Society of Pennsylvania, Philadelphia.

on phrenology in 1824, and its second edition just three years later had grown from 100 to 279 pages. Caldwell's vigorous proselytizing of the new doctrine earned him the title of the American Spurzheim.

A British visitor, Frances Trollope, heard Dr Caldwell's lectures on phrenology in Cincinnati in 1828, and she reported that Caldwell understood phrenology well enough, 'but neither his lectures nor his conversation had that delightful truth of genuine enthusiasm, which makes listening to Dr Spurzheim so great a treat'. Following the lectures series, however, 20 or 30 of the most erudite citizens of Cincinnati formed a phrenological society. According to Mrs Trollope, the second meeting of the society only had about half as many people in attendance but they managed to pass enough resolutions 'to have filled three folios'. Dues were to be collected at the third meeting, but only the treasurer attended, 'and so expired the Phrenological Society of Cincinnati' (Trollope, 1949).

After Gall died in 1828, Johann Spurzheim became the leading phrenologist of the world. His visit to the United States in 1832 stimulated new interest. After a few days in New York, he traveled to Boston, where he gave two series of lectures, one a popular series for the general public and another more advanced series for medical practitioners, both of which were well received. John Warren attended the advanced lectures and reported his inability to use phrenology as a guide to comparative anatomy. Spurzheim responded that phrenology was never meant to compare the brains of two different species. 'Phrenologists cannot compare the same organ in different species of animals', he said, 'nor even in different individuals of the same species; but must judge of each animal or man individually' (Spurzheim, 1832).

Before the lecture series was completed, Spurzheim developed typhoid fever and died. The collection of skulls that he had used in his lectures, as well as his own skull, were appropriated by the leading physicians of the city who founded the Boston Phrenological Society. The course of the Boston group followed the enthusiasm and decline of the Central Phrenological Society of Philadelphia, and the Boston Phrenological Society disbanded in the late 1830s (Walsh, 1972).

Physicians writing in the leading medical journals of America used the principles of phrenology in case reports to try to understand certain difficult clinical situations. As summarized in Table 1, early American academic physicians considered that the three clinical conditions that might best be understood using the principles of phrenology were (1) hereditary color-blindness, (2) self-abatement of epileptic seizures, and (3) the acquired loss of speech.

Color-blindness baffled contempoary medical science. Certain people discovered by accident that they saw only as gray what other people could differentiate as red and green. These surprised individuals, who were always males, had always considered their perception as

Table 1. Problems explained by the use of the phrenological doctrine of the brain in the 1820s and 1830s by American physicians.

Clinical problem	Phrenological explanation	Modern explanation
Color-blindness	Defect of the cerebral organ for color vision	Hereditary absence of retinal pigments
Behavioral seizure abatement	Shift of function from an abnormal (epileptic) cerebral organ	No real explanation
Loss of language	Damage of the cerebral organ for language	Damage to the language area of the left hemisphere

completely normal. Careful medical study revealed that except for this singular inability to differentiate red from green, their perception and intellect were truly normal. Study of family members, however, revealed that some male relatives also suffered from this enigmatic condtion. In 1824, the *Philadelphia Journal of Medical and Physical Sciences* summarized the hypothesis of a British physician that this problem resulted from the hereditary underdevelopment of the cerebral organ responsible for the appreciation of color (Butter, 1824). Treatments aimed at increasing the size of the organ by exercise, by intense study of red and green objects, were unsuccessful. The American psychiatrist, Pliny Earle (1809–1892), discussing the color-blindness within his own family, still considered the phrenological hypothesis as late as the 1840s (Earle, 1844). Phrenology offered an intriguing explanation for this disorder, even though it afforded no potential for therapy. Although we know today that hereditary color-blindness is a disorder of retinal pigments, damage to a very localized portion of the brain can produce a defect of color vision. Historical reviews of this brain disorder fail to recognize that phrenology had predicted its occurrence (Damasio et al., 1980; Zeki, 1990).

A second area where leading American physicians tried to use phrenology involved the complicated clinical situation now called self-abatement of epileptic seizures. A case report in the *American Journal ofMedical Sciences* described a young man who had suffered a skull fracture and subsequent frequent seizures. The patient had noted that his seizures were most likely to occur whenever he was making plans or arrangements; he learned to avoid a seizure by quickly shifting his mind to another subject whenever he felt an attack coming on. 'The phrenological physiologist explains this,' the author hypothesized, 'by the circumstances that causality, constructiveness, and ideality are in an injured state.' The cerebral organs that corresponded to these faculties were just below the site of the fracture.

When the patient was involved in planning, blood flow increased to these injured organs, triggering off epileptic discharges. 'But if the sufferer can turn his mind to other objects, that is, excite other faculties, the blood immediately takes its direction to the organs of these faculties, and is thus diverted from its former course' and the seizure is avoided (Epps, 1829).

Modern descriptions of the ability of some patients to prevent a seizure by some mental or physical activity avoid any attempt to explain the phenomenon in neurophysiological terms (Pritchard et al., 1985). Some modern neurologists have used special techniques to teach some patients how to avoid seizures; psychological interpretations are extensive but brain physiology is not mentioned (Forster, 1977).

A disorder of speech was the third symptom complex that stimulated early American physicians to invoke phrenology. This syndrome, which they called an amnesia for words, was most easily understood by assuming that the patients had suffered damage only to a small portion of the brain, that portion that was the phrenological organ of language. Three American reports in the late 1820s and early 1830s used phrenology to understand the clinical picture that today would be called aphasia.

In 1828, Samuel Jackson (1787–1872), a leading member of the medical faculty of the University of Pennsylvania, reported a patient who had lost the ability to speak (Jackson, 1828). 'I found my patient in bed,' he wrote, 'evidently in full possession of his senses, but incapable of uttering a word.' In trying to understand what had happened to this unfortunate person, Jackson first considered that the patient might have suffered paralysis of the muscles that generated speech. However, Jackson's examination showed that the tongue and other mouth muscles moved normally. Furthermore, the disorder involved all aspects of language, not just speech. 'When furnished with pen and paper,' Jackson described, 'he attempted to convey his meaning, but I saw that he could not recall the words and that he had written an unintelligible phrase.' The phrase written by the patient was: 'Diddoes doe the doe.'

Jackson utilized phrenology in his attempt to understand this patient's condition, and hypothesized that there was an interruption of blood supply to the brain, producing loss of language, 'the faculty of conveying ideas by words.' Since brain functions other than language were normal, including other intellectual capacities and the abilities to generate movements or appreciate sensations, Jackson concluded that the patient had not suffered generalized brain damage. Rather, the patient had experienced damage to only a small portion of the brain, a region that included the phrenological organ for the faculty of language.

Two years later, S. Henry Dickson (1798–1872) of the Medical College of South Carolina in Charleston reported a similar patient (Dickson, 1830). A 55-year-old man had a sudden onset of inability to speak most words. He could copy words but not write or speak them. Dickson offered the following hypothesis 'the *perception* [his italics, FRF] of objects seems to be irregularly defective,' not from any disorder of the sense organs or their nerves, 'but from something peculiar in the condition of the brain.' While Dickson did not invoke the doctrine of phrenology in his hypothesis, the editor of the *American Journal of Medical Sciences,* Nathanial Chapman, compared Dickson's patient to a patient reported in the British *Phrenological Journal* and suggested that the symptoms shown by Dickson's patient could only be understood by an analysis of phrenological principles (Chap-man, 1835).

Daniel Drake (1785–1852), the famous physician who wrote the major treatise on climate and disease in the upper Mississippi Valley, encountered a patient with loss of the ability to use language (Drake, 1834). Drake, like Jackson and Chapman, used phrenology to try to understand the patient's symptoms. The patient was a 45-year-old man with a complex problem involving a wound behind the left ear that may have become infected. He developed great difficulty in speaking or in writing nouns, especially proper names. Drake first thought that the disorder of speech might have been related to the paralysis of the facial muscles that follow damage to the seventh cranial nerve, 'which, as Mr. Bell thinks, regulates the function of articulation'. But the patient made errors in writing similar to his errors of speech, proving to Drake that the patient suffered a disorder of the language functions of the brain, not of the nervous control of the muscles of speech. As additional evidence for a brain disorder, Drake claimed that the patient experienced pain in his skull, overlying that portion of the brain 'which the phrenologist regards as the organ of language'.

In these three cases, as summarized in Table 2, leading American physicians looked to the doctrine of phrenology for an explanatory frame-work to understand how one brain function could be so seriously impaired while others remained unaffected. In none of these three reports did the authors accept the entire list of faculties or their precise locations as spelled

Table 2. Leading American physicians who used phrenology to explain a patient's inabiilty to use words.

Author	Date	Comment
Samuel Jackson	1828	Patient cannot speak and writes only: 'diddoes doe the doe'
S. Henry Dickson	1830	The journal editor, Nathaniel Chapman, suggests phrenology
Daniel Drake	1834	Compares the phrenological explanation to Bell's law (motor function of the seventh cranial nerve)

out by Spurzheim. Only Drake referred to that aspect of phrenology that involved skull shape. The authors never brought up the problem of hemisphere specialization; phrenological doctrine held that every cerebral organ existed in both cerebral hemispheres. In none of these American reports did the author mention if he thought that damage had occurred to both phrenological organs (either simultaneously or with the symptoms occurring only after the second lesion) or whether the loss of language occurred after damage to just one lesion. If they had ever hypothesized the latter, they might have asked, which hemisphere?

Samuel Jackson remembered the speech problem of his patient when he wrote his influential textbook of medicine. He tentatively accepted phrenology as the most likely explanation of brain function, but he thought existing phrenological maps were only a guide for further research. 'The brain is the instrument or organ of the intellect,' wrote Jackson, 'and it is more than probable, that the doctrine of Gall is correct, that assigns a particular organ of the brain to each faculty. The seat of these faculties is difficult to assign with precision, but it can be affirmed beyond a doubt, that the nobler and higher faculties are located in the anterior and superior parts of the brain.' (Jackson, 1832, p 32).

During this entire period of enthusiasm, many physicians had opposed phrenology. As early as 1824, John P. Harrison (1796–1849) of the University of Louisville had published a detailed critique. He argued against the division of the mind into separate categories, claiming that through introspection each of his readers could experience but a single stream of consciousness. Even if one accepted the existence of separate psychological faculties, their relationship to specific brain regions was problematical. Phrenological brain anatomy was intrinsically flawed, said Harrison, because no clear border delineated one phrenological organ from its neighbor (with one exception, the cerebellum). If the borders of the purported organs on the brain surface could not be rigorously specified, then precise measurements could not be made and correlations of organ size with psychological defects or abilities were not possible (Harrison, 1825).

The most detailed critique of phrenology was published by Thomas Sewall (1786–1845) of Columbian Medical College in Washington. He argued along the same lines as Harrison with additional special criticism concerning the most esoteric aspect of phrenology, the bumps on the skull. Sewall showed that there were variations in the thickness of the skull, variations from one individual specimen to another and from one location to another in any one person. Harrison performed measurements on human skull specimens, showing that one could not even predict the size of the brain from the volume of the intracranial space. This meant, therefore, that cranial measurements in the living subject could not determine overall brain volume or, even more obviously, could not calculate the size of supposed phrenological organs on the brain surface (Sewall, 1839).

By the late 1830s, the leaders of American medicine stopped using phrenological doctrine. Students and outlying practitioners occasionally mentioned phrenology. Four student theses at the University of Pennsylvania in the late 1830s were entitled 'Phrenology' or 'Craniosco-py'.[4] An obscure physician in rural New York reported a patient who had survived a gunshot wound to the head. The point of interest was the survival, but the doctor stated that the

---

[4] University of Pennsylvania medical theses:
    1835, F.M. Hamilton, 'Phrenology'.
    1837, J.M. Minor, 'Phrenology'.
    1839, T.D. English, 'Phrenology'.
    1841, E.J. Bee, 'Cranioscopy'

patient was mentally normal except, perhaps, for dysfunction in the phrenological organs in the path of the bullet. The editor of the journal in which the report was published asked for more detailed information concerning this point but none was forthcoming (Janson, 1840). A rural practitioner, Dr William B. Fahnestock, examined the skull of an executed murderer after the scalp had been removed, but performed no examination of the brain surface. By placing a candle within the skull, to determine the portions of the skull which transmitted the most light, he found, of course, that the phrenological organ of destructiveness had been enlarged (Fahnestock, 1840).

In the year 1838, two events seemed to promise a rejuvenation of American phrenology, but actually symbolized its demise as a science. The first of these events was the American tour of George Combe (1788–1858), a leading phrenologist from Scotland, who was not a physician, but a leading philosophical thinker. His book *The constitution of Man* was one of the major intellectual influences of the era. His brother Andrew (1797–1847) was a physician who had tried to use phrenology in clinical practice. Both had known Spurzheim when he lived in Edinburgh (Grant, 1965; Walsh, 1971).

During his visit, George Combe was treated as a leading philosopher, but his phrenology was ignored by leading medical practitioners. John Warren entertained Combe in his home but did not attend his phrenological lectures in Boston. Warren had given up on phrenology as an aid to his comparative neuroanatomy research and to his clinical practice. In Philadelphia, Combe met the anatomist Samuel George Morton (1799—1851), and added a phrenologic chapter to Morton's major work on the size and shape of human skulls, but Morton himself never discussed phrenology (Gould, 1978).

The second major phrenological event of 1838 was the publication of the *American Journal of Phrenology* by the Fowler brothers. The new periodical aimed initially at a scientific audience but within a year had become a magazine for popular consumption. The *Boston Medical and Surgical Journal* gave the first issue of the new phrenological publication a good review and wished the Fowlers success. After the appearance of a few issues, however, the *Boston Medical and SurgicalJournal* expressed the editorial opinion that the new phrenological magazine was so worthless and boring that it 'could put the inhabitants of a whole township to sleep' (Smith, 1840).

After 1840, phrenology in the United States was transformed into popular entertainment. It lost all scientific aspects and became a cross between a character reading and a humorous stage show (Stern, 1971). The new phrenologist had nothing to do with science or medicine; he had the appearance of the conman or entertainer who would define the personality of someone from the audience, 'a person completely unknown to me'. The entertainment level of the phrenology show grew with the development of phreno-mesmerism. A local was hypnotized and made to go through a series of antics when various organs were touched by the magnetic finger of the phreno-mesmerist. The dour librarian, for example, burst into boisterous laughter when the phrenological organ of mirth was touched (Fuller, 1982).

The involvement of the leading medical authorities in phrenology was gently forgotten. When Warren, Bell, Caldwell, and others died, their long obituaries did not mention phrenology. The leading opponent of phrenology, John Harrison, was scheduled to become the president of the American Medical Associaton but died before he could assume office. His long obituaries lamented his unexpected death and listed his many accomplishments; his successful opposition to phrenology, however, was not mentioned (Anonymous, 1850). When Charles Caldwell, the American Spurzheim, wrote his own autobiography, he listed Gall among the greats of the past who had been misunderstood, but the only appearance of the word phrenology was in the appended list of CaldweU's publications (Caldwell, 1855).

After phrenology had disappeared from most areas of medicine, it was still occasionally utilized in psychiatric explanation. Writing in the initial volume of the *AmericanJournal of Insanity*, published in 1844, Nathan S. Davis thought that 'some of the principles of phrenology' might be important in understanding insanity. He called for detailed post-mortem studies of the brain, with the attempt to link pathological changes in specific brain regions with specific forms of insanity. He did not mention the skull (Davis, 1844). Horace Buttolph, the superintendent of the State Lunatic Asylum at Trenton, New Jersey, thought that phrenology was 'the true science of the mind' (Buttolph, 1849). Lunacy could develop when one idea kept recurring over and over in the mind, thereby influencing the development of one phrenological organ to the detriment of others. Treatment involved individual effort by the patient to 'repress the over-strong'. Buttolph also did not mention anything about the skull (Carlson, 1959).

In the 1850s, the doctrine of phrenology was occasionally mentioned in American textbooks of medical physiology, but only to point out its weaknesses. John William Draper (1811–1882), professor of chemistry and physiology of the University of New York and one of the pioneers in the development of photography, mentioned phrenology in his textbook of physiology. He accepted that Gall and Spurzheim had been correct when they rejected the previous view that 'the brain acts as a unit'. But they had been incorrect in attributing faculties to specific regions. Draper spent a great deal of effort in examining the phrenological organ of amativeness, located in the region of the cerebellum. This was anatomically the most clearly defined phrenological organ because the cerebellum is clearly differentiated from surrounding portions of the brain. Draper pointed out that experiments in animals involving castration caused a decrease in amativeness but no change in the size of the cerebellum. Humans who had suffered brain damage that at autopsy was found to involve the cerebellum had experienced no decrease in amativeness. For these reasons, Draper rejected the doctrine of phrenology (Draper, 1856).

In the editions of his physiology textbook published in the 1830s, Robley Dunglison (1798–1869) of the University of Pennsylvania devoted 37 pages to phrenology but concluded that 'the views of Gall are by no means established'. In the editions of the same text published in the 1850s, Dunglison devoted less space to phrenology. He thought that scientific phrenology had suffered from the antics of popularizers. 'Indiscriminate divination from measurement of heads has been a sad detriment to phrenology as a branch of physiological science' (Dunglison, 1838, 1856).

In summary, phrenology was imported as a fully developed doctrine into the United States in 1822. Medical practitioners tried to use phrenology to understand their patients' clinical situations. When phrenology seemed to explain some symptoms, such as speechlessness, it was useful, but the doctrine was frozen in place and could not change. When it no longer seemed of any explanatory value, it was quietly dropped. In the United States, unlike in continental Europe, phrenology never faced any significant opposition because it attributed mental attributes to the activity of material, physical causes. Some historians have presented evidence that, in Britain, phrenology was used by younger medical practitioners to obtain authority and was opposed by established practitioners; such a situation did not occur in America (Cantor & Shapin, 1975).

Phrenology provided an explanatory framework for physicians to understand their patients' neurological and psychiatric symptoms, but did not lead to any significant treatment modalities. Phrenology declined when it no longer supplied a rationale for understanding brain function. Those individuals who had been the strongest proponents of the doctrine never confessed error, but simply stopped using phrenology.

Acknowledgements

This paper was presented at the Historical Conference on Brain Functions, Fort Myers, Florida, 4 January 1991. The author thanks Lynn Joy and Evan Melhado for their comments on earlier versions.

References

Anonymous (1850): Obituary. John P. Harrison. Am J Med Sci 19, 277–278.

Bell J (1822): On phrenology, or the study of the intellectual and moral nature of man. Philadelphia J Med Phys Sciences 4, 72–113.

Butter D (1824): Remarks on the faculty of perceiving colors. Philadelphia J Med Phys Sci 8, 198–199.

Buttolph HA (1849): The relation between phrenology and insanity. Am J Insanity 6, 127–136.

Caldwell C (1855): Autobiography. Philadelphia: J.B. Lippincott.

Cantor GN and Shapin S (1975): Phrenology in early nineteenth-century Edinburgh: an historigraphical discussion. Ann Sci 32, 195–218.

Carlson ET (1959): The influence of phrenology on early American psychiatric thought. Am J Psychiat 115, 535–538.

Chapman N (1835): Note to Grattan. Am J Med Sci 17, 467–469.

Cooter JR (1984): The cultural meaning of popular science: phrenology and the organization of consent in nineteenth-century Britain. Cambridge: Cambridge University Press.

Damasio A, Yamada T, Damasio H, Corbett J, McKee J (1980): Central achromatopsia: behavioral, anatomic, and physiologic aspects. Neurology 30, 1064–1071.

Davies JD (1955): Phrenology, fad and science: a nineteenth-century American crusade. New Haven: Yale University Press.

Davis NS (1844): Importance of a correct physiology of the brain. Am J Insanity 1, 235–243.

Dickson H (1830): Case of amnesia. Am J Med Sci 7, 359–360.

Drake D (1834): Case of partial amnesia in which the memory for proper names was lost. Am J Med Sci 15, 551–553.

Draper JW (1856): Human physiology, statical and dynamical; or, the conditions and course of the life of man. New York: Harper Brothers.

Dunglison R (1838): Human physiology, 3rd ed, Philadelphia: Blanchard and Lea. [8th ed, 1856.]

Earle P (1844): Memoirs. Boston: Damreil and Upham.

Epps J (1829): Affectation of the mind from injury of the brain (abstract). Am J Med Sci 5, 202.

Fahnestock WB (1840): Report on a series of experiments made by the medical faculty of Lancaster upon the body of Henry Cobler Moselmann, executed in the jail yard of Lancaster County, Pa, on the 20th of December, 1839. Am J Med Sci 26, 13–34.

Forster FM (1977): Reflex epilepsy, behavioral therapy, and conditional reflexes. Springfield, Illinois: Charles C. Thomas.

Fuller RC (1982): Mesmerism and the American cure of souls. Philadelphia: University of Pennsylvania Press.

Gould SJ (1978): Morton's ranking of races by cranial capacity. Science 200, 503–509.

Grant CA (1965): Combe on phrenology and free will. J Hist Ideas 26, 141–147.

Harrison JP (1825): Observations on Gall and Spurzheim's theory. Philadelphia J Med Phys Sci 11, 233–249.

Jackson S (1828): Case of amnesia. Am J Med Sci 3, 272–274.

Jackson S (1832): Principles of medicine. Philadelphia: Carey and Lea.

Janson H (1840): Case of gun-shot wound of the head and brain: recovery. Am J Med Sci 26, 248–249.

Parssinen TM (1974): Popular science and society: the phrenology movement in early Victorian Britain. J Social Hist 8, 1–20.
Pritchard PB III, Holmstrom VL, Giacinto J (1985): Self-abatement of complex partial seizures. Ann Neurol 18, 265–267.
Riegel RE (1933): The introduction of phrenology to the United States. Am Hist Rev 39, 73–78.
Sewall T (1839): An examination of phrenology in two lectures delivered to the students of Columbian College, District of Columbia. Boston: D.S. King.
Smith JVC (1840): An American phrenological journal. Boston Med Surg J 21, 245.
Spurzheim JG (1832): Outlines of phrenology. Boston: Marsh, Capen, and Lyon.
Spurzheim JG (1833): Phrenology in connexion with the study of physiognomy. Boston: March, Capen, and Lyon.
Stern MB (1971): Heads and headlines: the phrenological Fowlers. Norman: University of Oklahoma Press.
Temkin O (1947): Gall and the phrenological movement. Bull Hist Med 21, 275–321.
Trollope F (1949): Domestic manners of the Americans. New York: Alfred A. Knofp.
Walsh AA (1971): George Combe: a portrait of a heretofore generally unknown behaviorist. J Hist Behav Sci 7, 269–278.
Walsh AA (1972): The American tour of Dr Spurzheim. J Hist Med Allied Sci 27, 187–205.
Zeki S (1990): A century of cerebral achromatopsia. Brain 113, 1721–1777.

The promotion of phrenology in New South Wales, 1830–1850, at the Sydney Mechanics School of Arts

Catherine E. Storey

ABSTRACT

Sydney, New South Wales (NSW), Australia, began as a penal colony in 1788. British phrenologists would later show an intense interest in this new settlement, aroused by questions raised by convict transportation and indigenous assimilation into European culture. A more sinister engagement involved the scientific trafficking of Aboriginal skulls. This practice was seen, however, not as body snatching but as a meaningful contribution to the progress of science. In 1833, a group of educated, influential men formed the Sydney Mechanics School of Arts (SMSA). This organization was successful where previously learned societies had failed. These men aimed to see the diffusion of scientific and useful knowledge throughout the colony and to enhance the lot of the working man (mechanics). They planned to achieve this aim with lectures, demonstration classes, and the development of a library and museum. Phrenology fitted perfectly into their curriculum. From 1838 to the late 1840s, many of Sydney Town's prominent medical practitioners and other professionals delivered lectures promoting this "science." However, interest in the study of phrenology at the SMSA waned from the 1850s, when itinerant phrenologists turned the practice into a popular entertainment.

When Britain lost the American Colonies in the War of Independence, the British government also lost a convenient repository for its convict population. As the country's gaols filled and makeshift prison hulks overflowed, Thomas Townshend (1733–1800), First Viscount Sydney and Home Secretary in the Pitt Government, found the perfect solution—the eastern coast of "Terra Australis," charted by Captain James Cook (1728–1779) in 1770, where no European settlement had yet settled.

On May 13, 1787, Captain Arthur Philip (1738–1814) of the Royal Navy set sail from Plymouth in command of the First Fleet. Eleven ships carried more 1,500 persons, around half of whom were convicts, on a journey that would last eight months, to the other side of the globe. On January 26, 1788, Phillip landed his human cargo on the shores of a small inlet he named Sydney Cove, where he assumed the position as first governor of the Colony of New South Wales (NSW; Flannery 1999).

Free settlers began to arrive in the Colony from 1792. Although convict transportation did not cease in NSW until 1840, there was a steady increase in immigration of free settlers, and by 1820, the population had risen to 12,000. By this time, there was an established middle class consisting of government administrators, doctors, lawyers, churchmen, and teachers, all ready to improve their intellect and that of the community. The community was a mix of convicts, emancipists, free settlers, and the original indigenous population, now significantly reduced in numbers by disease and dispossession.

Flannery (1999, 5) has suggested that Europe at the time was extremely interested in the questions raised by this transplantation. "Could transportation redeem socially degraded felons? ... Could the Aborigines be brought into the European fold?" These were the questions that deeply engaged the early British phrenologists.

British phrenologists' acquisition of Aboriginal skulls

Private and institutional collectors were keen to acquire Aboriginal skulls. Sir Joseph Banks (1743–1820)—naturalist, botanist, and later president of the Royal Society—accompanied Captain James Cook on his first scientific voyage of discovery 1768–1771. It was during this voyage that Cook charted the eastern coast of Australia and provided Banks with the opportunity not only to examine the great diversity of new plant life but also to procure the skulls of these newly discovered Aborigines for European anatomists (Turnbull 2017).

With the rise in popularity of phrenology in Great Britain following the visit of Johann Spurzheim (1776–1832) to Edinburgh in 1816, British phrenologists were now in competition with the anatomists for these skulls.

After colonization, British phrenologists began to acquire these skulls from colonials. These men either plundered indigenous burial places or removed the heads from Aborigines killed in skirmishes with the military detachments sent to quell riots between traditional owners and new settlers (Turnbull 2017).

In 1820, Sir George Steuart Mackenzie (1780–1848)—Scottish phrenologist, a Spurzheim convert, and one of the founders of the Edinburgh Phrenology Society—published his assessments and deliberations on his collection in *Illustrations of Phrenology with Engravings*. In this work, Mackenzie examined the skull of Carnimbeigle, a "New Holland Chief" (one of several Aboriginal skulls in his collection). Mackenzie disclosed that this was

> the skull of Carnimbeigle, a chief of New South Wales, who was killed by a party of the 46th Regiment, in 1816. His skull is now in our possession, having been presented to us by Mr Hill, Surgeon, R.N. who received it from Lieutenant Parker of the 46th. (Mackenzie 1820, 233–34l; see Figure 1)

Mackenzie did concede from his observations that this chief possessed all the qualities of a leader—that is, confidence, courage, ambition, a strong sense of justice, and a talent for stratagem—but, he concluded,

> the progress of these people may be slow, and although their reasoning powers are not such as to lead us to think that their lower propensities can be under perfect controul (sic); still, by working on their love of approbation, the sense of justice, and veneration; and by exciting the organ of attachment by acts of kindness, much may be done for these miserable beings. (Mackenzie 1820, 235)

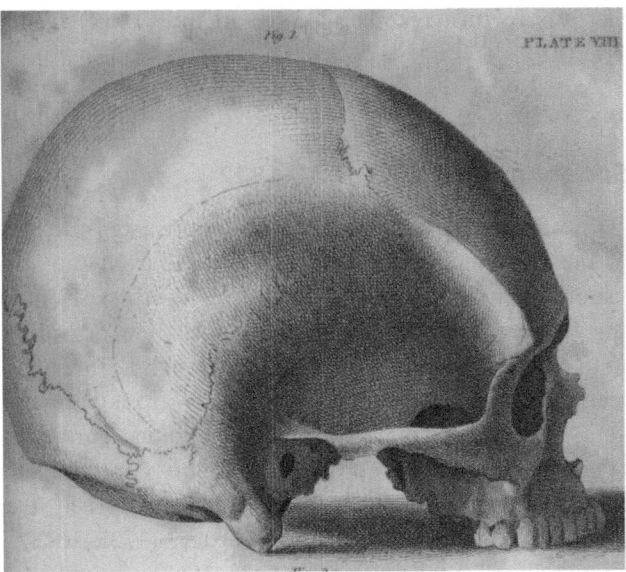

Figure 1. The skull of Carnimbeigle, New Holland Chief, from *Illustrations of Phrenology with Engravings* (1820).

A short report in the local Sydney press on July 7, 1826, confirmed that there was at least local knowledge of the practice of trafficking in human remains. "It is reported that Major M– took home a score or two of sculls (sic) of the Aborigines slaughtered in the late war" and later in the report confirmed that the phrenological societies of Europe were the likely recipients (Anon 1826, 2). Turnbull (2017) proposed that Major M–, is most likely to have been Major James Thomas Morriset (1780–1852) of the 48th Regiment, commander at Bathurst, NSW, in 1823. Morriset was known for his use of military force to quell resistance from the local Aboriginal people in defense of their lands (Parsons 1967). Turnbull made the case, however, that the number of skulls is likely to be an exaggeration.

In 1836, the French pathologist Francois Broussais (1772–1838) delivered a series of 20 lectures on phrenology at the University of Paris. All of the Broussais lectures were transcribed and republished in English in *The Lancet*. He considered that "the primitive state of man is a state of ignorance," a state found in the native of New Holland (Broussais 1836, 929). He also claimed that, due to a lack of written language and a less sophisticated range of spoken words, the "New Holland race has been left behind in the march of civilisation," and concluded "here is a specimen of their heads … he will never become civilised, because he lacks the cerebral organs necessary to become so" (Broussais 1836, 330).

Such analysis undoubtedly led to a damaging impression of the race, an opinion that lasted well into the twentieth century. Turnbull argued, however, that Mackenzie and George Combe (1788–1818), in their reviews of the indigenous skull, championed the development of "national phrenology," a branch of the science that aimed to achieve amicable assimilation. This advice was largely ignored by the Colonial authorities (Turnbull 2017, 157). There is no doubt, however, that the skulls sent from the colonies to the early European phrenologists were used to support the concept of racial inferiority, and that this belief was in turn readily accepted in the Colony from which the skulls originated (Poskett 2019).

Philosophical Society of Australia, 1821–1822

By 1820, the rising professional class of the Colony appreciated the need to improve the intellectual life of the community and to distance itself from the stigma of the penal colony. In 1821, some of these men formed the Colony's first scientific institution: the Philosophical Society of Australasia. They modeled their new organization on the Royal Society of London to encourage scientific endeavor, through lectures; the establishment of a lending library and museum; and, if the opportunity arose, support for the scientific expedition (Hoare 1974). The original group included Judge Barron Field (1786–1846), the first judge of the Australian Supreme Court; Dr Henry Grattan Douglas (1790–1865), an Irish medical practitioner and member of the Royal Irish Academy; Frederick Goulburn (1788–1837), colonial secretary; John Oxley, (1785–1828) surveyor general and inland explorer; Edward Wollstonecraft (1783–1882), successful businessman and cousin to Mary Wollstonecraft Shelley; John Bowman (1784–1846), principal colonial surgeon from 1823 to 1828; Patrick Hill, a surgeon-superintendent (and purveyor of the Carnimbeigle skull to Mackenzie); and Alexander Berry (1781–1873), a Scottish-born surgeon-turned-merchant and pastoralist. Major General Thomas Brisbane (1773–1860), governor of New South Wales from 1821 to 1825, accepted the position of patron of the society (Hoare 1974). These were all men of influence in the new Colony, and many were supporters of phrenology. Berry, a pastoralist in the region south of Sydney, and Oxley, during his explorations in search of pastoral land, both came into conflict with traditional owners of this land. Both Berry and Oxley were involved in the collection of skulls to pass on to British metropolitan institutions (Turnbull 2017).

Turnbull (2017) made the point, however, that men such as Berry who willingly collected these skulls saw themselves not as exploiters of human remains but as supporters of the advancement of science. Berry considered his sending of human skulls similar in kind to his donations of geological and natural specimens, which he dispatched to his old universities of St. Andrews and Edinburgh.

Barron Field, a judge of the Supreme Court, was an active contributor to the society. Early in 1822, he presented a lecture, *On the Aborigines of New Holland and Van Diemen's Land* (Field 1825). He concluded that Aborigines were not capable of embracing civilization. Reece, in a later analysis of this work, made the point that Field's contemporary assessment of the status of the Australian Aborigine was purely phrenologically based. He went on to conclude that, "Phrenology was certainly the most popular theory explaining Aboriginal inferiority" (Reece 1974, 87).

Despite all of the worthy intent and initial enthusiasm, the Philosophical Society ceased activity in just over a year. Although short-lived, Hoare (1974) made the point that this was a vital enterprise that raised the awareness of scientific endeavor and demonstrated what could potentially be achieved.

The Phrenological Society, 1828–1829

As early as 1825, a letter to the editor of the *Sydney Gazette* suggested the possibility of the establishment of a phrenological society

> Let me ask why should not we, in this rising capital, have a phrenological society, a philosophical society, and other learned bodies, commensurate with our importance. Allow me therefore, to propose … the establishment of a corporate body of phrenologists, and let them give their science from experimental induction, a grammatical form. It seems to me that it is a science which will become the algebra of mental philosophy. (Anon 1825, 3)

In 1826, many of the members of the original Philosophical Society, who had remained active in the promotion of intellectual activities, reformed to support a subscription library, and in this endeavor, Dr. William Bland (1789–1868) joined the group (Hoare 1974).

Bland had been a British naval surgeon when he killed a fellow officer in a duel. A military court convicted him of murder and sentenced him to seven years' in New South Wales. Bland arrived in the Colony in 1814 as a convicted felon, but was pardoned the following year. He went on to establish a successful private medical practice and a career as a state politician, and he was influential in the founding of many academic institutions in Sydney, including the Sydney Mechanics School of Arts and the University of Sydney (Cobley 1966).

The Australian Subscription Library flourished, and in June 1828, *The Sydney Gazette and NSW Advertiser* published this short notice:

> A great number of Gentlemen, most of whom are members of the Australian Library are on the eve of forming themselves into a body, to be called The Phrenological Society. About sixty members may be reckoned at present … it is conjectured that the number will be considerably augmented. (Anon 1828a, 2)

Within a few days, a correspondent who signed himself, "a phrenologist", wrote to the editor of *The Monitor* to announce the first meeting of the new Phrenological Society. Dr. Bland accepted the chair and delivered the first lecture, described as both elaborate and eloquent, on the nature of the science of phrenology and the advantages of this society. Reverend Dr. Dunmore Lang (1799–1878) accepted a position as committee members (Anon 1828b).

There was no further local news of the progress of the society in the local press until March 10, 1829 (Anon 1829a, 2). The *Sydney Gazette* published extracts from Bland's inaugural lecture in which he sought to rebuff any objections to the "science," calling out those who considered that "Phrenology, the branch of science we propose promoting, is itself a mere idle chimaera, engendered in the heated brains of certain German enthusiasts" most certainly "unfounded on reason" (Anon 1829a, 2).

The prospectus of the society dated June 11, 1828, accompanied the article (see Figure 2). This belated announcement triggered a prompt satirical reply from someone signed "Impromtu" to the editor

> To seek out skulls of every shape
> Bacon and Shakespeare, ass and ape,
> Phrenologists take pains;
> And in the search they're surely right
> For ne'er was system brought to light,
> So much in want of brains. (Anon 1829b, 2)

> June 11, 1828
> ## PROSPECTUS OF THE PROPOSED AUSTRALIAN PHRENOLOGICAL SOCIETY
> The objects of this Society are as follows:
> To give descriptive and explanatory lectures on the bones of the head, on the brain, and the mapping of the head, according to Gaul (sic) and Spurzheim's system.
> To collect casts of the heads of famous characters, of notorious characters of whatever description, and of the heads of different nations, or races of the human species.
> To form a museum illustrative of the comparative anatomy, particularly of the head in the brute creation.
> To collect specimens of malformations of the human head and form generally.
> In promotion of many of which objects, it is also proposed:
> To open correspondence with various Societies of a similar description both at home and abroad.
> To procure all the best works, both British and Foreign, on the subject of Phrenology.
> To criticise works on that subject; and finally
> To collect and preserve any other original papers on that subject, whether the productions of our own members, or of strangers

Figure 2. Prospectus of the proposed Australian Phrenological Society, June 11, 1828.

There was no further communication on the society's activities, and one can assume that the organization suffered the same fate as the Philosophical Society had in previous years (Hoare 1974).

From time to time, brief phrenologically based communications appeared in the local press, prompting one reader in 1830 to lament the space given to the promotion of phrenology. The author "Christian," a self-"determined enemy to Bumpologists and to Bumpimposition," considered "a science it never will be" and continued, that phrenology, "like our Australian diamond snake, is beautiful to the eye, but dangerous to meddle with. Its theory seems to be founded on the virtuous and wicked passions of man" (Anon 1830, 3).

Sydney Mechanics School of Arts (SMSA), 1833

The *Australian Dictionary of Biography* described the Scottish-born Reverend John Dunmore Lang as a Presbyterian clergyman, politician, educationalist, immigration organizer, historian, anthropologist, journalist, and gaol-bird (Baker 1967). Lang arrived in the Colony in 1823, where he became a powerful and at times controversial influence on Colonial society. In 1831, he returned to Scotland to secure both moral and financial support for his proposed school, the "Australian College," and recruit the skilled tradesmen (then known as mechanics) necessary for the build. Scotland was in the grip of a severe depression, and Lang had no difficulty in recruiting 52 emigrant mechanics to return to NSW. He also engaged several teachers, including the Reverend Henry Carmichael (1796–1862), a Presbyterian minister and passionate educator

(Nadel 1966). The group set sail on the *Stirling Castle*, which Lang charted for the return voyage. The journey lasted four months, and the group reached Sydney on October 13, 1831 (Anon 1831). During this long journey, Lang and Carmichael organized classes, five days per week, so as to "devote the time at sea to the moral and intellectual enlightenment of the mechanics" (Nadel 1957, 113). These mechanics were likely to be well acquainted with the Edinburgh School of Arts, established in 1821 (Wotherspoon 2013).

In February 1833, Sir Richard Bourke (1777–1855), then governor of NSW, wrote to Henry Carmichael, expressing a desire to speak with him about the possibility of establishing a mechanics' institute in Sydney (Wotherspoon 2013). On March 27, 1833, *The Sydney Monitor* reported the inaugural meeting of 200 men, "lovers of morals and intelligence," who met on March 22 to establish the Sydney Mechanics School of Arts (Anon 1833a, 2). Major Thomas Mitchell, surveyor-general, accepted the position as president, and Henry Carmichael became vice-president; but Dunmore Lang, who was present on the night, felt it unwise to take a place while embroiled in one of many public disputes.

On April 27, 1833, *The Sydney Gazette* announced that the SMSA, "has commenced its labours suspiciously" (Anon 1833b, 2). Carmichael delivered the introductory lecture (of two hours' duration), in which he outlined the circumstances that had led to its establishment, the means by which the institute would promote its aims and meet its objectives, and "to trace the bearing which the universal spread of knowledge is calculated to have upon the mental and moral relationship of social life" (Anon, 1833b, p2). The main aim of that institute was to facilitate "the diffusion of scientific and other useful knowledge as extensively as possible throughout the colony of New South Wales" (Nadel 1957, 116). A lending library, classes for mutual instruction, and the delivery of lectures on suitable topics of science and art would achieve these goals. The lecturers in these early years were some of the Colony's intellectually elite (Wotherspoon 2013).

In *The Colonist* of April 21, 1838, an article carried an account of the fifth annual report of the SMSA in which the report expressed the thanks of the Society to

> Dr Bland, for his handsome and very valuable donation of phrenological busts, no fewer than sixty in number, and we therefore hope, that with the aid of such materials for the elucidation of this interesting, and yet infant science, that Dr N, or some other member of the medical profession will come forward with a few lectures on the subject of phrenology. We consider it to be a science in some measure well founded, and one too, from which, when it comes to be properly understood, such practical good may be derived in regard to education and social intercourse. (Anon 1838a, 4)

The study of the "new science" of phrenology fit well with the aspirations of the institute. Here was a science that offered self-knowledge, an ability to identify one's failings and strengths, and (importantly) the means to allow the working man to reach his full potential. Bland's donation of the phrenological skulls appears to have renewed interest in phrenology, particularly at the SMSA. On November 23, 1837, *The Colonist* (a newspaper owned and edited by Dunmore Lang) recorded the success of lectures delivered in 1837, with a promise that next course of lectures at the SMSA would include phrenology (Anon 1837).

In 1838, Dr. Francis Lascelles Wallace (1811–1852), an Edinburgh-trained physician, began the first of many such lectures. *The Australian* reported the opening lecture on August 28, 1838, with Wallace, surrounded by "an imposing array of sculls(sic) of inhabitants of various nations, and casts of heads of persons familiar to the audience either by history or colonial notoriety." When it came to the examination of the Aboriginal skull, Wallace perpetuated the prevailing notion that these represented the "lowest in the scale of mortality and intellect" (Anon 1838b, 2). The report concluded, "the Mechanics Institute is flourishing to a degree that its most sanguine members could not have anticipated."

Interest was still keen in 1842, when a debate took place in Sydney over three nights. Robust, well-credentialed teams assembled for both sides of the argument to consider the question, "Is phrenology a true science?" When this mammoth session concluded, the affirmative won by an overwhelming majority (Anon 1842, 2). Interest in phrenology as a true science was still active.

Lectures on phrenology were part of the scientific programme offered at the SMSA until the late 1840s (see Figure 3). From 1846, there was a notable decline in such offerings, with the emergence of full-time itinerant public speakers who took phrenology to the community as a form of popular entertainment (Nadel 1957, 147).

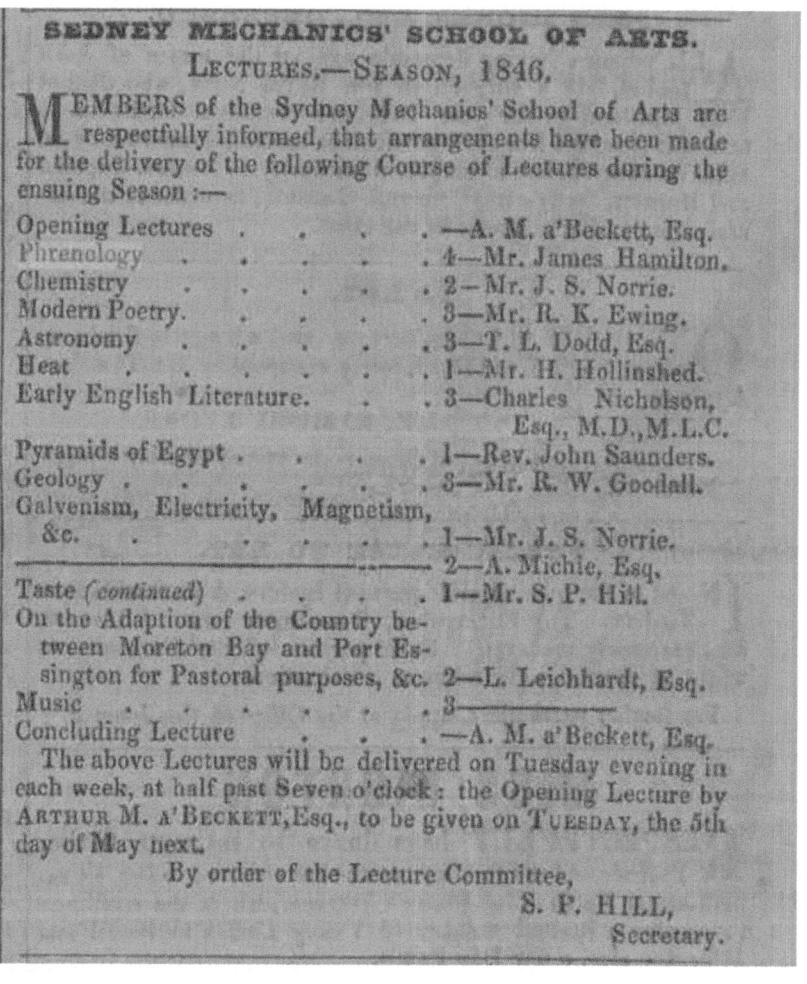

Figure 3. Lecture series at the SMSA, 1846.

Beyond the SMSA

One of many such itinerants was William David Cavanough (1834–1885). The local press frequently advertised and reported on his lectures. Cavanough had in his sights both politicians and lawmakers. The *Evening News* of December 19, 1877, reported that Cavanough lectured to an audience on the "intellectual and moral fitness of the new members of the present Parliament." He concluded that, although some of the legislators may have talent, the majority were of low intellect, ending his lecture with his opinion "that the present Parliament was not a true reflex of the intelligence of New South Wales," and said that "72 better men could be selected in a walk down the principal street of Sydney" (Anon 1877, 5). Before concluding the evening, Cavanough offered a phrenological examination of the heads of his audience.

By 1877, the intellectuals of the Colony who were once so involved in the promotion of phrenology at the SMSA, had long since disengaged from this subject. A comment by the prominent NSW medical practitioner Samuel Knaggs (1842–1921) summed up their position when he concluded:

> It must be admitted that the so-called science of Phrenology, as presented to us by many of those who make it a means of obtaining their livelihood—itinerant pretenders charlatans—is totally untenable in the present state of scientific advancement (Knaggs 1875, 351).

The "golden era" of phrenology, promoted by the intellectual community in the SMSA from 1838 to 1850, was long past.

Disclosure statement

No potential conflict of interest was reported by the author.

References

Anonymous. 1825. To the editor of the Sydney Gazette. *The Sydney Gazette and NSW Advertiser,* March 31: 3
Anonymous. 1826. *The Monitor,* July 7: 2
Anonymous. 1828a. *The Sydney Gazette and NSW Advertiser,* June 18: 2
Anonymous. 1828b. *The Monitor,* June 28: 5
Anonymous. 1829a. *The Sydney Gazette and NSW Advertiser,* March 10: 2
Anonymous. 1829b. *The Sydney Gazette and NSW Advertiser,* March 12: 2
Anonymous. 1830. *The Sydney Gazette and NSW Advertiser,* July 3: 3
Anonymous. 1831. *The Sydney Gazette and NSW Advertiser,* October 15
Anonymous. 1833a. *Sydney Monitor,* March 27: 2.
Anonymous. 1833b. *Sydney Gazette and New South Wales Advertiser,* April 27: 2.
Anonymous. 1837. *The Colonist,* November 23: 3
Anonymous. 1838a. Fifth annual report of the Sydney Mechanics School of Arts, for the year 1837. *The Colonist,* April 21: 4.
Anonymous. 1838b. *The Australian,* August 28: 2
Anonymous. 1842. *The Sydney Morning Herald,* October 10: 2
Anonymous. 1877. Character of the new members of parliament. *Evening News,* December 19: 5
Baker D. 1967. Lang, John Dunmore (1799-1878). In *Australian dictionary of biography.* National Centre of Biography, Australian National University. Accessed September 13, 2019. http://adb. anu.edu.au/bibliography/lang-john-dunmore-2326/text2953.

Broussais F. 1836. Lecture XX. – (The Last). Application of phrenology to history. *The Lancet* 26:929–44.

Cobley J. 1966. Bland, William (1789-1868). In *Australian dictionary of biography*. National Centre of Biography, Australian National University. Accessed September 13, 2019. http://adb.anu.edu.au/bibliography/bland-william-1793/text2027.

Field B. 1825. *Geographic memoirs on New South Wales; by various hands: Containing an account of the Surveyor General's late expedition to two new ports.* London: John Murray.

Flannery T. 1999. *The birth of Sydney*. Melbourne, Australia: Text Publishing.

Hoare M. 1974. Science and scientific associations in Eastern Australia, 1820-1860. PhD, Australian National University.

Knaggs S. T. 1875. The claims of phrenology to be classed as a science scientifically considered. *New South Wales Medical Gazette* 11:351–56.

Mackenzie G. S. 1820. *Illustrations of phrenology with engravings.* Edinburgh: A. Constable & Company.

Nadel G. H. 1957. *Australia's colonial culture: Ideas, men and institutions in mid nineteenth century eastern Australia.* Melbourne: F.W. Cheshire.

Nadel G. H. 1966. *Carmichael Henry (1796–1862).* In *Australian dictionary of biography*. National Centre of Biography, Australian National University. Accessed September 13, 2019. http://adb.anu.edu.au/biography/carmichael-henry-1881/text2209.

Parsons V. 1967. Morisset, James Thomas (1780–1852). In *Australian dictionary of biography*. National Centre of Biography, Australian National University. Accessed September 13, 2019. http://adb.anu.edu.au/biography/morisset-james-thomas-2482/text3313.

Poskett J. 2019. *Materials of mind: Phrenology, race, and the global history of science, 1815-1920.* Chicago: University of Chicago Press.

Reece R. H. W. 1974. *Aborigines and colonists. Aborigines and colonial society in New South Wales in the 1830s and 1840s.* Sydney: Sydney University Press.

Turnbull P. 2017. *Science, museums and collecting the indigenous dead in Colonial Australia.* Cham, Switzerland: Palgrave Macmillan. Palgrave Studies in Pacific History.

Wotherspoon G. 2013. *The Sydney Mechanics' School of Arts: A history.* Sydney: Sydney Mechanics' School of Arts.

Lampooning Phrenology

The phrenological illustrations of George Cruickshank (1792–1878): A satire on phrenology or human nature?

Gül A. Russell

ABSTRACT

For a brief period in1826, George Cruickshank (1798–1878), already an established artist in political satire and book illustration, turned to phrenology. He produced one initial print (*Bumpology*), followed by a collection of six plates of 33 engravings, linked by an explanatory preface, under the title, *Phrenological Illustrations or an Artist's View of the Craniological System of Doctors Gall and Spurzheim*. It was published during what is regarded as "the phrenological craze" in Britain. The illustrations were also produced at the height of Cruickshank's staggering creative productivity. In 1873, as phrenology was making its exit from scientific credibility into history, Cruickshank's phrenological illustrations were reissued by popular demand. Yet in contrast to his other works, these illustrations have received little attention in modern scholarship. The ways and the extent to which his caricatures constitute a contribution to the history of phrenology deserve to be studied. Here they are analyzed together with his descriptions in the prefaces to both the 1826 and 1873 editions. They reveal a surprising knowledge of phrenology in relation to Spurzheim and Gall. Furthermore, their uniquely innovative features will be identified in the context of other contemporary caricatures, and the fundamental significance of Cruickshank's achievement and its impact will be evaluated.

Introduction

In 1874, the *British Medical Journal* published the following note:

> Dr. Ferrier, who is as little like Falstaff as any man we know, had certainly no intention of being humorous when he published in the last volume of the *West Riding Reports*, and subsequently reported at the British Medical Association at Bradford, his famous localization of the functions of the brain nor is it probable that he expected to be the source of humor of others. These researches of Dr. Ferrier, and their bearing on phrenology, afford, however, to the veteran George Cruickshank an excuse for reissuing his droll *Phrenological Illustrations*; and after enlightening himself by reading Dr. Ferrier's essay, we do not know that a man could do better than amuse himself by turning to Cruickshank's volumes of grotesque and mirth-moving caricatures. (Vol. 1, 679, Jan. 3, pp. 21–22)

This comment was made almost half-a-century after the first appearance of George Cruickshank's *Phrenological Illustrations, or an Artist's View of the Craniological System of Doctors Gall and Spurzheim* in 1826 (see Figure 1). Although phrenology was being given the *coup de grâce* as a "science," Cruickshank's phrenological caricatures, which had gone out of

Figure 1. George Cruickshank by Alfred Croquis [Daniel Maclise (1806–1870)]. London: James Fraser (ca. 1833). Etching, August 1833 (Patten, 386).

print, were being reissued by popular demand (Cruickshank 1873). Yet Cruickshank had turned to phrenology for a brief period between February and August 1826 to leave us (with the exception of a separate print, *Bumpology*) a brilliant collection of 33 phrenological Illustrations in six plates (Cruickshank 1826a, 1826b).

In quantity, these constitute only a tiny fraction of Cruickshank's staggering output of more than 12,000 prints during the period of 1818 to 1829 (Jerrold 1882; Patten 1992). They have received little attention, largely mentioned with a passing reference (Buchanan-Brown 1980, 11–44) or with only one or two of the prints selectively considered (Finger and Eling 2019; Patten 1992). Yet they were enthusiastically reviewed in the periodicals of his time (Hone 1827; North and Wilson 1826; Thackeray 1840) and represent a significant aspect of the response to phrenology in Britain during the early decades of the nineteenth century.

The extent to which Cruickshank's caricatures constitute a contribution in the graphic arts, to the history of the reception of phrenology in Britain, and in mirroring and possibly even shaping social attitudes through "comical" vignettes deserves to be explored. Moreover, there is an interesting parallel between the rise and decline of phrenology and Cruickshank's career, which appears to have "virtually ended after 1826" (Patten 1974). But why did Cruickshank turn to phrenology at all for a brief period in 1826, when phrenology had been spreading in Britain from the beginning of his career as an artist? Already viewed as a subject of amusement, it was also exploited in caricatures by leading graphic artists, such as Thomas Rowlandson (Figure 2). Furthermore, his *Phrenological Illustrations* demonstrates a level of knowledge that raises the question of how and from what kind of sources Cruickshank acquired this information. A brief overview of the reception of phrenology in Britain helps to provide us with some answers.

Background: The reception of phrenology

On you we rest. to check the encroaching sway
This outré science gains from day to day;
Investigation's blood hound scent employ its force,
On themes more worthy of our scrutiny;
Rob this attractive magnet of its force,
And check this torrent's inundating course. (Cruickshank, Preface, *Illustrations*, 1826)

Phrenology was introduced into Britain in 1802, covering Franz Joseph Gall's (1758–1828) ideas and lecture tours on the continent, not only in English medical journals but also in widely read popular press. What may have been "a collective yawn" (Finger and Eling 2019, 391) initially had become "a craze" from London to Edinburgh within two decades (Patten 1992). Between 1823 and 1826, the number of phrenological societies had risen from one to 24, with 64,250 publications on phrenology, including 57 books and pamphlets (Parssinen 1974).

The diffusion of phrenology has largely been attributed to the London visit of Johann Gaspar Spurzheim (1776–1832) in 1814. His lecture demonstrations, followed by the *Physiognomical System of Drs. Gall and Spurzheim* (1815), resulted in intense debates, reviews, and synopses in periodicals (Anonymous 1826, 1829; Finger and Eling 2019; Gordon 1815; Wyhe 2004, 2007). Shifting the emphasis of the "organology" of Gall, with whom he had worked for more than a decade, Spurzheim had turned from cerebral anatomy to the psychosocial potentials of the "new science." By 1818, he was promoting what came to be known as "phrenology," a term he did not coin (Finger and Eling 2019). Gall, too, visited London in 1823. Although his lectures had less impact than those of Spurzheim, George Combe (1788–1858), in Edinburgh, further reinforced the interest across all social classes and the controversy with his treatises, *Outline* or *Elements of Phrenology* (1824) and *Constitution of Man* (1828), especially after the second's 1836 edition (Parssinen 1974; Secord 2014).

The way with which a wide range of people could practice phrenology contributed to both its popularization through cheap literature and to the controversy surrounding its use and interpretation. The scathing reviews of the *Physiognomical System of Drs. Gall and Spurzheim* (1815) helped to make phrenology fair game for editorialists, other writers, and graphic artists (Jeffrey 1826; Jeffrey and Gordon 1817/1818).

Figure 2. Thomas Rowlandson: Gall demonstrating on a skull, to a group of five men, surrounded by his extensive collection of busts and skulls, human and animal, on shelves, 1808. Etching. Wellcome Collection.

"Craniology" (palpation of the protuberances of the skull) or "cranioscopy" inspired "amusement" (Anonymous 1815; Cooter 1984; Finger and Eling 2019). There were attempts under the guise of aiming "to explain its leading principles, to expose its absurdities and innocent raillery, to invalidate its positions with popular arguments" (*Craniad* 1818, *Preface*, p. iv), as well as with limericks.

> [O]h Spurzheim! … Be thou our patron saint …
> Thus have we proved what ne'er was proved before
> But which, once proved, can ne'er be doubted more,
> That every faculty is born and bred,
> And rear'd to full perfection in the head.

That mind depends on brains we've clearly shown,
For when the brains are out, the mind is flown;
That skulls contain the laws of human life,
Which often are with human laws at strife. (Wyhe 2004)

Imitated versions of "scientific lectures" had also become a form of entertainment, tailored to reach beyond the middle classes to a wider audience, including the working classes, and courses were advertised in local newspapers. Spurzheim gave two courses of 18 lectures in 1825 at the large Crown and Anchor Tavern that were published in the *Lancet* (Spurzheim 1825b), and again in 1826 at the London Institution to a full house (Review 1826; Spurzheim 1825a, 1825c). At the same time, self-proclaimed itinerant practitioners were already exploiting phrenology as a business for profit (Cooter 1984; Finger and Eling 2019; Parssinen 1974). By 1826, when Cruickshank embarked on his *Illustrations*, the "craniological mania" was said to have "spread like a plague, possess[ing] every gradation of society from the kitchen to the garret, as a kind of "a species of intellectual mushroom" (Cooter 1984, 1989, 135).

Biographical note

The diffusion of phrenology in Britain coincided with Cruikshank's long career, which began at the age of eight, when he apprenticed to his father, Isaac Cruikshank (1764–1811). His older brother, Isaac Robert (1789–1856), was also a caricaturist and illustrator, with whom he intermittently collaborated (Patten 2004/2007; Patten 1992). His father was one of the three leading caricaturists of the late 1790s, together with James Gilray (1756/1757–1815), who had a major influence on Cruickshank, and Thomas Rowlandson (1756–1827). In their hands, political satire had become "a most feared of social weapons by unmasking pretensions and killing by ridicule" (Gombrich and Kris 1938, 1940; Patten 1983).

Cruickshank gained notoriety initially with his satirical prints, continuing the propaganda war against Napoleon by the British (see Figure 3) and attacking not only leading politicians but also the royal family (Buchanan Brown 1980, 11; Cooter 1984). In the early 1820s, however, he turned to social satire, campaigning for legal reform and temperance, having witnessed the destruction of both Gilray and his father by alcohol (Altick 1978; Patten 1992). Then, in 1823, beginning with *Grimm's Tales*, he became the foremost illustrator of literary classics, including Scott's novels and Dickens' *Oliver Twist* (Kitton 1899/1903; Patten 1992).

This change of direction has been attributed to the rather substantial royal bribe of £100 in 1820 for a pledge "not to caricature His Majesty 'George IV' in any immoral situation."

To take this offer was infinitely preferable. For example, Philipon, the editor of *La Caricature* in France, was accused of libel for the portrayal of Louis Philippe, the Bourgeois King (1830–1848), as a "pear" (with the double meaning of *poivre*, as in "fat head"; Davis 2018, 5–9; Gombrich and Kris 1938). Additionally, public interest was shifting from individual personalities to social comedy and good humor (Patten 1992). In 1826, when Cruickshank turned to phrenology, despite the mounting craze, there was no evidence of any prior interest on his part.

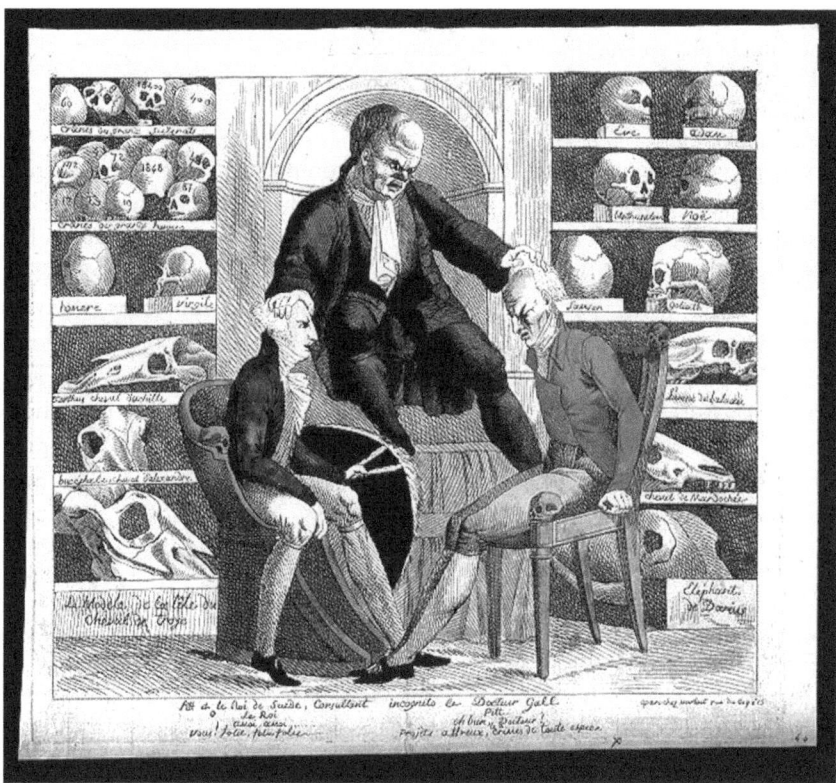

Figure 3. French response to the savage British satires of Napoleon. Anonymous: King of Sweden [Gustav Adolf VI (1778–1837)] and Mr. Pitt [the Younger (1759–1806)] "incognito," consulting Dr. Gall, 1806. Musée de la Ville de Paris.

Cruickshank's initial attempt: *Bumpology*

In fact, the subject appears to have been suggested to him by John Doyle (1797–1868)—also a political caricaturist, painter, and lithographer—who had gained recognition with the exhibition of his *Turning Out the Stag* at the Royal Academy in 1825. Doyle is considered the founder of the style subsequently made famous by *Punch* (Baker 2004). This may explain why his pen name—H. T. D. B.—shows on Cruickshank's first phrenological caricature, *Bumpology* (see Figure 4), etched for his print seller, William Humphrey, on February 24, 1826.

This print portrays a phrenologist with an exaggerated nose and chin, his profile repeated on one of the numerous heads. He is feeling the scalp of a bumptious-looking young man. His assistant, standing behind, has recorded, "Very large wit no. 32," while his mother looks on with a sheepish grin. In the background are a large bookcase and framed phrenological drawings. On the desk, a paper ironically reads, "Thurtle [the infamous murderer] Shown to Be Craniologically an Excellent Character" (Anonymous 1824). An added limerick further lampoons, "The prognostication of the phrenologist":

> Pore o'er the Cranial Map with learned eyes,
> Each ring hill and bumpy knoll descries,
> Here secret fires, and there deep mines of sense
> His touch detects beneath each prominence.

George Cruikshank, after H.T.D.R.   BUMPOLOGY.   *February 24, 1826.*
" Pores o'er the Cranial Map with learned eyes."

Figure 4. George Cruickshank: *Bumpology (A very large wit, No.32)*, Feb. 24, 1826. Etching.

The figure in the print is identified as James De Ville (1777–1846), a self-proclaimed specialist who had become well known in the 1820s for the plaster casts of phrenological "heads" on which the protuberances of the organs of Gall and Spurzheim were exactly mapped. He published treatises to explain them and examined skulls in his private consulting rooms on the Strand (De Ville 1824, 1826). Although regarded as illiterate by some, having started in lighthouse fittings and plaster casting, De Ville was a contributor to the *Phrenological Journal*. In 1825, he had organized Spurzheim's lectures in the Crown and Anchor Tavern (Browne 1846; Cooter 1984).

Neither the depiction of the scene nor the limerick is particularly distinctive. For example, the earlier print by Cocking represents what would become a generic scene of a craniological examination in a private consulting room (Figure 5). It is on the frontispiece of the anonymous book, *Craniology Burlesqued, in Three Serio-comic Lectures, Humbly dedicated to the patronage of Drs. Gall and Spurzheim, by a friend to Common Sense* (1816/1818).

Figure 5. R. Cocking, *A Phrenological Examination* (Spurzheim?), 1816. Etching. Wellcome Collection.

A foreign-looking phrenologist, dressed in black frock coat, frills, and buckled shoes (Spurzheim?), is looking at what is revealed by his calipers and ruler on the scalp of a large man seated in an arm chair, his left boot resting on a pile of books, with a bust on its side. The astonishment on his face is mirrored by the expression and gesture of his assistant in the doorway. The satire is enhanced by adding to the "shape" the "size" of the bump as a measure of the power of the organ. There are books on shelves and on the floor, numerous phrenological busts with idealized profiles or various expressions, along with a skull in the shadows. A partly open cupboard reveals a calf's head on a dish—possibly a reference to the controversial similarities of the propensities of humans and animals. This setting is subsequently repeated with slight variations. By 1825, caricatures were portraying the phrenologist or the examination of bumps.

The *Phrenological Illustrations*

> There is a great deal humor in these whimsical caricatures, and they are, perhaps, the best and the only answer which the bumpological reveries of Drs. Gall and Spurzheim deserve, or should be treated with. (Advertisement, *Illustrations* 1826)

Six months after *Bumpology*, Cruickshank turned away from the tradition of depicting superficial examinations of bumps to create an entirely new style with the publication in August 1826 of his *Phrenological Illustrations, or An Artist's View of the Craniological System of Doctors Gall and Spurzheim* (Figure 6).

The title refered to both Gall and Spurzheim, but Cruikshank used the term "phrenology," which Gall detested and Spurzheim (1818) popularized, to distinguish his own "improved" version of the doctrine from that of Gall. (Spurzheim provided relatively

PHRENOLOGICAL ILLUSTRATIONS,

OR

An Artist's View

OF THE

CRANIOLOGICAL SYSTEM OF DOCTORS GALL AND SPURZHEIM.

BY

GEORGE CRUIKSHANK.

LONDON:

PUBLISHED BY GEORGE CRUIKSHANK, MYDDELTON TERRACE, PENTONVILLE; AND SOLD BY J. ROBINS AND CO. IVY LANE, PATERNOSTER ROW;
S. KNIGHTS, SWEETING'S ALLEY, ROYAL EXCHANGE; AND G. HUMPHREY, 24, ST. JAMES'S STREET.

MDCCCXXVII.

Figure 6. The title page of the *Phrenological Illustrations*, 1826. Woodcut. Wellcome Collection.

minor variations in the logic and basic tenets, but reclassified, renamed, and increased the number of faculties.) At the same time, Cruickshank referred to their "craniological system," a term shunned by Gall (whose "organology" focused on the brain, not the skull), whose primary focus was the brain. The title may be a reflection of the public perception of the two men, preaching what is essentially the same cranially based doctrine (Finger and Eling 2019). Upon the publication of the *Illustrations*, as one reviewer rightly put it: "Be the name right or wrong, this method of illuminating our understanding is the most entertaining, perhaps … the most rational and useful, that has yet appeared" (Hone 1827).

Beneath the title is a wood engraving of three faces—with front, side, and back views of their partially bald scalps; roman numerals signify the phrenological organs, but there are no bumps. Here, Cruickshank's conception was already in stark contrast to existing phrenological heads as represented elsewhere. In the *Physiognomical System of Drs. Gall and Spurzheim* (1815), a full-page illustration (Figure 7) shows three views of an idealized head with Gall's cranial markers for his underlying "organs of mind" (Finger and Eling 2019). Various versions, including anatomical textbook style of skulls, had appeared in numerous publications. For example, a copy of the engraving from Spurzheim was placed at the end of *Craniology Burlesqued* (1816). A modified profile of the face (Figure 8)

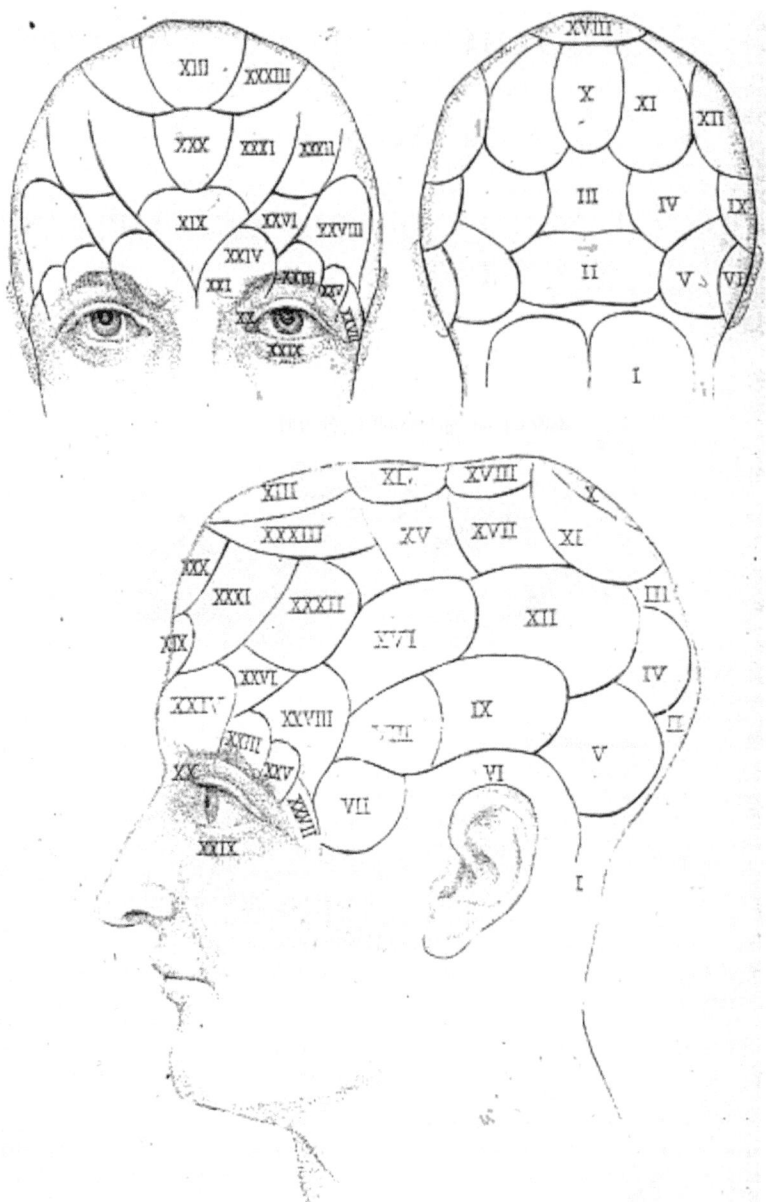

Figure 7. The phrenological organs as mapped on the human head, in the *Physiognomical System of Drs. Gall and Spurzheim*, (1815). Etching.

appeared in the *Craniad* (1818). A top view was added in the frontispiece of Combe's *Elements of Man* (1825).

But whereas others directly copied or vulgarized the original phrenological head, Cruickshank created his own unique version. He translated the mapping of a theoretical doctrine into amusing real portraits, hair partly shaved, revealing an incongruously demarcated scalp with Roman numerals and jovial, laughing faces, as if in the midst of a conversation—perhaps laughing at the subject itself. The gentle humor and subtle wit

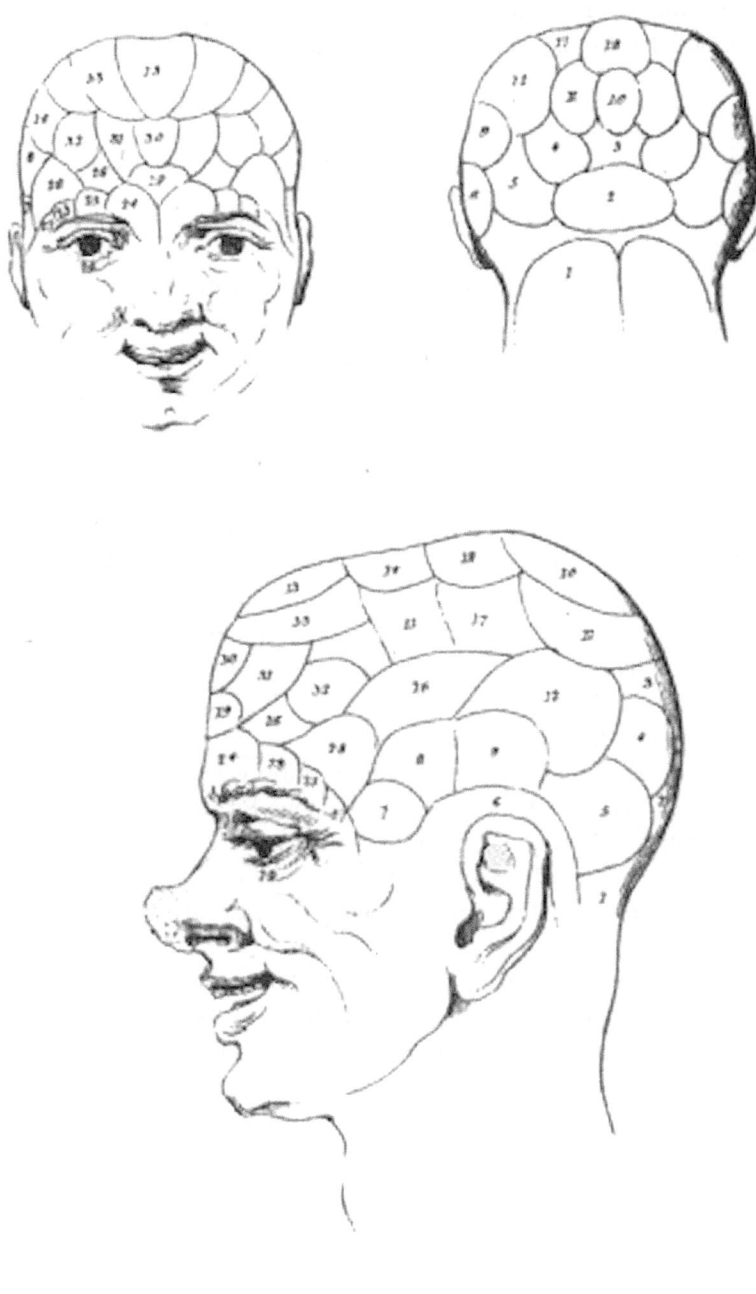

Figure 8. Frontispiece. *The Craniad; or Spurzheim Illustrated.* Edinburgh, 1817.

effectively set the tone for the whole collection. This was not going to be the usual "crude" satire on "bumps," despite the contemporary public perception of phrenology.

The preface

Any interpretation of Cruickshank's *Phrenological Illustrations* needs to start with the preface, which reveals the underlying seriousness with which Cruickshank is approaching the subject. The opening statement, although brief, conveys his understanding of phrenology as "a doctrine proposed to elucidate the character, the passions, the morals, and the interior faculties of man, from certain appearances exhibited by the brain and skull" (Preface 1826) This in itself is different from the usual attempts proposing to give a "true" explanation of the system that are actually facades aimed at satirizing craniology.

Despite his homage to both Gall and Spurzheim, Cruickshank largely followed the 33 faculties (not Gall's 27) listed in Spurzheim's *Physiognomical System* (1815, Table, 579–580; Review 1826; Spurzheim 1825c). Accordingly, he categorized the organs into two orders: (I) feelings, to which belong propensities; sentiments; and (2) understanding, which includes knowing and reflecting intellectual faculties. At the same time, he comparatively referred to Gall and Spurzheim, bringing out some of the differences in their definitions to humorous effect. He then indicated how he would endeavor to represent the individual faculties on each plate—through their manifestations in human behavior in real terms. By deliberately creating incongruities between the faculties as defined and his graphic representations of them, he injected further irony and humor. Thus, the preface serves to reveal the intent of the artist, setting the tone for what he called his "illusions" and, more importantly, to provide a coherent structure for his numerous illustrations. At the same time, he provided just enough information about the subject and the figures in the illustrations for the viewer to identify the faculties with no further need for explanatory captions.

Cruickshank's sketchbooks reveal that he was exploring alternative designs in numerous trial drawings in pencil. Each plate portrayed one large scene at the center of the long sheet, representing an "organ of faculty," and four smaller but related scenes at the corners. Technically, he adapted the existing practices from two contemporary sources: the caricature prints of folding screens decorated with pasted cutouts and highly popular scrapbooks with etched designs in the margins of prints (as first introduced by Henry Alken in his *Sporting Scrap Book of 1824*; Patten 2004/2007). Doubtless, he was also bringing his vast experience in book illustration into caricature. In the process, he introduced an entirely novel style to illustrate the basic phrenological concepts in each of his plates.

The result is a serious endeavor in caricature to first define and then translate the function of phrenological "organs" into a series of amusing vignettes of human behavior within varying social contexts. His approach was remarkably consistent with the fundamental role in the doctrine of both Gall and Spurzheim, of "empirical and naturalistic studies of humans in society and of species in nature," in determining the "location of brain function" (Cooter 1984, 2–3).

First, in organizing his plates, he took liberties with Spurzheim's "order" to combine propensities and intellectual faculties as he saw fit. Accordingly:

(1) Plate I—centered on the propensity of philoprogenitiveness—includes, with amativeness and self-love, the knowing faculties of individuality and number (Figure 9).

(2) Plate II is all propensities, centered on adhesiveness, with inhabitiveness, constructiveness, combativeness, and destructiveness (Figure 10).

(3) Plate III consists of color, form, space, and order, all intellectual faculties of knowing, associated with drawing as the large image (Figure 11).

(4) Plate IV adds to the propensities of covetousness, secretiveness, size, and firmness, the knowing faculties of tune, time, and weight (Figure 12).

(5) Plate V brings together ideality, and approbation with the knowing faculty of language in the center, and the reflecting faculties of wit, comparison, and imitation (Figure 13).

(6) Plate VI ends on hope as the large image, adding to the propensities of conscientiousness, veneration, cautiousness, and benevolence the reflecting faculty of causality (Figure 14).

Although linked by the explanatory preface, the faculties are selectively defined, but only to enhance the artist's purpose. In the process of designing their manifestations (or "the activity of their powers"), Cruickshank developed innovative features that uniquely characterize his phrenological illustrations. The essence of caricature is a pictorial likeness in a few simple lines, sometimes exaggerated or disfigured, as in satirical portraits, yet tapping into our

Figure 9. Cruickshank, *Phrenological Illustrations*, 1826. Etching, Plate I. Wellcome collection. Representing amativeness (physical love), self-love, philoprogenitiveness, individuality, number.

Figure 10. Cruickshank, *Phrenological Illustrations*, 1826. Etching, Plate II. Wellcome Collection. Representing inhabitiveness, constructiveness, adhesiveness, combativeness, destructiveness.

perceptual abilities to recognize objects and faces by minimally defined critical features. In his phrenological illustrations with complex scenes, however, Cruickshank created the amusing context by associating his subjects with culturally familiar and at times sensational fairground figures of public attraction, which his viewers, irrespective of social class, could immediately identify in the shop windows of the print sellers where they were displayed.

For example, Cruickshank suggested that, "To verify seat of the organ of *Combativeness*, that Dr. Gall ascertained by examination of the head of a pugnacious tribe of boys from the streets, he should visit *Donnybrook Fair*." This was the infamous, the most notorious site in Dublin, which had come "to exhibit such continuous scenes of riot, bloodshed, debauchery, and brutality, as only the coarsest taste and the most hardened heart could witness without painful emotion … This was by day; 'the orgies of the night' may better be imagined than described" (*Parliamentary Gazetteer*, 1845). His subsequent illustration of combativeness (Plate I) gains its impact by the association of the depicted scene of violence with the actual place.

Similarly, for the representation of the intellectual faculty of number (Plate I), Cruickshank indicated, "Among the individuals remarkable for their great perfection of this organ, our *sapient friend Toby* deserves a distinguished place" (Preface, Plate I). Toby was the generic

Figure 11. Cruickshank, *Phrenological Illustrations*, 1826. Etching, Plate III. Wellcome Collection. Representing color, form, drawing, space, order.

name for performing pigs that displayed a skill "never heard of before to be exhibited by an animal of the swine race" and, most remarkably, "discover a person's thoughts." Brought into the limelight by an illusionist in 1817 at the famed Royal Promenade Rooms in the Spring Gardens, Toby remained a major attraction at the Bartholomew Fair in London between 1818 and 1823 (Altick 1978). Among other tricks, he would correctly pick up the called-out number, as in Cruickshank's illustration, that spelled "vittles," then "with a joyful grunt, left to have some" (Bentley 1982, 88–104). Since the eighteenth century, he had inspired satirical comments, verses, and popular songs (Tegg 1831). Toby was used to parody meretricious celebrity (as "a far greater object of admiration to the English nation than ever was Sir Isaac Newton"; Cattrell 2006, 42–43), Literary figures mentioned him, including Charles Dickens, whose books Cruickshank had illustrated. He had even published an autobiography, *The life and adventures of Toby, the sapient pig: with his opinions on men and manners. Written by himself* (or, rather, by N. Hoare).

Moreover, Toby was already featured in Thomas Rowlandson's satirical caricature, as *The Wonderful Pig … displaying his erudition to a crowd of amazed ladies and gentlemen* and *The Surprising PIG well versed in all Languages, perfect Arethmatician Mathematician & Composer of Musick (1785)* (Figure 15). Thus the "pig" in illustration, although simply labeled *number*, would have been a thoroughly familiar figure with rich associations in addition to the total incongruity with the phrenological faculty it represents.

P. 4

Figure 12. Cruickshank, *Phrenological Illustrations*, 1826. Etching, Plate IV. Wellcome Collection. Representing covetiveness, secretiveness, the organ of tune, size, weight, firmness.

After describing that "individuality is particularly conspicuous in superficial people," Cruickshank added, "The French Gentleman, lately exhibited, being all superficies is surely entitled to distinction." Here he was playing on the words superficial and superficies, the visible external surface of a body, outward part, and appearance. The reference is to another well-known fairground character: "Seurat, the Living Skeleton," who was born in 1798 in Troyes, Champaigne, France, and brought to London. His "form" was "unparalleled in the annals of humanity and in every respect entitled to the living Spectre" (Seurat 1825, 20). He was "5ft, 7 and a half inches" with no substance beneath the surface. All of his hair was "shaved off for the purpose of exposing the formation of the skull to the scrutiny of scientific observers." He wore "a wig … when not called upon to bare his cranium. In regard to the construction of the back of the head, … the organ, denominated by the disciples of Spurzheim, *Philoprogenitiveness* is altogether wanting" (Seurat 1825, 9). Cruickshank's illustration is modified from Seurat's *Autobiography*, which was illustrated by his brother, Robert.

In the same way, size (Plate IV) is represented by Daniel Lambert (1770–1809), the fattest man in British history, weighing 739 pounds at his death. His coffin required 112 square feet of wood, built with wheels to allow for easy transport (Figure 16). Despite a sloping approach dug to the grave, it took 20 men almost half an hour to drag his casket into the trench, in a newly opened burial ground to the rear of St Martin's Church (Altick 1978). Lambert

Figure 13. Cruickshank, *Phrenological Illustrations*, 1826. Etching, Plate V. Wellcome Collection. Representing ideality, wit, language, imitation & approbation, comparison.

became an object of display and the subject of paintings and caricatures. These figures were so familiar that, without any detailed description, their evoked associations greatly enriched the meaning of the image and enhanced the humor.

By a brilliant stroke of genius, Cruickshank turned Spurzheim's faculties into objects of "mirth" by showing how they could be manifest in multiple ways in human behavior, with which the viewer could both empathize and, witnessing the frailties in others, laugh at. Cruickshank appeared to be exploiting in his caricatures what has been described as the appeal of phrenology, responding to "the desire for knowledge about the characteristics of oneself, and especially of others" (Wyhe 2007, 42).

Cruickshank also designed dramatic scenes that were not only a carryover from his illustrations of Dickens' works but also derived from William Hogarth (1697–1764), who "treated his subjects in his etchings in the manner of a dramatist, considering the picture as a stage and his figures as players" (Hogarth, quoted in Moore 1948). Cruickshank was thoroughly familiar with Hogarth's prints, and he also attending the theater regularly (Patten 1992; Paulson 1989). In two instances, Cruickshank actually incorporated into his illustrations a character from a well-known play, as well as a comic stage actor to impersonate the phrenologist.

For Spurzheim, causality is ranked as the highest intellectual knowing faculty together with comparison, which Cruickshank equated with that of inquisitiveness. In his representation, he reduced it to mere curiosity. He used Paul Pry, the central character in

Figure 14. Cruickshank, *Phrenological Illustrations*, 1826. Etching, Plate VI. Wellcome Collection. Representing hope, conscientiousness, veneration, cautiousness, benevolence, causality (inquisitiveness).

a popular farce in three acts that had opened at the Haymarket Theatre in London on September 13, 1825 (Anonymous 1910, 780). Consumed with curiosity and unable to mind his own business, Pry would conveniently leave behind an umbrella everywhere he went as a convenient excuse to return and eavesdrop. Cruickshank's portrayal of Pry holding his umbrella and saying, "I hope I am not intruding," is a perfect example of the ingenuity of his application of Spurzheim's definition of causality: As "we cannot invent or create causes, we can observe the circumstances under which events take place, observing the succession of events or regularity with which they occur as causes" (Lectures, June 4, 1825). At the time, a comic song based on Paul Pry's character was also circulating in 1826 (*The Universal Songster*, Vol. 3, p. 97). On occasion, when Cruickshank was referring to the protuberance of a specific "organ," he presented evidence supportive of the activity relative to its function. Hence, "Our amusing friend Paul Pry, has a most special development, in a hemispherical form on the superior part of his forehead" (Preface).

He sustained the humor by further drawing attention to the discrepancy in the definitions of Spurzheim and Gall, as exemplified by inhabitiveness. He pointed out that, "to this organ is attributed in man, self-love, … in animals, physical height" and that, "according to Dr. Gall, what is physical in other animals is moral in man." He then illustrates it with a snail (Plate II).

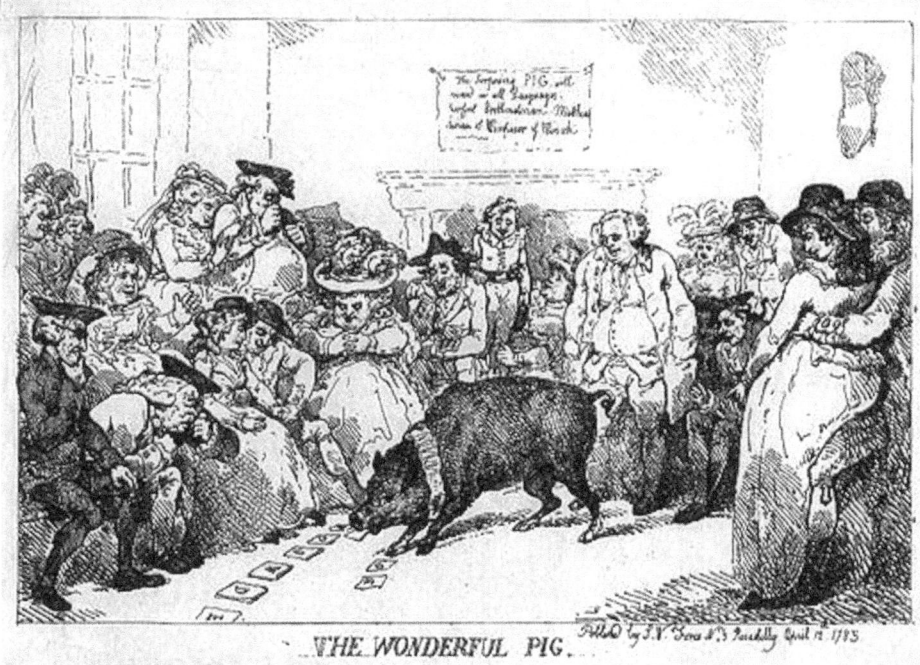

Figure 15. Thomas Rowlandson, *The Wonderful Pig ... well versed in all Languages, perfect Arethmatician Mathematician & Composer of Musick, 1785. Etching.* National Gallery of Art (Rosenwald Collection).

Similarly, amativeness (Plate I), he pointed out, is "placed at the head of all the others by Drs Gall & Spurzheim. Man, it appears, has a great brain (cerebrum) and a little brain (cerebellum) to which the organ of physical love is attributed." Then he endeavored to give a literal delineation, in which a lusty apothecary is on his knees before a maid. Interestingly, his depiction is closer to Gall's original organ of sexual gratification than Spurzheim's rendering "to meet trends of polite society" (Cooter 1984, 78; Finger and Eling 2019).

Philoprogenitiveness (literally, love of offspring) was also drawing considerable attention (Patten 1992), and Cruickshank described it as "most conspicuous in females, but certain males are said to excel in this inclination" (Preface). He then depicted both in one "huge and happy family" that would "make a cynic laugh" (Hone 1827). Victorian novelist William Makepiece Thackeray asked in his *An Essay on the Genius of George Cruikshank* (1840), "Did we not forego tarts, in order to buy his prints?" To purchase the entire plates, they had formed a "joint stock company of boys, each drawing lots afterwards for the separate prints," for the right to pick one of the sheets, which they then colored themselves. Drawing the first lot, he seized upon *Philoprogenitiveness* and conveyed the full exuberance of the scene which is worth noting in full (Figure 17):

> The composition writhes and twists about like the Kermes of Rubens. No less than seven little men and women in nightcaps, in frocks, in bibs, in breeches, are clambering about the head, knees, and arms of the man with the nose; their noses, too, are preternaturally developed; the twins in the cradle. ...The second daughter, who is watching them; the youngest but two, ... squalling in a certain wicker chair; the eldest son, ... yawning; the eldest daughter, ... preparing with the gravy of two mutton chops, a savory dish of Yorkshire pudding for eighteen persons; the youths ... examining her operations (one a literary gentleman, in a remarkably neat nightcap and

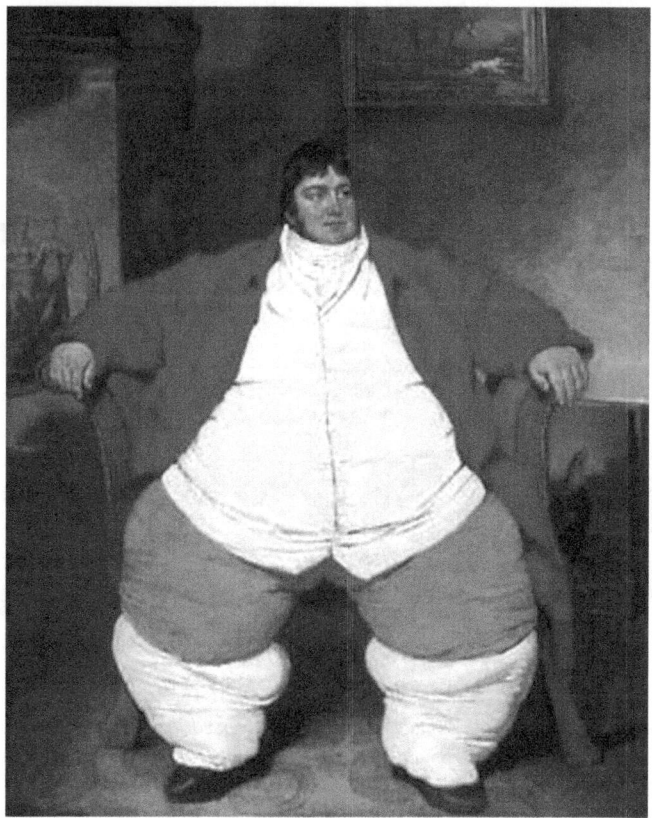

Figure 16. Portrait of Daniel Lambert by A. van Assen (after Joseph Parry 1744–1826, 1804. Etching. Wellcome Collection.

pinafore, who has just had his finger in the pudding); the genius … at work on the slate, and the two honest lads … the good humored washerwoman, their mother, all, all, save, this worthy woman, have noses of the largest size. Not handsome certainly are they, and yet everybody must be charmed with the picture. It is full of grotesque beauty. … He loves children in his heart; every one of those he has drawn is perfectly happy, and jovial, and affectionate, and as innocent as possible. He makes them with large noses, but he loves them, … The smiling mother reconciles one with all the hideous family: they have all something of the mother in them something kind, and generous, and tender. (Thackeray 1840, 5–6)

Thackeray added his own touch of humor by suggesting, "The artist has at the back of his own skull, we are certain, a huge bump of philoprogenitiveness." A comparison with other interpretations of *Philoprogenitiveness* shows Cruickshank's "ingenuity *and jovial humor in this* marvelous print, *full of* grotesque beauty" (Thackeray 1840, 6).

One example is the print *An Old Maid's Skull Phrenologized* (1830) by E. F. Lambert (Figure 18; Bynum 1968). It is a crude satirization of the view, attributed to Dr. S (Spurzheim) that "philoprogenitiveness" is most prominent in women and animals. The humor derives from gross exaggeration by representing an ugly, obviously barren "old

Philoprogenitiveness —

Figure 17. Philoprogenitiveness from Plate III, *Phrenological Illustrations*, 1826.

maid," contrary to the size of her "bump," and her dog, curled up in her wig on a chair, and equating them by implication.

On self love (Plate I): Cruickshank explained that it was "first discovered in a beggar who accused his pride as the cause of his mendacity, and considered himself too important to acquire any business." As "This faculty is said to give us a great opinion of our own person," he illustrated it with a dandy holding up a mirror to admire his reflection.

On adhesiveness (Plate II), phrenologists implied friendship as a propensity that "applies to all that we possess, animate and inanimate." His faithful illustration is a couple of rather unattractive individuals thrown out of a gig into a pond, and sticking in the mud (Figure 19), which Cruickshank viewed as the human condition, "common to many" (Preface 1873).

Constructiveness (Plate II) is illustrated by a spider's web "in no instance perpetually exercised than by the humble insect." Destructiveness is a bull in a china shop (Plate II).

Cruickshank pointed out the differences between Gall and Spurzheim to comic effect: For Spurzheim, "*Conscientiousness* is the organ of *Righteousness*, and Gall thinks there is no organ of conscience." He then portrayed a Dickensian Fagin-like character paying a pittance for a whole bundle of clothing.

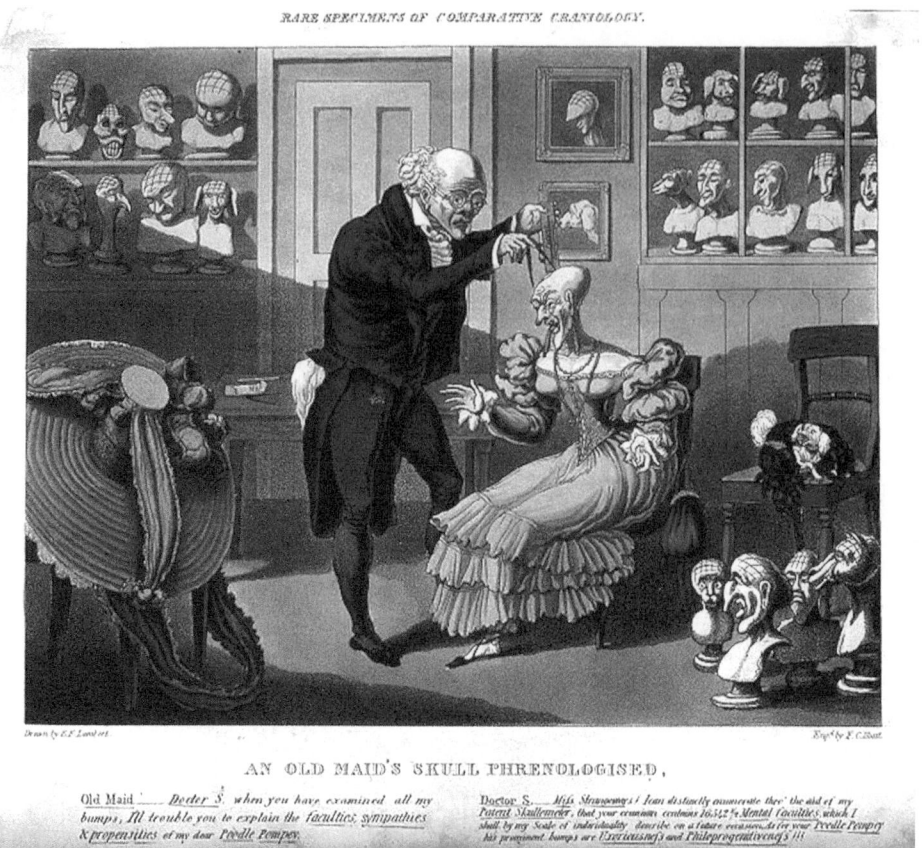

Figure 18. E. F. Lambert (c1790–1846), *An old maid's skull phrenologised*, 1830. Etching. Science Museum Group.

On veneration: "Dr. Gall has observed this organ in persons chiefly with bald heads … common to all classes and description of persons." Then Cruickshank illustrated a portly man salivating before a huge side of beef (Plate VI).

On hope: "Dr. Spurzheim considers this a particular sentiment; but Dr. Gall regards it as belonging to every organ." Yet the "poor dumb animal" in Cruickshan's example (Plate VI) appears to have little hope: a starved dog wistfully looking at his master gnawing a bone.

Benevolence (Plate VI) was defined by Spurzheim as the organ that "Directs our actions in relation to others, disposes us to be kind and agreeable and assist those in suffering (Lectures, June 4, 1825.) Cruickshank, after ironically commenting that "Timely chastisement has saved many a wretch from the gallows," represented a procession in which a man is being flogged, tied to a wagon, with the House of Correction on one side and a gibbet dominating a hill in the background.

In Plate IV he deviated from his usual pattern and placed seven vignettes, centered on tune, which is illustrated by an organ-grinder. Playing on Spurzheim's definition of time with reference to "duration and succession of events" (Lectures 1825), Cruickshank placed a tiny clock face "familiar to everyone" surmounted by a winged hourglass above, and weight "a crown resting on a cushion" below, as "no loyal man will offer any objection."

Figure 19. Adhesiveness from Plate II, *Phrenological Illustrations*, 1826.

"Dr. Spurzheim [had] most conveniently placed *Time … Between* the organs of individuality, *space, order, tune,* and *cause,*" Cruickshank instead brought together covetiveness, ("the metropolis unfortunately affords too many scenes of this propensity") represented by "a familiar one" of a pick-pocket, and firmness, represented by a *pavior* with his rammer ("a character now being consigned to oblivion"). After stating, that "an essential difference is said to coexist between size and form," he used Daniel Lambert, the fattest man in Britain to illustrate size, and a dandy admiring the shape of his boot for form.

On secretiveness, Cruickshank pointed out that, according to Gall, "its sphere of activity is considerable" with "cunning, prudence, the *savoir faire*, the capacity of finding means to succeed, hypocrisy, lies, intrigues, dissimulation, duplicity, falsehood, in poets … dramatic. plots; and slyness in animals!" He then exemplified "one of the advantages," showing a woman hiding a letter in the fireplace.

He deliberately downplayed the intellectual seriousness of organs categorized under faculties of knowing or reflecting by representing them with concrete physical objects, with "humble creatures" (such as a snail or a spider), or by making both the protuberance as described, and the supporting evidence as given, appear ridiculous and absurd. For example, he stated that

[A]n elevation in the midst of the superior part of the forehead in the shape of a 'reversed pyramid' is sure to possess the organ of comparison and analogy. Dr. Gall has found the heads of two Jesuits to be remarkably distinguished. The Artist trusts, he adds, that "it is not necessary to be a Jesuit in order to possess this faculty on the example given in Pl. V."

He contrasted an exceedingly thin and tall man with a long umbrella tucked under his arm walking out of Loud Acre into Little St. Martin's Lane with a fat, dumpy woman, a third of his height, coming toward him.

Language he depicted in the large image of Irishwomen aggressively haggling at Billingsgate Fish Market. As for the reflecting faculties of wit, comparison, and imitation, he stated that the essence of wit "is said to consist in its peculiar manner of comparing, which always excites gaiety and laughter." Then he represented a terrified woman in a churchyard at night who almost drops her lantern, seeing on the gravestone in front of her: "Alas Poor Yorick" (Plate V), reminding us of Hamlet's soliloquy on the skull of the court jester.

Approbation and imitation were brought together in one representation (Figure 20). Cruickshank explained that approbation was considered as "essential to society, exciting other faculties, and producing emulation to the point of honour," and ironically added that his illustration is "in a manner consonant to the wishes of the admirers of the Phrenological system." As imitation (Plate V) *sui generis* belongs to none but acts upon them all, and as the individuals possessing it like to be Actors, The Artist has taken the liberty of giving his illusion of it in the exhibition of the Phrenologist himself." The term "illusion" is significant, as Cruickshank chose for his illustration of it a comic actor, Charles Mathews (1773–1835), well-known for his exceptional ability of impersonation and for his "At Home" series in London (Andrews 2006, 111–126; Barnet 1965). Here the phrenologist is on a stage, demonstrating a skull (with ambiguous-looking skulls on the table in front of him) to a packed middle-class audience, applauding him. He is holding up a skull he cannot see, as he is wearing dark glasses and appears to be blind.

Thus, the phrenologist is not what he seems. Is the actor the phrenologist or the phrenologist merely an actor? The deception is also perpetrated on a gullible audience, which is giving approbation to a mimic notably mistaken for the model. Is the approbation given to the phrenologist (which would be false) or to the mimic, the stage actor for his impersonation? Cruikshank has created a brilliantly provocative illusion of imitation.

The levels of deception could well derive from the controversy on the definition of "mimicry" and caricature that had centered on Charles Mathews. He was defended against the charge of mimicry, in the sense of mechanical imitation versus natural impersonation, or the assumption of being someone else (*Blackwood Magazine* 1820). The discussion also involved whether "a caricaturist exaggerates and distorts," and the question, "what are they [caricatures] but just representations of individual character and *habit* under peculiar circumstances?" (Andrews 2006, 115). The subtle complexity of Cruickshank's representation becomes all the more significant in light of the emerging new understanding of "deception" and "mimicry" in appearance, as will be developed in biology.

To Spurzheim's phrenological faculties, Cruickshank added the organ of drawing which, he claimed, was "not noticed by the learned advocates of Science of Phrenology, but which in the Artist's opinion is highly deserving of a place." Here he either ignored or was unaware of the extensive discussion by Gall and Spurzheim on artists and the arts, including painting (Finger and Eling 2019; Spurzheim, Lectures, 1815). He mischievously confessed that he had not yet been "able to satisfy himself as to the precise seat of this organ, or as to the extent of its specter of activity; but he has attempted an illusion of it" (Plate III). Then, in a remarkable series of verbal and visual puns, he showed in one central vignette a carter drawing a cart; a drawing academy; a child drawing a toy behind, staring at a dentist drawing a tooth; a publican at one window drawing a cork; a woman at another drawing beer; and a caged bird drawing up water from a tiny bucket. Cruickshank was at his best playing on words and images in multiple variations. At the same time, he demonstrated the inventiveness of the faculty he introduced,

Figure 20. Imitation & approbation from Plate V, *Phrenological Illustrations*, 1826.

The related organs of space, order, rorm, and color, essential to an artist, were placed around drawing. For space, Cruickshank suggested an alternative to the protuberance at the eyebrows toward the middle lien of the forehead "as the more positive seat of this faculty," and he portrayed a man with a huge, bulging stomach. Order was represented by the English school tradition of achieving it with the "birch." For the "blending of *colours*," he chose an African slave.

Phrenological caricatures by others

The originality, subtle wit, suggestive amusement, and gentle humor in Cruickshank's illustrations stand out all the more when we consider other contemporary phrenological illustrations both prior and subsequent to Cruickshank's prints. These appear to follow the recurring pattern of the phrenologist, either examining exaggerated "bumps" in the privacy of his consulting room or giving a public lecture. The depicted "phrenologists" are usually identified as Gall, Spurzheim, Combe, or De Ville. To ensure the satire is not missed, an added lengthy textual caption typically repeats what was already obvious.

For example, in the print published in September 1826, one month after Cruickshank's *Illustrations*, a phrenologist, his bald head deformed with red bumps, addresses a mixed audience of people also conscious of their own protuberances. The bookcases display rows of indiscriminately juxtaposed subjects, Latin and English titles of the classic and famed contemporary writers next to the "Memoirs" of Harriette Wilson (the notorious courtesan patronized by Wellington and Byron), and a plaster bust of Gall next to the sensational murderer, Thurtle (Fraser 2004). The title, *The Calves' Heads and Brains or a Phrenological Lecture*, is not quite represented. The names on the print are given as J. Lump for the artist and L. Bump for the engraver. The print is actually attributed to Henry Thomas Alken (1785–1851), a caricaturist and illustrator of sporting subjects (Figure 21). A concluding caption further satirizes the gullibility of the public: "On the Craniums of this highly gifted and scientific Audience, the Organ of Implicit faith Under Evident Contradictions, Stands beautifully develop'd to a Surprising and Prominent degree." The phrenologist here is identified as George Combe (1788–1858), whose second edition of the *Elements of Phrenology* (1824) had just been reviewed (September 1826) in the *Edinburgh Review* (George 1938, 1967; Gibbon 1878).

To view Cruickshank's *Illustrations* as simply another satirical portrayal of phrenology and to place it with the "grotesque caricatures in the style of Hogarth's *Physiognomists* drawn by E. F. Lambert and R. Cocking, among others" (Cooter 1984, 23), totally misses the profound significance of his unique treatment.

Cruickshank's contemporaries set him on a level with Hogarth, hoping "that posterity would delight in him as one of the foremost venerated old masters" (Jerrold 1882, 218). The invocation of Hogarth that had become a touchstone with which to judge effective comic acting (Davis 2018, 9) was conferred on him. It was not only his humor justly perceived as "his highest gift" but also his "honesty of purpose" as a humorist that was compared to Hogarth's. "Is any man more remarkable," asked Thackeray (1840, 52), "than

Figure 21. Henry Thomas Alken (1784–1851): *Calves' Heads and Brains or a Phrenological Lecture 1826.* On the print, the given names are J. Lump, Artist, and L Bump, Engraver. The print is attributed to Alken, a popular sporting illustrator. Lithograph.

our artist for telling the truth after his own manner?" As such, the *Phrenological Illustrations* may "have furnished material for the first attempt at illustrated journalism" (Jerrold 1980, 91).

Despite his extraordinary appeal across all classes, popularity, and warm praise, Cruikshank did not gain much income from the sale of his prints. They were sold by the hundreds, not by the thousands, with reorders of a dozen or so sets to be colored on demand (Patten 1992). He published the *Phrenological Illustrations* himself to bypass the print sellers, with the "Engravers' Copyright Act" (also known as "Hogarth's Act") still in the future (June 25, 1735) to legally recognize the authorial rights of individual artists.

The reissue of the *Phrenological Illustrations* in 1873

The Preface of the 1826 edition of Cruikshank's book is preceded by a quotation from Johann Gottfried von Herder (1744–1803):

> Where is the hand that shall grasp that which resides beneath the skull of man? Who shall approach the surface of the now tranquil, now tempestuous abyss? We shudder at contemplating the powers contained in such small a circumference, by which a world may be illuminated or a world destroyed.

The quote can be found in Johann Caspar Lavater's (1741–1802) *Essays on Physiognomy, Calculated to Extend the Knowledge and the Love of Mankind* (1797, 28). Cruickshank had a copy (Patten 1992). His inclusion of the quote on a separate dedicatory page is highly significant. The quote appeared in a review of the *Phrenological Illustrations* in *The Atheneum* (1827, Vol. 20, 72–73) by William Hone, his friend and bookseller. Hone used the first part of Herder's question ("Where is the hand that shall grasp that which resides beneath the skull of man?") to give Cruickshank as the answer. It was an astute judgment as Cruickshank dealt exactly with the manifestations of "what lies beneath the skull": "the organs of the brain" and not the bumps on the surface.

Furthermore, he impresses as being fully aware of the developments in phrenology. In the 1826 preface, he fairly accurately conformed to Spurzheim's table of classification and number of organs, as well identifying some of his differences from Gall. He even mentioned "Mr. Foster" with reference to his alternative definition of "ideality." Thomas Forster had used the term "phrenology" before Spurzheim (although he did not coin the word), and the two men were close friends (Forster 1815).

In his new preface to the reissue of the *Phrenological Illustrations* in 1873, Cruickshank named some of the major critics of phrenology. Lord Francis Jeffrey's review of Combe's *A System of Phrenology* in 1826 had delivered a severe blow to the phrenologists by amassing almost all of the previously applied logical, theological/ethical, and anatomical arguments against it (*Edinburgh Review*, 1826). Furthermore, William B. Carpenter, a physiologist, had seriously challenged phrenology's credibility in 1846. Cruickshank then referred to the scientific meetings and the research of David Ferrier, the leading contemporary figure on localization of brain function (Clarke 1971; Ferrier 1873, 1874; Finger 2004; Pearce 2003; Report 1874; Young 1968, 1970/1990). His inimitable style of "punning" is worth presenting in full.

> As the subject of phrenology has been brought before the public by the "British Association," at their meeting held at Bradford, on the 20th of Sept. this year (1873). I have thought it

desirable (for reasons which I will explain) to republish my Illustrations of that subject, first published in 1826 nearly fifty years back—when Drs. Gall & Spurzheim and Brothers Combes, were looked upon as great "Nobs" in the matter, when their doctrines were fully believed in, acted upon, and were exceedingly popular; although opposed at the time by some men of high intellect, such as Lord Jeffrey, and others, and at a later date—in 1846—by Drs. Carpenter, and now it appears that Dr. Ferrier, with Dr. Carpenter and other Doctors and Professors have given Gall & Spurzheim's Phrenology a most terrible "Knock on the Head," so much so that by their hard thumps on the "Bumps," they have "Turned the Brain" of Phrenology completely round, so that the guiding power, like the helm of a ship, is at the stern, or back, instead of, as it was formerly believed to be, in the fore-part of the front; the eyes of course, the "Lookout" or "Watch." (Preface 1873)

How Cruickshank came to know of these figures may be more of a reflection of the high publicity given to such meetings and discussions—as, for example, with Ferrier's research, reported in the *Times* (September 22 and 24, 1873)—than of extensive reading on his part.

His biographers emphasize that he had no formal education. Presumably, he picked up his knowledge from the streets, conversations, and popular entertainment (Jerrold 1980; Patten 1992).

Conclusions

Does Cruickshank's *Phrenological Illustrations*, one may ask, constitute a satire on phrenology as such, or a brilliant use of phrenology to convey the *comédie humaine*, perceived without passing judgment with insight, gentle humor, and compassion by an acute observer of human nature and social behavior?

First of all, Cruickshank's illustrations, in rendering the abstract graphically tangible, captured what has been described as the unique capacity of phrenology: "Never again was there to be anything like phrenology—at least not with the capacity to, at the same time, render the intangible tangible, to explain comprehensibly the human mind, and to predictively and optimistically act upon and inform the human condition" (Cooter 1989, 270–271).

Moreover, as Cruickshank himself wrote:

> Now the observer will see, in looking over my *Illustrations of Phrenology*, that the alterations in the supposed operations of the "Organs of the Brain," do not in any way alter the facts and features of my illustrations. For instance, the "organ of Tune" which I have represented was grinding in 1826 has been grinding ever since, and will go on grinding as long as the organ lasts; and "Self-love" and "Amativeness" must continue as long as the world shall last; and with respect to "Adhesiveness"—unfortunately many persons will still be found" sticking to the mud:" and indeed I may say that all the organs represented will still hold their place. (Preface 1873)

Cruickshank was prescient in the timeless appeal of his *Phrenological Illustrations*. What his unassuming "little *jeu d'esprit*" graphically achieved was no less than conveying the *comédie humaine*—the "ambitious vision" subsequently attempted by such literary figures as Honoré de Balzac (1799–1850).

Disclosure statement

No potential conflict of interest was reported by the author.

References

Altick R. D. 1978. *The shows of London.* Cambridge: Harvard University Press.

Andrews M. 2006. *Charles Dickens and his performing selves: Dickens and the public readings.* London: Oxford University Press.

Anonymous. 1815. Essay on Drs. Gall and a system of craniology, &c. *The Gentleman's Magazine* XXXV: 440–442.

Anonymous. 1816. *Craniology burlesqued: In three serio-comic lectures, humbly recommended to the patronage of Drs. Gall and Spurzheim, by a friend to common sense.* 2nd ed., 1818. London: Effingham Wilson.

Anonymous. 1824. Further particulars of Thurtell &c. *Examiner*, January 18, 40–41.

Anonymous. [Review]. 1826. Spurzheim's principles of phrenology. *The Phrenological Journal and Miscellany* 3 (October):XII: 635.

Anonymous. 1829. Phrenology. *The Literary Gazette: A Weekly Journal of Literature, Science, and the Fine Arts*, September, 599–600.

Anonymous. 1910. Paul Pry. In *Encyclopaedia Britannica*, 780. 11th ed. New York: The Encyclopaedia Britannica Co.

Baker K. 2004. Doyle, John [HB] (1797–1868). *Dictionary of National Biography*.

Barnet M. C. 1965. Mathew Allen (1783–1845). *Medical History* 9:16–28. doi:10.1017/s0025727300030106.

Bentley G. E. 1982. The freaks of learning. *Colby Quarterly* 18 (2):Art. 3:88–104.

Browne J. P. 1846. Memoir of the late Mr. James De Ville. *Phrenological Journal* 19:329–44.

Buchanan-Brown J. 1980. *The book of illustrations of George Cruickshank.* London: Charles Tuttle Co.

Bynum W. F. 1968. An old maid's skull phrenologised. *Journal of the History of Medicine* 23:386.

Cattrell V. A. C. 2006. *City of laughter: Sex and Satire in eighteenth-century London.* London: Atlantic Books.

Clarke E. 1971. David Ferrier. In *Dictionary of scientific biography*, ed. C. C. Gillespie, vol. 4, 593–95. New York: Charles Scribner.

Cooter R. J. 1984. *The cultural meaning of popular science; phrenology and the organization of consent in nineteenth century Britain.* Cambridge: Cambridge University Press.

Cooter R. J. 1989. *Phrenology in the British isles: An annotated, historical bibliography and index.* Metuchen: The Scarecrow Press, Inc.

Cruickshank G. 1826a. *Phrenological illustrations.* Six plates. Price 8s. plain; 12s. colored. London: Robins and Company.

Cruickshank G. 1826b. *Phrenological illustrations, or an artist's view of the phrenological system of doctors Gall and Spurzheim.* London: George Cruickshank, Myddelton Terrace, Pentonville. Sold by J. Robins and Co. Ivy Lane, Paternoster Row.

Cruickshank G. 1873. *Phrenological illustrations, or an artist's view of the phrenological system of doctors Gall and Spurzheim.* London. Republished for the artist. Frederick Arnold, 86 Fleet St.

Davis I. 2018. *Comic acting and portraiture in Late-Georgian and regency England*, 5–9. London: Cambridge University Press.

De Ville J. 1824. *Outlines of phrenology, as an accompaniment to the phrenological bust.* London: J. De Ville, 367 Strand.

De Ville J. 1826. *Manual of phrenology as an accompaniment to the phrenological bust.* London: J. De Ville, 367 Strand.

Ferrier D. 1873. Experimental researches in cerebral physiology and pathology. *West Riding Lunatic Asylum Medical Reports* 3:30–96. Illus., p. 42.

Ferrier D. 1874. On the localization of the functions of the brain. *The British Medical Journal* 1:679.

Finger S. 2004. *Minds behind the brain. A History of the pioneers and their discoveries.* London: Oxford University Press.

Finger S., and P. Eling. 2019. *Franz Joseph Gall—Naturalist of the mind, visionary of the brain.* New York: Oxford University Press.

Forster T. 1815. XII observations on a new system of phrenology, or the anatomy and phsiology of the brain, of Drs. Gall and Spurzheim. *The Philosophical Magazine* 45 (201):44–50. doi:10.1080/14786441508638384.

Fraser A. 2004. Thurtell John (1794–1824). In *Oxford dictionary of national biography*, ed. H. C. G. Matthew and B. Harrison. Oxford: Oxford University Press.

George M. D. 1938. *Catalogue of the political and personal satires preserved in the Department of Prints and Drawings in the British Museum*, vol. VI, 1784–92. pp. xxvi, 197, 259. (1952): Vol. X. London: British Museum.

George M. D. 1967. *Hogarth to Cruickshank: Social change and graphic satire*. London: Walker & Co.

Gibbon C. 1878. *Life of George Combe*. London: Macmillan & co.

Gombrich E. H., and E. Kris. 1938. The principles of caricature. *British Journal of Medical Psychology* 17:319–42. doi:10.1111/j.2044-8341.1938.tb00301.x.

Gombrich E. H., and E. Kris. 1940. *Caricature*. Harmondsworth: Penguin Books.

Gordon J. 1815. The doctrines of Gall and Spurzheim. *Edinburgh Review* 25 (June):227–68.

Hone W. 1827. On George Cruickshank. *The Atheneum* 20:72–73.

Jeffrey F. 1826. A system of phrenology. *Edinburgh Review* 44 (September):253–318.

Jeffrey F., and J. Gordon. 1817/1818. *Craniad, or Spurzheim illustrated*. Edinburgh: W. Blackwood; London: T. Cadell & W. Davies.

Jerrold B. 1882. *The life of George Cruickshank*. In *Two Epochs*, vol. 2, London: Chatto & Windus.

Kitton F. G. 1899/1903. *Dickens and His Illustrators: Cruickshank*. Amsterdam: S. Emmering.

Moore R. E. 1948. *Hogarth's literary relationships*. Minneapolis: University of Minnesota Press.

North C., and J. Wilson. 1826. Noctes Ambrosiana, No xxix. *Blackwood's Edinburgh Magazine* 20 119:770–792, 782. November 1.

Parssinen T. M. 1974. Popular science and society: The phrenology movement in early Victorian Britain. *Journal of Social History* 8 (1) Autumn:1–20, 39. doi:10.1353/jsh/8.1.1.

Patten R. 1992. *George Cruikshank's life, times, and art: 1792–1835*, vol. 1. New Brunswick: Rutgers University Press.

Patten R. L. 1974. *George Cruickshank: A revaluation*. Ed. R. Patten, vol. 35. Princeton: Princeton University Library. Issues 1–2. [NR 1992.]

Patten R. L. 1983. Conventions of Georgian caricature. *Art Journal* 43 (4):331–38. doi:10.1080/00043249.1983.10792251.

Patten R. L. 2004/2007. George Cruickshank. In *Oxford dictionary of national biography*, ed. H. C. G. Matthew and B. Harrison. Oxford: Oxford University Press.

Paulson R. 1989. *Hogarth's graphic works*. 3rd ed. London: The Print Room.

Pearce J. M. S. 2003. The West Riding Lunatic Asylum. *Journal of Neurology Neurosurgery & Psychiatry* 74 (8):1141. September. doi:10.1136/jnnp.74.8.1141.

Review. 1826. Spurzheim's principles of phrenology. *The Phrenological Journal and Miscellany* 3. Edinburgh (August 1825-October 1826); [Also: Review of *Phrenological Illustrations*] Vol. IV, Art. XII: 635–39.

Report. 1874. *The 43rd meeting of the British association for the advancement of science*. Held at Bradford 20 September, vol. 2. London: John Murray.

Secord J. A. 2014. *Visions & science*. Oxford: Oxford University Press.

Seurat C. 1825. A living skeleton. authentic memoir of that singular human prodigy. In *Illustrated with three engravings of the extraordinary human being, accurately drawn from the life by Robert Cruickshank*, ed. C. Seurat. London: John Fairburn, Broadway, Ludgate Hill.

Spurzheim J. 1825a. Lectures. *Atheneum*, June 4, 38.

Spurzheim J. 1825b. Lectures on Phrenology. *Lancet* viii (2):270–75. June 1. (July 16): 33–38. doi:10.1016/S0140-6736(02)83754-0.

Spurzheim J. 1825c. *Phrenology, or, the doctrine of the mind*. 3rd ed. London: H. Knight, Pall Mall East.

Tegg T. 1831. The fortune teller. In *The vocal annual, or singer's own book for 1831: A collection of the newest and most popular songs of the day*, 275–77. London: Thomas Tegg.

Thackeray W. M. 1840. Essay on the genius of George Cruickshank. *Westminster Review* 34 (June):1–60,5–6.

Wyhe J. V. 2004. Was phrenology a reform science? Towards a new generalization for phrenology? *History of Science* xlii:313–31.

Wyhe J. V. 2007. *Phrenology and the origins of Victorian scientific naturalism.* Farnham: Ashgate.

Young R. M. 1968. The functions of the brain: Gall to Ferrier (1808–1886). *ISIS* 59 (3):250–68. doi:10.1086/350395.

Young R. M. 1970/1990. *Mind, brain and adaptation in the nineteenth century: Cerebral localization and its biological context from Gall to Ferrier.* Oxford: Oxford University Press.

Mark Twain's phrenological experiment: Three renditions of his "small test"

Stanley Finger

ABSTRACT

Samuel Langhorne Clemens (1835–1910), the American humorist and author better known as Mark Twain, was skeptical about clairvoyance, supernatural entities, palm reading, and certain medical fads, including phrenology. During the early 1870s, he set forth to test phrenology—and, more specifically, its reliance on craniology—by undergoing two head examinations with Lorenzo Fowler, an American phrenologist with an institute in London. Twain hid his identity during his first visit, but not when he returned as a new customer three months later, only to receive a very different report about his humor, courage, and so on. He described his experiences in a short letter written in 1906 to a correspondent in London, in humorous detail in a chapter that appeared in a posthumous edition of his autobiography, and in *The Secret History of Eddypus, the World Empire*, a work of fiction involving time travel, which he began to write around 1901 but never completed. All three versions of Twain's phrenological ploy are presented here with commentary to put his descriptions in perspective.

Physiognomy, the idea that character can be determined by examining the physical features of the body, has roots in antiquity but has continued to draw supporters well into the modern era. During the closing years of the eighteenth century, Viennese physician Franz Joseph Gall (1758–1828) considered his doctrine a variant of physiognomy (Finger and Eling 2019). He contended that, rather than there being just a few basic faculties of mind, such as perception, cognition, and memory, there are a great many independent faculties. He ultimately settled on 27 for humans, including organs for tune, mathematics, and wit, writing that, although we share 19 of these organs with other animals, eight are unique to humans. He further attempted to show that each faculty is dependent on a specific part of the cortex.

Gall identified the sites of his chosen cortical organs by studying and comparing heads and skulls from the extremes of society (e.g., thieves, artistic geniuses, great mathematicians), as well as by studying various animals, comparing genders, and following development. His basic assumption was that the growing brain shapes the skull, and he argued that physiologically large and active organs could be correlated with tell-tale bumps, whereas abnormally small organs would reveal themselves by specific

---

Color versions of one or more of the figures in the article can be found online at www.tandfonline.com/njhn.

deficiencies and locus-specific cranial depressions. Thus, cranioscopy was Gall's primary method, although he also studied individuals with brain damage, conducted human autopsies, dissected animals, correlated emerging traits with brain region growth, and the like.

Gall presented his revolutionary doctrine in his *Anatomie et Physiologie du Système Nerveux en Général, et du Cerveau en Particulier*, a four-volume work with an accompanying atlas from 1810–1819, and soon after in his less-expensive *Sur les Fonctions du Cerveau et sur Celles de Chacune de ses Parties* of 1825, translated a decade later as *On the Functions of the Brain and Each of Its Parts* (Gall 1825, 1835; Gall and Spurzheim 1810–1819). He never called his life's work "phrenology." This term was coined by American physician Benjamin Rush in 1805 and was popularized by Johann Spurzheim (1776–1832), who had assisted Gall before setting off on his own in about 1813 (Finger and Eling 2019; Noel and Carlson 1970; Rush 1811). After abandoning the term "physiognomy," which he used while still in Vienna, Gall preferred *Schädellehre* when presenting his doctrine in German and *organologie* when writing in French, sometimes also using other terms for it. Nonetheless, his doctrine, Spurzheim's modifications of it, and other variations of the doctrine linking faculties of mind to specific cortical areas were soon subsumed under phrenology.

To say the least, phrenology was controversial from the start. For conservative clerics and their followers, associating the mind with the brain smacked of materialism and seemed to preclude free will. Aware of this, Gall devoted substantial parts of his lectures and writings to show otherwise. Other critics were more focused on whether his claims were verifiable and replicable, as befitting a true science. Some individuals observed and palpated skulls on their own, and were unable to confirm what Gall and his followers were contending. Others set forth to conduct autopsies and dissections, with some anatomists contending that the external table of the skull does not faithfully follow finer brain morphology—this dissociation being particularly notable for the smaller organs Gall had located behind the (obscuring) frontal sinuses (e.g., Anonymous [Gordon] 1815; Hamilton 1831, 1845). To the phrenologists starting with Gall, however, these critics were simply poor observers. Had they been more skilled, they would have been able to confirm nature's real "facts."

The phrenologists were also accused of selecting only cases consistent with their *a priori* beliefs and biases, while showing either a blatant disregard for contradictory cases or finding dubious reasons for dismissing more troublesome cases. Peter Mark Roget (1779–1869), the British physician best remembered for his *Thesaurus*, was not lost for words when he wrote that, with "such convenient logic … it would be easy to prove anything," and for this very reason, "they will be rejected as having proved nothing" (Roget 1824, 437; Kruger and Finger 2013)

Despite these criticisms against phrenology as a real science, phrenology continued to thrive in the United States during the second quartile of the nineteenth century. As noted by historian John Davies (1955), few American physicians conducted revealing experiments on the subject, and the great majority of Americans were not particularly interested in delving into the scientific evidence for or against the doctrine. The inhabitants of the New World were, however, pragmatic, and what they found in phrenology was an easily understood way to know themselves and others, and a useful guide for living a healthier life, making education and career choices, choosing a spouse, and so on.

Two enterprising young men—Orson Squire Fowler (1809–1887) and his younger brother, Lorenzo Niles Fowler (1811–1896)—later joined by Samuel Roberts Wells (1820– 1875), were quick to turn phrenology into a highly profitable business on American soil. Starting in Boston in 1835, they and the men they trained set forth to read skulls across the expanding nation, in addition to writing phrenology books, publishing a specialized phrenological journal, selling skulls, casts, and porcelain busts, and offering courses, charts, and other materials for would-be practitioners anxious to capitalize on the growing fad (Stern 1971).

Mark Twain and the allure of phrenology

Samuel Langhorne Clemens (1835–1910), America's most famous writer and humorist, grew up in the small Missouri town of Hannibal in this *zeitgeist* (see Figure 1). He would later adopt the pen name Mark Twain (a riverboat term signifying deep enough water for passage) and reflect on the industry created by "Fowler and Wells." In a recent edition of his *Autobiography*, we find:

> In America, forty or fifty years ago, Fowler and Wells stood at the head of the phrenological industry, and the firm's name was familiar in all ears. Their publications had a wide currency and were read and studied and discussed by truth-seekers and converts all over the land. (Twain 2013, 335)

Young Clemens was an acute observer, and he speculated that the itinerant phrenologist was less than honest when claiming to be reading heads, even if the residents of

Figure 1. American writer and humorist Samuel Langhorne Clemens (Mark Twain) in 1909, the year before he died. (Photograph courtesy of Google Images.)

Hannibal were not suspicious. He was particularly bothered by how the phrenologist always made his clients seem like George Washington or some other esteemed person, a practice that would, of course, bring forth more paying customers.

> It is not at all likely, I think, that the traveling expert ever got any villager's character quite right, but it is a safe guess that he was always wise enough to furnish his clients character-charts that would compare favorably with George Washington's. … This general and close approach to perfection ought to have roused suspicion, perhaps, but I do not remember that it did. It is my impression that the people admired phrenology and believed in it and that the voice of the doubter was not heard in the land. (Twain 2013, 335)

Twain left Hannibal in his late teens, and he read a phrenology book while living and working in St. Louis in 1855 (Weaver 1852). Intrigued, he even copied parts of it into one of his notebooks (Twain 1855/1976). During this time and over the next 17 years as he traveled the country, he appeared to think fairly positively about phrenology as a quick means for assessing character. But with the passage of time, he had additional doubts about the men claiming to be proficient at reading skulls. Driven to see things with his own eyes, he decided to put phrenology to the test during the early 1870s, while in London.

What he would later call his "small test" would involve a simple deception. He would have one of the most respected phrenologists of the time provide him with a head reading, while doing everything he could to conceal his own identity. Then, after some time had passed, he would return as Mark Twain to have his head reexamined, not informing the practitioner that this was, in fact, a return visit.

Twain's ploy was both brilliant and practical. After all, if the before-and-after charts revealed the same features, he would have new respect for phrenology as a means for assessing character and a useful tool for his character portrayals. And should the two readings differ in significant ways, he would have ample subject matter to lampoon in his lectures, novels, and travelogues, which were now providing him with the income he needed to maintain his growing family along with their lavish home in Hartford, Connecticut.

Twain's chosen phrenologist was Lorenzo Fowler, who, along with his older brother Orson, had left the family's publishing house in 1863, leaving its management to other family members. Lorenzo subsequently emigrated to London, where he founded the Fowler Institute on Fleet Street (Stern 1969, 1971). Twain could not have found a better target, as Lorenzo was extremely well known on both sides of the Atlantic.

For reasons unknown, Twain did not convey what transpired in a public way immediately after seeing Fowler. In fact, he revealed nothing about his London head readings prior to the new century. He did, however, begin to ridicule phrenology, the phrenologists, and what they represented shortly after visiting the Fowler Institute. He did this in his two greatest works, *Tom Sawyer* (Twain 1876/1911) and *Huckleberry Finn* (Twain 1884/2003), and in several other well-known writings (Finger 2019). No longer intrigued by phrenology as a doctrine that might have some validity, or merely suspicious of phrenologists always giving positive head readings, or left pondering about what cues these men really used, Gall's so-called "new science" of man with its many variants had become fair game for a humorist with a sharp pen that could function like a loaded gun.

The brief letter version

Near the end of his life, Twain received a letter from England. It was written on December 6, 1906, by Frederic Whyte (1867–1941), a Reuters correspondent and author/translator who had become an editor at Cassell and Company. His objective was to get Twain to write an opinion piece about phrenology for a popular London periodical.

Whyte had read Alfred Russel Wallace's (1823–1913) *The Wonderful Century*, in which the British naturalist, anthropologist, explorer, and evolutionist argued for a serious study of phrenology—asking what it encompassed, the evidence for it, and its utility (Wallace 1898). He explained to Twain that he had convinced the editors of the *Daily Graphic* to have a symposium about phrenology in this periodical, one in which the opinions of leading scientists and other famous people would be presented. He mentioned the names of some of the prominent individuals he had invited, before writing, "I am most anxious to have a few lines from you."

In his return letter, Twain referred to Whyte as a "Gentleman" from England, who "wonders why phrenology has apparently never interested me enough to move me to write about it" (Twain 2013, 334). Whyte was not entirely, correct. True, Twain had not published anything exploring, endorsing, or exposing phrenology, although he had used phrenological terms and included phrenological fragments in some of his literary pieces. But Whyte was clearly mistaken in thinking that Twain had never pursued phrenology. Although Twain rightfully informed him in his return letter, dated December 18, 1906, that he did not publish on the subject, because "I never did *profoundly* study phrenology," he was, in fact, fairly knowledgeable about the subject and its purveyors.

While still in St. Louis, he had read the Reverend George Sumner Weaver's (1818–1908) *Lectures on Mental Science According to the Philosophy of Phrenology* (Weaver 1852) and had even copied parts of it (including its diagram of a head showing faculty groupings) into a notebook. Later on, he conversed with other phrenologists, including Frederick Coombs (1803–1874), the eccentric author of *Popular Phrenology* (Coombs 1841; see Cowan, Bancroft, and Ballou 1964, 4–7). And by the time of Whyte's letter, he had had his head "read" by at least three leading phrenologists: Lorenzo Fowler in 1872 or 1873; Edgar C. Beall (fl. 1890s) in 1885; and Lorenzo's enterprising daughter, Jessie Fowler (1856–1932), in 1901.

In his return letter, Twain explained why he had been ridiculing phrenology since the early 1870s. He did this after stating that he "never did *profoundly* study phrenology," and while contending, "therefore I am neither qualified to express an opinion about it nor entitled to do so" [italics added]. In his words:

> In London, 33 or 34 years ago, I made a small test of phrenology for my better information. I went to Fowler under an assumed name and he examined my elevations and depressions and gave me a chart which I carried home to the Langham Hotel and studied with great interest and amusement—the same interest and amusement which I should have found in the chart of an impostor who had been passing himself off as me and did not resemble me in a single sharply defined detail. I waited 3 months and went to Mr. Fowler again heralding my arrival with a card bearing both my name and *nom de guerre*. Again I carried away an elaborate chart. It contained several sharply defined details of my character but it bore no recognizable resemblance to the earlier chart. These experiences gave me a prejudice against phrenology which has lasted until now. I am aware that the prejudice should have been

against Fowler, instead of against the art; but I am human and that is not the way prejudices act. (UCLC 07599; also Twain 1958, 64; 2013, 334)

Whyte printed Twain's response January 12, 1907, in London's *Daily Graphic* under the heading "Bumps and Brains." It was titled "Mark Twain's 'Small Test': Why He Is Prejudiced Against Phrenology" (Twain 1907a). British scientist Francis Galton (1822–1911) and authors G. K. Chesterton (1874–1936) and George Bernard Shaw (1856–1950) also contributed to the symposium.

Whyte sent a note of thanks to Twain on January 18, 1907 (Mark Twain Papers/Berkeley UCLC 35853), and he included a copy of Twain's now-published letter. Rather than being pleased, Twain reacted furiously. Not fully comprehending what Whyte was requesting, he assumed he had provided him with a personal letter, not a document for everyone to read. On January 29, 1907, he had his secretary fire off a scathing letter to Whyte, telling him, "there was nothing about your first letter that indicated you would use his private answer in a public way," further asserting that what was done violated Twain's contract with his publisher (UCLC 07632).

There are no known subsequent letters from Whyte to Twain or Twain to Whyte. But Twain was so incensed that he decided to attach a brief but vitriolic footnote to his autobiographical piece dated February 10, 1907. In his words, "The English gentleman was not really a gentleman: he sold my private letter to a newspaper" (Twain 2013, 336).

The autobiographical version

The best-known version of Twain's ploy was dictated shortly after he wrote his letter to Whyte (Finger 2019). Nonetheless, this longer, more humorous portrayal of what transpired in London was not published until Charles Neider included it in his 1958 edition of Twain's autobiography (Twain 1958). It can also be found in the second volume of the newer "complete and authoritative edition" of his autobiography, the one quoted here (Twain 2013).

After writing that it was his "impression that the people [of his childhood] admired phrenology and believed in it, and that the voice of the doubter was never heard in the land" (p. 335), Twain continued:

I was reared in this atmosphere of faith and belief and trust, and I think its influence was still upon me, so many years afterward, when I encountered Fowler's advertisements in London. I was glad to see his name and glad of an opportunity to personally test his art. The fact that I went to him under a fictitious name is an indication that not the whole bulk of the faith of my boyhood was still with me; it looks like circumstantial evidence that in some way my faith had suffered impairment in the course of the years. I found Fowler on duty, in the midst of the impressive symbols of his trade. On brackets, on tables, on shelves, all about the room, stood marble-white busts, hairless, every inch of the skull occupied by a shallow bump, and every bump labeled with its imposing name, in black letters.

Fowler received me with indifference, fingered my head in an uninterested way, and named and estimated my qualities in a bored and monotonous voice. He said I possessed amazing courage, an abnormal spirit of daring, a pluck, a stern will, a fearlessness that were without limit. I was astonished at this, and gratified too; I had not suspected it before; but then he foraged over on the other side of my skull and found a hump there which he called "caution." This hump was so tall, so mountainous, that it reduced my courage-bump to a mere hillock by comparison, although the courage-bump had been so prominent up to that time—

according to his description of it—that it ought to have been a capable thing to hang my hat on; but it amounted to nothing, now, in the presence of that Matterhorn which he called my Caution. He explained that if that Matterhorn had been left out of my scheme of character I would have been one of the bravest men that ever lived—possibly the bravest—but that my cautiousness was so prodigiously superior to it that it abolished my courage and made me almost spectacularly timid. He continued his discoveries, with the result that I came out safe and sound, at the end, with a hundred great and shining qualities; but which lost their value and amounted to nothing because each of the hundred was coupled up with an opposing defect which took the effectiveness out of it. However, he found a *cavity*, in one place; a cavity where a bump would have been in anybody's else's skull. That cavity, he said was all alone, all by itself, occupying a solitude, and had no opposing bump, however slight in elevation, to modify and ameliorate its perfect completeness and isolation. He startled me by saying that that cavity represented the total absence of the sense of humor! He now became almost interested. Some of his indifference disappeared. He said he often found bumps of humor which were so small that they were hardly noticeable, but that in his long experience this was the first time he had ever come across a *cavity* where a bump ought to be.

I was hurt, humiliated, resentful, but I kept these feelings to myself; at bottom I believed his diagnosis was wrong, but I was not certain. In order to make sure, I thought I would wait until he should have forgotten my face and the peculiarities of my skull, and then come back and try again and see if he had really known what he had been talking about, or had only been guessing. (Twain 2013, 335–36)

And with regard to his return visit, Twain wrote:

After three months I went to him again, but under my own names this time. Once more he made a striking discovery—the cavity [for Humor] was gone, and in its place was a Mount Everest—figuratively speaking—thirty-one thousand feet high, the loftiest bump of humor he had ever encountered in his life-long experience! I went from his presence prejudiced against phrenology, but it may be, as I have said to the English gentleman [Whyte], that I ought to have conferred the prejudice upon Fowler and not upon the art which he was exploiting. (Twain 2013, 336)

This was Mark Twain at his best, the humorist readers and audiences around the world adored for his tongue-in-cheek humor, color, fanciful allusions, perfectly placed pauses, and comical understatements. The piece was also classic Mark Twain because it carried an important message about the gullibility of humankind for half-baked ideas presented as facts. As such, it would be the version cited in books and articles briefly noting Twain's rejection of phrenology (e.g., Finger 2019; Finger and Eling 2019; Lopez 2002; Stern 1969, 1971; Stone 2003).

The *Eddypus* version

The least-quoted version of Twain's deception can be found in a short story, *The Secret History of Eddypus, the World Empire*. Of the three extant pieces, this rendition was actually the first one written and the only one presented as fiction or fantasy. Indeed, were it not for Twain's published letter to Whyte and his more detailed autobiographical piece, this version, with its commentator looking back from a time in the far distant future, would probably go unrecognized as a whimsical takeoff on what had been a real-life experience.

Twain never finished *Eddypus* and he was not anxious to see it published in his lifetime. Rather, he wanted its publication to be postponed until well after his death, because he knew it would offend some individuals and religious groups. Adding to the intrigue, the precise dating of *Eddypus*, and especially the part based on his phrenological deception, is less than certain.

Based on a notebook entry from February 3, 1901, it would seem that Twain had just commenced working on *Eddypus* at this time. It has been suggested that his phrenological chapter was written shortly after he had his head "read" by Jessie Fowler, Lorenzo's surviving daughter, on March 7, 1901, at the firm's American Phrenological Institute in New York City (Tuckey 1972, 20). While there, he either purchased or was gifted a copy of Orson and Lorenzo Fowler's *New Illustrated Self-Instructor in Phrenology and Physiology* (Fowler and Fowler 1859). His personal copy was signed "Clemens, 1901," and he based the boring lecture on phrenology in *Eddypus* on it.

The phrenological chapter with its two different head readings gives the impression of being a later "add-on." John S. Tuckey (1972, 24) wrote, it "is not well integrated with what precedes and follows," and surmised, "it appears to have been thrown into *Eddypus* for no better reason than that Clemens was interested in phrenology at the time." Robert Hirst, currently the head of the Mark Twain Papers at Berkeley, also described it as "clearly a kind of insertion," when he showed it to me in 2018.

Twain's youngest daughter, Jean Clemens (1880–1909), appeared to be the person responsible for typing most of the penned text (including the phrenological chapter). But she did not type last three chapters of Book 2, which were probably written in 1902. Thereafter, Twain seemed to have shelved this project (Tuckey 1972, 24–27).

The historical backdrop for the *Eddypus* story can be traced to Twain's decision to join an ongoing war against Mary Baker Eddy (1821–1910). During the 1860s, Eddy began to promote her version of religious-based faith healing, and in 1879, she established the Church of Christ, Scientist, which succeeded in attracting a considerable following. Then, in 1899, she gave a "millennial" talk about her movement's progress and future, a speech covered by many reporters (Gill 1998). Although lauded by her followers, what she said irritated many people, leading some of her antagonists to use the media to portray her as insane, a fraud, or both mad and a charlatan—actions intended to slow her influence (Squires 2017). Twain joined these forces with his powerful pen because, as he astutely commented when dealing with some of his human targets, murder was illegal.

Twain never set forth to denigrate faith healing. He understood the power of the mind and had witnessed enough personally—including with his mother, wife, and children—to realize that faith healing could, under some circumstances, function as an adjunct to traditional medicine (Ober 2003, 207–22; Squires 2017, 89–117). In fact, it seemed to work unusually well with "believers," whose problems originated from misguided or "sick" minds. Instead, Twain went after "Mother Eddy" as a "shameless old swindler," "the queen of frauds and hypocrites," and a demagogue intent on amassing massive institutional power. He portrayed her as an abusive tyrant bilking the masses, and as a leader unable to tolerate dissent—hence, a threat to democracy (Tuckey 1972, 21; Twain 1972, 323). Furthermore, how Eddy was dogmatically promoting the idea that *everything* is in the mind was, to an intellect like Twain's, akin to taking a giant step backward into the Dark Ages.

Twain did his best to expose Eddy in a series of satirical articles in the *North American Review* and *Cosmopolitan* from 1899 to 1903, as well as in a later compilation titled *Christian Science* (Twain 1907b). He also wrote letters, some directly to Eddy. *Eddypus* was a part of his assault. Indeed, in selecting this title for his piece, he combined Eddy's name with "pus," as he likened what she was spreading to a nasty infection.

In this fictional work, which has us looking back from 1,000 years in the future to Twain's own time, Twain appears as the "sometimes Bishop of New Jersey." As put by the narrator to a secret correspondent, Bishop Twain was "a revered priest of the earlier faith," described as being before Christian Science and Roman Catholicism joined together to establish a despotic stranglehold that threw the world into profound darkness. Branded a heretic for his opposition to how history was being rewritten and for how theocracy was replacing democracy, he was hanged in "A.D. 1912 = A.M. 47." The ancient book that had just been discovered was written 14 years before Bishop Twain's demise.

In Chapter 1 of Book 2 of *Eddypus*, we are told that the Bishop, despite his impressive learning, had a notable defect, "his lack of the sense of humor"! The narrator even provided examples to show how he could have been funnier (Twain 1972, 339). That he lacked a sense of humor was supposedly what Lorenzo Fowler had told Twain during his first visit to his London establishment. Clearly, Twain was still unable to stop thinking about what the phrenologist had told him during the early 1870s.

In Chapter 2 of Book 2, the narrator provided a "curious character sketch" of the Bishop. And this is where Twain seemed to paste the London ploy into this fictional work.

The narrator began:

One perceives that a poet had paid the historian [Bishop Twain] a majestic compliment; that it had produced a physical change in his skull, in the nature of an enlargement, that he had hopes that this might mean a corresponding enlargement of his mental equipment, and also additions to the graces of his character. To satisfy himself as to these matters he went to a magician to get enlightenment. He calls this person a "phrenologist." He nowhere explains, except figuratively, who or what the phrenologists were, and it seems probable he was not able to classify them quite definitely; for whereas in the beginning of his third chapter he twice speaks of them as "those unerring diviners of the human mind and the human character," in later chapters he always refers to them briefly and without ornament as "those damned asses."

In this place I will insert the first division of the fragmentary character-sketch; and, with diffidence, I will add a suggestion: Might not the historian have been mistaken concerning the poem? It does not mention him by name; may it not have been an apostrophe to his country, instead of to him? (Twain 1972, 348)

The Bishop's own words followed:

It was in London — April 1st, 1900. In the morning mail came a Harper's Weekly, and on one of its pages I found a noble and beautiful poem, fenced around with a broad blue-pencil stripe. I copy it here.

THE PARTING OF THE WAYS

Untrammeled Giant of the West,
With all of Nature's gifts endowed,
With all of Heaven's mercies blessed,
Nor of thy power unduly proud —
Peerless in courage, force, and skill,
And godlike in thy strength of will, —

Before thy feet the ways divide:
One path leads up to heights sublime;
Downward the other slopes, where bide

The refuse and the wrecks of Time.
Choose then, nor falter at the start,
O choose the nobler path and part!

Be thou the guardian of the weak,
Of the unfriended, thou the friend;
No guerdon for thy valor seek,
No end beyond the avow'd end.
Wouldst thou thy godlike power preserve,
Be godlike in the will to serve!
                                        —Joseph B. Gilder.

It made me blush to the eyes. But I resolved that I would do it, let it cost me what it might. I believe I was never so happy before.

My head began to swell. I could feel it swell. This was a surprise to me, for I had always taken the common phrase about swell-head as being merely a figurative expression with no foundation in physical fact. But it had been a mistake; my head was really swelling. Already—say within an hour—the sutures had come apart to such a degree that there was a ditch running from my forehead back over to my neck, and another one running over from ear to ear, and my hair was sagging into these ditches and tickling my brains.

I wondered if this enlargement would enlarge my mental capacities and make a corresponding aggrandizement in my character. I thought it must surely have that effect, and indeed I hoped it would. There was a way to find out. I knew what my mental calibre had been before the change, and I also knew what my disposition and character had been: I could go to a phrenologist, and if his diagnosis showed a change, I could detect it. So I made ready for this errand. I had no hat that would go on, but I made a turban, after a plan which I had learned in India, and shut myself up in a four-wheeler and drove down Piccadilly, watching out for a sign which I had several times noticed in the neighborhood of New Bond street. I found it without trouble

    —"BRIGGS AND POLLARD AMERICAN PHRENOLOGISTS." (Twain 1972, 348–50)

Bishop Twain explained why he did not want to reveal his identity or pertinent personal information during his first visit to the office of the two men.

What I desired was the exact truth. If I gave my real name and quality, these people would know all about me: might that not influence their diagnosis? might they not be afraid to be frank with me? might they not conceal my defects, in case such seemed to be found, and exaggerate what some call the great features of my mind and character? in a word, might they not dishonorably try to curry favor with me in their own selfish interest instead of doing their simple and honest duty by me? Indeed this might all happen; therefore I resolved to take measures to hide my identity; I would protect myself from possible deception, and at the same time protect these poor people from sin. (Twain 1972, 350)

His rendition continued:

Briggs and Pollard were on hand up stairs. There were bald-headed busts all around, checkered off like township maps, and printed heads on the walls, marked in the same way. Briggs and Pollard had been drinking, but I judged that the difference between a phrenologist drunk and a phrenologist sober was probably too small to materially influence results. I unwound the turban and took a seat, and Briggs stood up behind me and began to squeeze my head between his hands, paw it here and there, and thump it in spots—all in impressive silence. Pollard got his note-book and pencil, and made ready to take down Briggs's observations in short-hand. Briggs asked my name; I told him it was Johnson. Age? I told him another one. Occupation? Broker, I said—in Wall Street—when at home. How long a broker? Five feet eight and a half [this being Twain's actual height]. Question misunderstood, said Briggs: how long in the broking *business*?

Always. Politics? Answer reserved. He got other information out of me, but nothing valuable. I was standing to my purpose to get an estimate straight from the bat and the bumps, not a fancy scheme guessed out of the facts of my career. Briggs used a tape-measure on me, and Pollard wrote down the figures:

"Circumference, 46 inches. Scott! this ain't a human head, it's a prize pumpkin, escaped out of the country fair."

It seemed an unkind remark, but I did not say anything, for allowances must be made for a man when his beverages are working.

"Most remarkable craniological development, this is," mumbled Briggs, still fumbling; "has valleys in it." He drifted into what sounded like a lecture; not something fresh, I thought, but a flux of flatulent phrases staled by use and age. "Seven is high-water mark on the brain-chart of the science; the bump that reaches that altitude can no further go. Seven stands for A_1, *ultima thule*—that is to say, very large; organ marked by 7 is sovereign in its influence over character and conduct, and, combining with organs marked 6 (called large), direct and control feeling and action; 5 (called full) plays a subordinate part; it and 6 and 7 press the smaller ones into their service; 4 (called average) have only a medium influence; 3 (called moderate) below *par*; medium influence, more potential than apparent; 2 (called deficient) leaves the possessor weak and faulty in character and should be assiduously cultivated; while organs marked 1 are very small, and render their possessor almost idiotic in the region where they predominate." (Twain 1972, 350–51)

And turning to what they revealed about the Bishop's true character:

"In the present subject we find some interesting combinations. Combativeness 7, Destructiveness 7, Cautiousness 7, Calculation 7, Firmness 0. Thus he has stupendous courage and destructiveness, and at first glance would seem to be the most daring and formidable fighter of modern times; but at a second glance we perceive that these desperate qualities are kept from breaking loose by those two guardians which hold them in their iron grip day and night,—Cautiousness and Calculation. Whenever this bloody-minded fiend would crave and slash and destroy, he stops to calculate the consequences; then he quits frothing at the mouth and puts up his gun; at this point his total destitution of Firmness surges to the front and he gets down in the dirt and apologizes. This is the low-downest poltroon I've ever struck." (Twain 1972, 351)

The Bishop of New Jersey was stunned.

This ungracious speech hurt me deeply, and I came near to striking him dead before I could restrain myself; but I reflected that on account of drink he was not properly responsible for his acts, and also was probably the sole support of his family, if he had one, so I thought better of it and spared him for their sake; in case he had one. Pollard had a hatchet by him; I was not armed.

Amativeness 6. Probably keeps a harem. No; spirituality, 7. That knocks it out. A broker with spirituality! oh, call me early mother, call me early mother dear! Veneration 7. My! can that be a mistake? No—7 it is. Oh, I see—here's the solution: self-esteem 7. Worships himself! Acquisitiveness, 7; secretiveness, 7; conscientiousness, 0. A fine combination, sir, a noble combination." I heard him [Briggs] mutter to himself, "Born for a thief."

"Veracity? Good land, a *socket* where the bump ought to be! And as for— (Twain 1972, 351–52)

The narrator stepped back in at this point, commenting:

There the first division breaks off. The Bishop makes no comment, but leaves it so. This silence is to me full of pathos; it is eloquent of a hurt heart, I think; I feel it, and am moved by it, after the lapse of ten centuries; centuries which have swept away thrones, obliterated dynasties and the very names they bore, turned cities to dust, made the destruction of all grandeurs their province, and have not suffered defeat till now, when this little, little thing rises up and mocks them with its immortality—the unvoiced cry of a wounded spirit!

The Bishop did not rest there. He had come to believe that the phrenologists were merely guessers, nothing more, and that they could rightly guess a man only when they knew his history. He resolved to test this theory. He waited several months, then went back to those experts clothed in his ecclesiastical splendors, with his chaplain and servants preceding and announcing him, and submitted his mentalities and his character to examination once more. His "regimentals," as he calls them, disguised him, and the magicians were not aware that they had seen him before. This is all set down in the seventh chapter of Volume 4 and forms the first paragraph of the second division of the fragmentary character-sketch. The Bishop then summarizes the results of his two visits, under the head of "Remarks of the Charlatan Briggs—with Verdicts." Thus:

OBSCURE STRANGER.	RENOWNED BISHOP
"Not a head—a prize pumpkin."	"A noble head—sublime!"
"Low-down poltroon."	"Lion of the tribe of Judah!"
"Bloody-minded fiend"	"Heart of an angel!"
"Probably keeps a harem"	"Others are dirt in presence of this purity!"
Worships himself."	"Here we have divine humility!"
"Born for a thief."	"This is the very temple of honor!"
"Veracity? Good land!"	"This soul is the golden palace of truth!"

The fragment closes with this acrid comment: "Phrenology is the 'science' which extracts character from clothes" (Twain 1972, 352–53; see Figure 2).

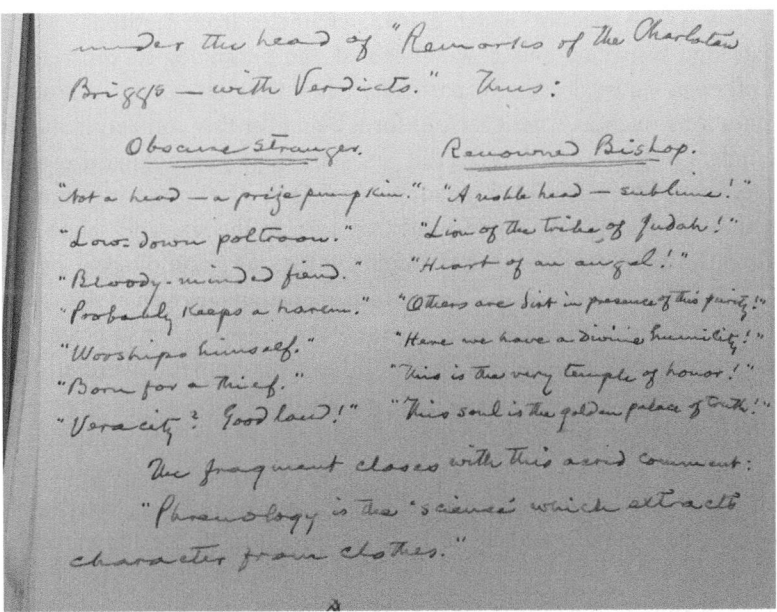

Figure 2. The page from *Eddypus* on which Twain penned, "Phrenology is the 'science' which extracts character from clothes." (Photograph taken by the author at the Bancroft Library, University of California, Berkeley, with permission of Robert Hirst, Head, Mark Twain Papers.)

Discussion

Mark Twain's ploy involving two trips to a leading phrenologist, the first hiding his identity and the second identifying himself, tells us much about both the man and the status of phrenology at the time (for more, see Finger 2019; Gribben 1972). But although the "new science" was already three-quarters of a century old and no longer quite as faddish during the 1870s, when his deception took place, what is uncertain is whether he was mimicking others who might have tried the same sort of trickery or whether he had dreamed up the deception on his own.

There were earlier attempts to hide identities from phrenologists, some even from before 1828, when Gall died. For example, after leaving Vienna for Berlin, where he lectured and visited asylums during the spring of 1805, Gall made his way to nearby Potsdam, where he had been invited to lecture by Queen Louise (1776–1810). He spoke before an audience of 230 people on May 3, 1805, but unlike his queen, King Friedrich Wilhelm III (1770–1840) remained skeptical. Hence, the monarch invited him to a banquet with prisoners dressed as army officers. Gall supposedly recognized bumps signifying *angriffslust* (hostility) and *zerstörungswut* (destruction) on one of the "soldier's" skulls, and so impressed the king with his assessments that the ruler revealed his ruse and presented him with a ring containing precious stones (Oehler Klein 1990, 50, note 120; Finger and Eling 2019).

Twain's deception in London, however, ended differently than Gall's in Potsdam. This time the leading phrenologist of the day emerged looking like a charlatan and a fool. But of greater importance is what Twain gleaned from Fowler's two very different head readings. He concluded that, rather than determining basic character traits by reading skulls, the phrenologists were, in fact, guided largely or exclusively by noncranial cues. Twain damned the profession in a tongue-in-cheek way in *Eddypus*, when he concluded that phrenology is "the 'science' which extracts character from clothes."

As he had long suspected and now confirmed, the phrenologists must also have been relying on other noncranial cues. In particular, they were attending to how their clients answered questions, such as what they did for a living. In this context, it should be noted that the founders of Fowler and Wells did not start their phrenological examinations by observing, feeling, and measuring a person's head. Only after assessing "the constitution, temperament, or make-up" of a subject did they turn to the size and shape of the skull— and only then did they move on to assessing the various phrenological organs of mind (Stern 1971, 17). "Few people today," Twain historian Alan Gribben (1972, 47) wrote, "are aware that the conformation of the skull by itself was considered insufficient for complete phrenological knowledge of a person, and that the temperaments were an indispensable component of the discipline worked out by Spurzheim."

George Combe (1788–1858), who was mentored by Spurzheim, further emphasized knowing a person's temperament, which he associated with four different systems (sanguine, bilious, nervous, lymphatic), albeit allowing for mixtures (e.g., nervous-lymphatic). And following what Spurzheim did in 1832, Combe introduced his variation of Gall's doctrine to welcoming Americans in 1838–1840. The now-modified doctrine, with its emphasis on assessing temperament, was readily incorporated into phrenology texts such as George Weaver's *Lectures on Mental Science According to the Philosophy of Phrenology* —the book Samuel ("Sammy") Clemens read in St. Louis in 1855.

Thus, Twain emerged from his two head readings fully aware of the fact that phrenologists were gleaning valuable information not just by studying crania but by attending to how a person dressed, his or her answers to pertinent questions, facial expressions, and so on. That he had anticipated this finding is suggested by how he dressed, responded to questions, and conducted himself during his first session with Lorenzo Fowler. His deception was, in fact, based on first feeding Fowler false information.

Twain recognized that having access to personal information could make any phrenologist (or medium or palmist) appear like a master of the trade and, more generally, human nature (Quirk 2007). In *Eddypus*, he even wrote about how Bishop Twain "had come to believe that the phrenologists were merely guessers ... and that they could rightly guess a man only when they knew only when they knew his history." He even brought the importance of having pertinent information to the fore well before starting on *Eddypus*.

In 1884, in his *Adventures of Huckleberry Finn*, Twain described how Huck allows two tricksters to join him and Jim (a runaway slave) on their raft as they make their way down the dangerous Mississippi River. One of the frauds claims to be the lost son of the deceased Duke of Bridgewater, and therefore the rightful inheritor of his title, whereas the other, not to be outdone, passes himself off as the "pore disappeared Dauphin, Looy the Seventeen"—that is, the King of France! When Huck inquires what they do for a living, the "Duke" states he does "a little in patent medicines; theatre-actor ... ; take a turn at mesmerism and *phrenology* ... ; sling a lecture sometimes; ... most anything that comes handy," whereupon the "King" says he can cure cancer and paralyses by "layin' on o' hands," preaches, and is able to "tell a fortune pretty good, *when I've got somebody along to find out the facts for me*" (Twain 1884/2003, 121; italics added).

Still, the question that remains is whether the detailed information Twain presented about his two trips to the phrenologist in his 1907 autobiographical account can be accepted as factual. Earlier historians have raised this important question, and all seem to agree that Twain took his share of liberties to make his autobiographical accounts more exciting, colorful, and appealing. For example, Madeline Stern doubted that Lorenzo Fowler, whom she described as an "authority on mnemonics," could "have forgotten Mark Twain's unmistakable features in a matter of three months," further noting that "bumps" and "cavities" "were rarely part of the Fowler phrenological vocabulary" (Stern 1971, 183). Alan Gribben also expressed some concerns, writing, "Mark Twain's efforts at writing truthful autobiography frequently fell into the comic pattern of his fiction" (Gribben 1972, 63).

Neither Stern nor Gribben mentioned the striking asymmetries Fowler supposedly discovered on Twain's skull, which, if real, would have been surprising for a person as mentally and physically healthy as Twain had been. Large cranial asymmetries are more characteristic of mentally impaired people, individuals with brain damage. Also telling is that the phrenological books and charts sold by Fowler and Wells treated the cortical organs as symmetrical entities, providing a place for a single number from 1 to 7 for each organ's power. Unfortunately, the charts Twain received have been lost and there are no surviving records of what Fowler actually wrote or dictated about Twain's bumps and phrenological organs.

For these and other reasons, questions will continue to be raised about precisely what transpired in London and, more to the point, the extent to which Twain took artistic liberties, inadvertently presented false memories, or allowed his fertile imagination to run unchecked when describing his phrenological examinations. After all, Twain expressed no regrets about mixing storytelling with facts in his *Autobiography*. "I don't believe these details are right but I don't care a rap," he once stated, continuing, "They will do just as well as the facts" (Twain 1958, xiv). And with this mindset, he did not hesitate to stretch the truth and embellish his life story in ways that amused him—ways he knew would elicit smiles from his readers.

What remains highly likely is that Twain did make two trips to Lorenzo Fowler's institute in London during the early-1870s, that he did try to hide his identity on the first but not his second visit, and that he was provided with different "head" readings. Additionally, after previously showing interest in phrenology as a means of character assessment, he returned to America intent on lampooning the phrenologists and how they really garnered information in his writings. Further suggesting that his later descriptions of what transpired with Lorenzo Fowler were not entirely made up is a document showing that he had planned to carry out the same sort of two-visit deception on a New Orleans fortune teller a decade earlier (Paine 1917, 51).

In retrospect, Twain's ploy in its three versions (i.e., his brief letter to Whyte, his autobiographical dictation, and his futuristic novel) is revealing of a man with many interests, sometimes a slash-and-burn agenda, and always a fertile mind. At the same time, what he wrote also reveals a lot about phrenology and how it was faring over his lifetime (for more, see Finger 2019). For all intents and purposes, and as reflected in Twain writings, phrenology as a scientific endeavor to understand the mind and brain had died with Gall in 1828. But it was still a way for ambitious itinerants and well-heeled businessmen to make a living in America before, during, and even after its disruptive Civil War of the 1860s. Phrenology was a movement with significant social features and, for writers, a fad with literary appeal and a fabulous vocabulary of its own.

The *zeitgeist* was clearly changing when Twain took it upon himself to walk into Lorenzo Fowler's emporium on what might have been a sunny or perhaps a foggy or rainy day in London. Forward-looking physicians (e.g., Paul Broca in France) and experimentalists (e.g., Gustav Fritsch and Eduard Hitzig in Prussia; David Ferrier in England) were now busily assessing patients with brain damage and studying animals subjected to lesions or electrical brain stimulation (Finger 1994, 2000; Young 1970). These new pioneers of the brain rejected craniology yet, in accordance with what Gall was the first to claim publicly, they were now confirming that specialized parts of the cerebral cortex are, in fact, associated with different functions, such as speech and voluntary movements. Also consistent with what the phrenologists had long been claiming, their research was showing that our most noble faculties (e.g., speech, controlled attention, higher ideation) are dependent on cortical territories in the front of the brain.

In many ways, Twain, a brash American with little formal education who boldly stepped forth to test phrenology when it was already losing favor, was in tune with the spirit of the times. Joining the vanguard during the 1870s, he came prepared to dismiss and ridicule the craniological part of phrenology as only he could with his wit and his pen. In so doing, he provided vivid descriptions of how phrenologists were swindling people, while also lacing his verbal portraits with phrenological terms to convey the situation even more clearly. For instance, he not hesitate to associate an adversary's more primitive propensities with the growth of the posterior region of his brain, as he did with riverboat captain Tom Ballou in *Life on the Mississippi*. Twain wrote that Ballou "had more selfish organs than any seven men in the world—all packed in the stern-sheets of his skull, of course, where they belonged," adding as only he could, that they "weighed down the back of his head so that it made his nose tilt up in the air" (Twain 1883/ 2012, 171).

Also worthy of note in this context is how Twain, especially during his final decade, seemed unable to shake the sobering thought that we are more machine-like than we might think, implying that free will is merely an illusion (e.g., in *The Mysterious Stranger*; Twain 1916). These ideas can be associated with the shift from metaphysically oriented to brain-based theories of mind that was taking place at this time.

Many of Twain's best writings shed light on the rise and fall of what French physiologist François Magendie (1783–1855) labeled a "pseudo-science" (Magendie 1834, 89). What Magendie did not realize, but what Twain just might have understood, is that, while phrenology suffered from faulty methodologies, it still put into play many seminal ideas about the mind (e.g., a great number of inherent faculties) and brain (cortical localization of function) that certainly merited attention. These positive features of phrenology, which Twain never challenged when assailing cranioscopy, would become foundation stones for the modern neurosciences.

Acknowledgments

I would like to thank Paul Eling for his help on this manuscript, which was stimulated by the research for our recently published book on Franz Joseph Gall and the origins of phrenology. I am also thankful to Robert Hirst of the Mark Twain Project (Berkeley) for all of his help, including directing me to important resources, showing me original copies of *Eddypus*, and commenting on an earlier draft of this article.

Disclosure statement

No potential conflict of interest was reported by the author.

References

Anonymous. [Gordon, J.]. 1815. The doctrines of Gall and Spurzheim. *Edinburgh Review* 25: 227–68.
Coombs F. 1841. *Coomb's popular phrenology*. Boston: F. Coombs.
Cowan R. E., A. Bancroft, and A. L. Ballou. 1964. *The forgotten characters of old San Francisco*. San Francisco: Ward Richie Press.
Davies J. D. 1955. *Phrenology fad and science: A 19th-Century American crusade*. New Haven, CT: Yale University Press.
Finger S. 1994. *Origins of neuroscience: A history of explorations into brain function*. New York: Oxford University Press.
Finger S. 2000. *Minds behind the brain: A history of the pioneers and their discoveries*. New York: Oxford University Press.
Finger S. 2019. Mark Twain's fascination with phrenology. *Journal of the History of the Behavioral Sciences* 55: 99–121. doi: 10.1002/jhbs.21960.
Finger S., and P. Eling. 2019. *Franz Joseph Gall: Naturalist of the mind, visionary of the brain*. New York: Oxford University Press.
Fowler O., and L. Fowler. 1859. *New illustrated self-instructor in phrenology and physiology*. New York: Fowler and Wells.

Gall F. J. 1825. *Sur les Fonctions du Cerveau et sur Celles de Chacune de ses Parties.* Vol. 6. Paris: J.-B. Baillière.

Gall F. J. 1835. *On the functions of the brain and each of its parts: With observations on the possibility of determining the instincts, propensities, and talents, or the moral and intellectual dispositions of men and animals, by the configuration of the brain and head.* Vol. 6. Ed. N., Trans. W. Lewis. Boston: Marsh, Capen and Lyon.

Gall F. J., and J. G. Spurzheim. 1810-19. *Anatomie et Physiologie du Système Nerveux en Général, et du Cerveau en Particulier.* Vol. 4 and atlas. Paris: F. Schoell.

Gill G. 1998. *Mary Baker Eddy.* New York: Perseus Books.

Gribben A. 1972. Mark Twain, phrenology and the "temperaments": A study of pseudoscientific influence. *American Quarterly* 24:45–68. doi:10.2307/2711914.

Hamilton W. 1831. An account of experiments on the weight and relative proportions of the brain, cerebellum, and tuber annulare, in man and animals, under the various circumstances of age, sex, country, etc. In *The anatomy of the brain, with some observations on its functions,* ed. A. Munro, 4–8. Edinburgh: J. Carfrae.

Hamilton W. 1845. Researches on the frontal sciences, with observations on their bearings on the dogmas of phrenology. *Medical Times* 12:159–60, 177–79, 371, 379.

Kruger L., and S. Finger. 2013. Peter Mark Roget: Physician, scientist, systematist; his *Thesaurus* and his impact on 19th-Century neuroscience. In *Literature, neurology, and neuroscience: Historical and literary connections,* ed. S. Finger, A. Stiles, and F. Boller, 173–95. Oxford: Elsevier. Progress in Brain Research 203.

Lopez D. J. 2002. Snaring the Fowler: Mark Twain debunks phrenology. *Skeptical Inquirer* 26:33–36.

Magendie F. 1834. *Précis Élémentaire de Physiologie.* 4th ed. Brussels: H. Dumont.

Noel P. S., and E. T. Carlson. 1970. Origins of the word "Phrenology". *American Journal of Psychiatry* 127:154–57. doi:10.1176/ajp.127.5.694.

Ober K. P. 2003. *Mark Twain and medicine: Any mummery will cure.* Columbia: University of Missouri Press.

Oehler Klein S. 1990. *Die Schädellehre Franz Joseph Galls in Literatur und Kritik des 19. Jahrhunderts.* Stuttgart: Fischer Verlag.

Paine A. B. 1917. *Mark Twain's letters.* Vol. 1. New York: Harper & Brothers.

Quirk T. 2007. *Mark Twain and human nature.* Columbia: University of Missouri Press.

Roget P. M. 1824. Cranioscopy. In *Supplement to the fourth, fifth, and sixth editions of the Encyclopaedia Britannica,* vol. 3, 419–37. Edinburgh: Printed for A. Constable and Co. [Originally published 1818.].

Rush B. 1811. *Sixteen introductory lectures to courses of lectures upon the institutes and practice of medicine.* Philadelphia, PA: Bradford and Innskeep.

Squires L. A. 2017. *Healing the nation: Literature, progress, and christian science.* Bloomington: Indiana University Press.

Stern M. B. 1969. Mark Twain had his head examined. *American Literature* 41:207–18. doi:10.2307/2923950.

Stern M. B. 1971. *Heads and headlines: The phrenological fowlers.* Norman: University of Oklahoma Press.

Stone J. L. 2003. Mark Twain on phrenology. *Neurosurgery* 53:1414–16. doi:10.1227/01.NEU.0000093429.94129.F0.

Tuckey J. S. 1972. Introduction. In *Mark Twain's Fables of man,* ed. M. Twain, 1–29. Berkeley: University of California Press.

Twain M. 1855/1976. *Mark Train's notebooks & journals, 1855-1873. Vol. 1: What I was at 19–20.* Berkeley: University of California Press.

Twain M. 1876/1911. *Adventures of Tom Sawyer.* Montgomery, AL: NewSouth Books.

Twain M. 1883/2012. *Life on the Mississippi.* Ware, UK: Wordsworth Editions.

Twain M. 1884/2003. *Adventures of Huckleberry Finn.* New York: Bantam Dell.

Twain M. 1907a. Mark Twain's "small test." Why he is prejudiced against phrenology. *The Daily Graphic* (London), January 12. [pages unnumbered].

Twain M. 1907b. *Christian science.* New York: Harper and Brothers.

Twain M. 1916. *The mysterious stranger.* New York: Harper & Brothers.
Twain M. 1958. *The autobiography of Mark Twain, including chapters now published for the first time,* ed. C. Neider. New York: Harper and Brothers.
Twain M. 1972. The secret history of Eddypus, the world empire. In *Unpublished manuscript in Mark Twain's Fables of man,* ed. J. S. Tuckey, 315–85. Berkeley: University of California Press.
Twain M. 2013. *Autobiography of Mark Twain.* Vol. 2. Ed. B. Griffin and H. E. Smith. Berkeley: University of California Press.
Wallace A. R. 1898. *The wonderful century: Its successes and its failures.* London: Swan, Sonnenschien, & Co.
Weaver G. S. 1852. *Lectures on mental science according to the philosophy of phrenology.* New York: Fowler and Wells.
Young R. M. 1970. *Mind, brain and adaptation in the 19th Century.* Oxford, UK: Clarendon Press.

Dr. Oliver Wendell Holmes on phrenology: Debunking a fad

Stanley Finger

ABSTRACT

Oliver Wendell Holmes, Sr. (1809–1894) was a Boston physician, a professor of medicine at the Harvard Medical School, and a writer of prose and poetry for general audiences. He was also one of the most famous American wits of the nineteenth century and a celebrity not bashful about exposing costly, absurd, and potentially harmful medical fads. One of his targets was phrenology, and the current article examines how he learned about phrenology during the 1830s as a medical student in Boston and Paris, and his head-reading with Lorenzo Fowler in 1858. It then turns to what he told readers of the *Atlantic Monthly* (in 1859) and Harvard medical students (in 1861) about phrenology being a pseudoscience and how phrenologists were duping clients. By looking at what Holmes was stating about cranioscopy and practitioners of phrenology in both humorous and more serious ways, historians can more fully appreciate the "bumpy" trajectory of one of the most significant medical and scientific fads of the nineteenth century.

Dr. Oliver Wendell Holmes (1809–1894)—frequently confused with his son, Supreme Court Justice Oliver Wendell Holmes, Jr. (1841–1935)—was a popular figure in the United States during the nineteenth century (for biographies, see Morse 1896; Howe 1939; Hoyt 1979; Tilton 1947). He achieved prominence as a physician, by writing poetry, essays, serials, and novels, and as an entertaining speaker at various events. Holmes (Figure 1) might have been America's best-known physician and most celebrated wit, "a man of reason and an irrepressible humorist," throughout much of the century (Gibian 2001, 1).

As a physician-scientist, Holmes did his best to expose worthless and sometimes harmful medical practices. And with his highly creative mind, he did not hesitate to merge his exceptional abilities to speak, write, and make people laugh with his thoughts about futile medical practices. Along with homeopathy and heroic therapies (e.g., bleeding, purging), phrenology was one of his targets. Nonetheless, Holmes's biographers and others writing about his forays into science and medicine have provided little beyond a few lines from one source showing how Holmes ridiculed the phrenologists and some of their claims prior to the American Civil War, when phrenology as a science was on the decline, but when the laity were flocking to have their heads "read" (e.g., Dowling 2006; Gibian 2001; Hoyt 1979; Podolsky and Bryan 2009; Tilton 1947).

With the hope of providing a better picture of what Holmes wrote about phrenology and why, the present article will begin with some basic information about phrenology in

Figure 1. Dr. Oliver Wendell Holmes (1809–1894), *circa* 1879.

Boston and Paris during the 1820s and 1830s, while Holmes was a medical student. Mention will then be made of his desire to modernize American medicine with its many follies. Following how he had his own head read in 1858, two pieces showing what he concluded about phrenology will be examined: one from the *Atlantic Monthly* in 1859, and the other from an 1861 published lecture delivered at Harvard, where he was a fixture on the medical faculty. Commentary and closing thoughts about Holmes, his agenda, and his impact will be provided.

Holmes's Boston

Oliver Wendell Holmes was born in Cambridge, Massachusetts, close to the Harvard campus, in 1879. His father, a Calvinist minister, enrolled him in Phillips Academy when he was 15 years old. A year later, in 1825, he matriculated at Harvard College, graduating in 1829 with a reputation for being bright and quick witted, but also as a nonstop talker.

Holmes spent his next year studying law at Harvard's Dane Law School, only to find that it was not for him. He did, however, achieve fame at this time with a short poem called "Old Ironsides," published in 1830, which helped save the decommissioned American frigate (*The Constitution*) from destruction.

Having no interest in the ministry or business, and knowing poetry could not be a career choice, he next tried medicine. He enrolled in Boston's Medical College, founded in 1810. Although staffed by only five physicians, it was in many ways an extension of the Harvard Medical School, then called the Massachusetts Medical College, where some of his professors also taught.

During this time, at least half of Boston's medical profession endorsed phrenology, which intrigued both physicians and laity in the region (Walsh 1976). Anatomist and surgeon John Collins Warren (1778–1856), one of his teachers, was one of phrenology's first supporters, having learned about Gall and his doctrine while in Europe in 1801–1802 (Warren 1860, 1921). In 1830, he began giving annual "dissertations" on the subject, and in 1832, when Johann Gaspard Spurzheim (1776–1832) visited Boston, he sought out Warren, who invited him to lecture at the Medical College. Spurzheim also gave public lectures at Boston's Atheneum Hall and at Harvard, and when he became ill and died that November, Warren helped with his autopsy. Quoting one of Holmes's biographers, "Many of those who had sat in on the autopsy on Spurzheim had also attended his public lectures, among them young Holmes" (Tilton 1947, 74).

Holmes might also have learned about phrenology from anatomist and surgeon Winslow Lewis, Jr. (1799–1875), another of his teachers. Lewis removed Spurzheim's head at the autopsy, and he was also a founding member of the Boston Phrenological Society, established later in 1832.

Physician James Jackson (1777–1867) also merits attention in this context. He treated Spurzheim during his final illness. Holmes felt close to Jackson, and when Jackson lectured on Spurzheim's autopsy, Holmes recorded what he had to say in a notebook labeled *Lectures on the Theory and Practice of Medicine*.

Thus, Holmes met, knew, and interacted with several highly regarded Boston-area physicians who met Spurzheim and had an understanding of phrenology. Without question, he would also have garnered additional information by reading about Gall and Spurzheim's neuroanatomical research and phrenological ideas in books, journals, and even the local newspapers. Unfortunately, what Holmes thought about the doctrine at this moment in time is unknown.

Medicine and phrenology in Paris

Encouraged by James Jackson, Holmes continued his medical education (prior to getting his diploma) in Paris. This was where Gall and Spurzheim had settled in 1807 and wrote their books, after leaving Vienna in 1805 (Finger and Eling 2019). With many forward-looking teachers, numerous hospitals, free lectures, and unmatched opportunities to observe patients and conduct autopsies, Paris had more to offer aspiring medical students than any other city at the time (e.g., Ackerknecht 1967; Dowling 2006; Jones 1973, 1978; Shyrock 1947, 1960; Tilton 1947; Warner 1998).

Pierre Charles Alexandre Louis (1787–1872) was the physician who had the greatest effect on Holmes. A modest man, Louis worked at La Pitié, a hospital on the Left Bank of the River Seine. He favored evidence-based medicine, meaning medicine supported by detailed case notes, extensive observations, tests, autopsies, large samples, and statistics. Louis eschewed general theories of disease, as well as time-honored "heroic" measures, which he personally found ineffective at the bedside (Louis 1828). With Louis serving as his guide, Holmes's stated goal was to modernize American medicine, in part by making it more scientific and in part by purging it of worthless medicines and exposing frauds.

As was the case in Boston, Holmes attended lectures by a number of physicians and surgeons in Paris known to favor phrenology. Several were active in the *Société Phrénologique*, founded in 1831 (Ackerknecht 1956; Laugée 2014; Renneville 1998). Gabriel Andral (1797–1876), who Holmes admired, was the organization's president, and François-Joseph-Victor Broussais (1772–1838), whose ideas Holmes considered old fashioned, was its secretary. He also mentioned Jean-Baptiste Bouillaud (1796–1882) in his 1882 reminiscences of his mentors,[1] although not in the context of this *Société*.

Holmes left Paris shortly before the *Musée de Phrenologie* was opened to the public in January 1836. This was also just before Broussais began giving his popular lecture series on phrenology and had his first phrenology book published (Broussais 1836). Still, Holmes probably visited a nearby phrenology shop, where John Collins Warren's son, Jonathan "Mason" Warren (1811–1867), a fellow medical student in Paris, went to buy items for his father's phrenology collection. In addition, he reminisced about seeing Baron Dominique Jean Larrey (1766–1842), Napoleon Bonaparte's (1769–1821) favorite surgeon, who had lectured on phrenology and supplied Gall with some of his most famous cases (Gall 1835, 5, 16–18).

Who Holmes did not mention in his 1882 lecture on his early teachers is also interesting. He said nothing about Joseph Vimont, a member of the *Faculté de Médecine* who lectured on phrenology (in 1829) and was currently publishing his *Traité de Phrénologie Humaine et Comparée* (Vimont 1832–1835). Nor did he bring up Jean-Baptiste Mège (1791–1866) or Félix Voison (1794–1882), who were also promoting Gall's new science in Paris at this time.

With regard to phrenology's opponents, he mentioned having attended a lecture by François Magendie (1783–1855), but said nothing about him lecturing on phrenology or being the first physician to label it a pseudoscience (Magendie 1834, 89; Finger and Eling 2019, 480–481). As for Jean Pierre Flourens (1794–1867), who was then emerging as Gall's and Spurzheim's most visible opponent, there is nothing about him or his mentor, Georges Cuvier (1769–1832), in Holmes's letters home or reminiscences (Finger and Eling 2019, 454–463; Eling and Finger 2019).

Interestingly, Holmes took back two skeletons and some skulls on his return to Boston late in 1835. But how he was feeling about phrenology is again hard to judge, given that he had written no surviving letters about it, other than one informing his parents (in 1834) that he was "delighted to hear that they had a slight row [fight] in the prints about phrenology" (Morse 1896, 1, 125–126).

Medical quackery

For Holmes, speaking out against medical quackery or "humbugs" began in earnest in 1840, one year before he joined Harvard's medical faculty. In that year, he gave four public lectures titled *The Natural Diet of Man, Astrology and Alchemy, Medical Delusions of the Past*, and *Homeopathy and Its Kindred Delusions* (see Holmes 1892). He did his best to dismember Samuel Hahnemann's (1755–1843) homeopathy ("A mingled mass of perverse ingenuity, of tinsel erudition, of imbecile credulity, and of artful misrepresentation";

Holmes 1892, IX, 99), and chided physicians still relentlessly bleeding, purging, and blistering their patients.

Phrenology might also have been on his mind at this time. After all, Edinburgh phrenologist George Combe (1788–1858) had recently lectured in Boston (in 1838). And although phrenology was losing its luster among physicians, his teacher, John Collins Warren, still was continuing to collect skulls and casts, even opining in 1847, "the study of the forms of the crania enable us in some measure to understand the degree of intellectual power possessed by individuals" (Warren 1860, 13).

Holmes invited Warren to a lecture he delivered on phrenology at Harvard in 1850, but what he told his audience is far from clear. As for his book sources, one might have been Gall and Spurzheim's (1810–1819) "great work" in French. It was among the books in his private library (this collection is now at Harvard's Countway Library).

Whereas members of the medical profession were turning away from phrenology, interest in it as a practical tool for self-betterment and character assessment was increasing among the laity (Davies 1955). Phrenology's ascent to fad status in the United States was largely due to Lorenzo Niles Fowler (1811–1896) and his younger brother, Orson Squier Fowler (1809–1887), who opened their phrenology business in New York in 1836. Joined by Samuel Wells (1820–1875) a few years later, these businessmen reaped great profits reading heads, lecturing, giving courses, selling skulls and casts, and publishing charts, journals, and books (Stern 1971). By mid-century, even people in remote towns were lining up to have their heads read by the masters, their trainees, and numerous itinerant "professors" traveling from town to town. For Oliver Wendell Holmes, physician-scientist, phrenology's decline from an interesting (even if flawed) science to a popular amusement that told people what they wanted to believe about themselves had to be horrifying.

Holmes's head reading

Holmes decided to have his own head read in 1858 by Lorenzo Fowler, then in New York City. What we know about his head reading comes largely from a copy of the Fowlers' *The Illustrated Self-Instructor in Phrenology* (Fowler and Fowler 1859) containing the "Chart and Character of O. W. Holmes" that he once owned, and a penned document in the *Oliver Wendell Holmes Memorabilia Scrapbook*, which he might have started before others took it over (Lokensgard 1940, 714–715; Stern 1971, 130–131, 299).[2]

Fowler gave Holmes high ratings (sevens and sixes) for mental temperament and for his faculties of cautiousness, approbativeness, conscientiousness, benevolence, mirthfulness, locality, language, eventuality, causality, comparison, human nature, agreeableness, conjugality, combativeness, secretiveness, firmness, hope, constructiveness, sublimity, imitation, form, size, and order. But he only scored fours on continuity, veneration, color, calculation, and time, and he received an even lower score for self-esteem. How Fowler could have found so many distinct cortical organs in his head reading could have registered on Holmes before he even finished.

After stating that Holmes had a "nervous temperament," E. R. Gardiner, the recorder of the document in the Holmes scrapbook, wrote:

> The desire to distinguish yourself, and to be all that education & circumstances can possibly make you, is a very prominent trait. … You have excessive Conscientiousness & adhere very

rigidly to what you think is right. ... You have a great love of Fun & and a quick & ready sense of the ridiculous. ... You have very distinct powers of observation ... a good memory of facts. ... Language is very large; & you are seldom at a loss for words to communicate your ideas. .... You are well qualified to analyse, describe, illustrate & ... are firm and fixed in your purposes. ... You are combative & fond of debate, love opposition ... [and] would succeed best in some literary pursuit; in teaching some branch or branches of natural science.

The Professor at the Breakfast-Table

Holmes attacked the phrenologists and their cranioscopy in public just one month after having his head read. His venue was the *Atlantic Monthly*, a new periodical he helped start and had just named (Gibian 2009; Scudder 1901). In 1857, he published *The Autocrat of the Breakfast-Table* in monthly installments (Holmes 1892, 1). It featured a medically knowledgeable protagonist who started conversations on a wide range of topics, yet had to endure unpredictable and occasionally hilarious interruptions from other boarders. Readers begged for more and with newly found star status Holmes began a second series in 1859, *The Professor at the Breakfast Table* (Holmes 1892, II) Twelve years later he would publish still another set of conversations, *The Poet at the Breakfast Table* (Holmes 1892, III), which, like the others, would also be published as books.

In the August installment in the second series, the Professor sets forth to debunk phrenology. In its entirety, this segment reads as follows:

Having been photographed, and stereographed, and chromatographed, or done in colors, it only remained to be phrenologized. A polite note from Mssrs. Bumpus and Crane [as in cranium], requesting our attendance at their Physiological Emporium, was too tempting to be resisted. We repaired to that Scientific Golgotha.

Mssrs. Bumpus and Crane are arranged on the plan of the man and the woman in the toy called a "weather house," both on the same wooden arm suspended on a pivot,—so that when one comes to the door, the other retires backwards, and vice versa. The more particular specialty of one is to lubricate your entrance and exit,—that of the other to polish you off phrenologically in the recesses of the establishment. Suppose yourself in a room full of casts and pictures, before a counterful of books with taking titles. I wonder if the picture of the brain is there, "approved" by a noted Phrenologist, which was copied from *my*, the Professor's, folio plate, in the work of Gall and Spurzheim. An extra convolution, No. 9, *Destructiveness*, according to the list beneath, which was not to be seen in the plate, itself a copy of Nature, was very liberally supplied by the artist, to meet the wants of the catalogue of "organs." Professor Bumbus is seated in front of a row of [rich] women,—horn-combers and gold-beaders, or somewhere about that range of life,—looking so credulous, that, if any Second Advent Miller or Joe Smith should come along, he could string the whole lot of them on his cheapest lie, as a boy strings a dozen "shiners" [bait fish] on a stripped twig of willow.

The Professor (meaning ourselves) is in a hurry, as usual; let the horn-combers wait,—he shall be bumbed without inspecting the antechamber.

Tape round the head,—22 inches. (Come on, old 23 inches, if you think you are the better man!)

Feels thorax and arm, and muzzles round among muscles as those horrid old women poke their fingers into the salt-meat on the provision-stalls at the Quincy Market. Vitality, No. 5 or 6 [Ascending numbers from 1–10 signify how well developed faculties are], or something or other. *Victuality*, (organ of epigastrium) some other number equally significant.

Mild champooing [shampooing] of the head now commences. 'Extraordinary revelations! Cupidiphilous, 6! Hymeniphilous, 6+! Paediphilous, 5! Deipniphilous, 6! Gelasmiphilous, 6!

Musikiphilous, 5! Uraniphilous, 5! Glossiphilous, 8!! and so on. Meant for a linguist.— Invaluable information. Will invest in grammars and dictionaries immediately.—I have nothing against the grand total of my phrenological endowments.

I never set great store by my head, and did not think Mssrs. Bumpus and Crane would give me so good a lot of organs as they did, especially considering that I was a *dead*-head on that occasion. Much obliged to them for their politeness. They have been useful in their way by calling attention to important physiological facts. (this concession is due to our immense bump of Candor.)

Directly after this first part, Professor entered into "A short lecture on Phrenology, read to the Boarders at our Breakfast-Table."

I shall begin, my friends, with the definition of a Pseudo-science. A *Pseudo-science* consists of *nomenclature*, with a self-adjusting arrangement, by which all positive evidence, or such as favors its doctrines, is admitted, and all negative evidence, or such as tells against it, is excluded. It is invariably connected with some lucrative practical application, its professors and practitioners are usually shrewd people; they are very serious with the public, but wink and laugh a good deal among themselves. The believing multitude consist of women of both sexes, feeble minded inquirers, poetical optimists, people who always get cheated in buying horses, philanthropists who insist on hurrying up the millennium, and others of this class, with here and there a clergyman, less frequently a lawyer, very rarely a physician, and almost never a horse-jockey or a member of the detective police.—I do not say that Phrenology was one of the Pseudo-sciences.

A Pseudo-science does not necessarily consist wholly of lies. It may contain many truths, and even valuable ones. The rottenest bank starts with a little specie. It puts on a thousand promises to pay on the strength of a single dollar, but the dollar is very commonly a good one. The practitioners of the Pseudo-science know that common minds, after they have been baited with a real fact or two, will jump at the merest rag of a lie, or even at the bare hook. When we have one fact found us, we are very apt to supply the next out of our imagination. (How many persons can read Judges xv. 16 correctly the first time?) The Pseudo-sciences take advantage of this,—I did not say that it was so with Phrenology.

I have rarely met a sensible man who would not allow that there was *something* in Phrenology. A broad, high forehead, it is commonly agreed, promises intellect; one that is "villanous low" and has a huge hind-head back of it, is wont to mark an animal nature. I have rarely met an unbiased and sensible man who really believed in the bumps. It is observed, however, that persons with what the Phrenologists call "good heads" are more prone than others toward plenary belief in the doctrine.

It is so hard to prove a negative, that, if a man should assert that the moon was in truth a green cheese, formed by the coagulable substance of the Milky Way, and challenge me to prove the contrary, I might be puzzled. But if he offer[s] to sell me a ton of this lunar cheese, I call on him to prove the truth of the Gaseous nature of our satellite, before I purchase.

It is not necessary to prove the falsity of the phrenological statement. It is only necessary to show that its truth is not proved, and cannot be, by the common course of argument. The walls of the head are double with a great air-chamber between them, over the smallest and most closely crowded "organs." Can you tell me how much money there is in a safe, which also has thick double walls, by kneading its knobs with your fingers? So when a man fumbles about my forehead, and talks about the organs of *Individuality, Size*, etc., I trust him as much as I should if he felt of the outside of my strong-box and told me that there was a five-dollar or a ten-dollar-bill under this or that particular rivet. Perhaps there is; *only he doesn't know anything about it*. But this is a point that I, the Professor, understand, my friends, or ought to, certainly, better than you do. The next argument you will all appreciate.

I proceed, therefore, to explain the self-adjusting mechanism of Phrenology, which is *very similar* to that of the Pseudo-sciences. An example will show it most conveniently.

A. is a notorious thief. Mssrs. Bumpus and Crane examine him and find a good-sized organ of Acquisitiveness. Positive fact for Phrenology. Casts and drawings of A. are multiplied, and the bump *does not lose* in the act of copying.—I did not say it gained.—What do you look for? (to the boarders.)

Presently B. turns up, a bigger thief than A. But B. has no bump at all over Acquisitiveness. Negative fact; goes against Phrenology.—Not a bit of it. Don't you see how small Conscientiousness is? *That's* the reason B. stole.

And then comes C., ten times as much a thief as either A. or B.,—used to steal before he was weaned, and would pick one of his own pockets and put its contents in another, if he could find no other way of committing petty larceny. Unfortunately, C. has a *hollow*, instead of a bump, over Acquisitiveness. Ah, but just look and see what a bump of Alimentiveness! Did not C. buy nuts and gingerbread, when a boy, with the money he stole? Of course you see why he is a thief, and how his example confirms our noble science.

At last comes along a case which is apparently a *settler*, for there is a little brain with vast and varied powers,—a case like that of Byron, for instance. Then comes out the grand reserve-reason which covers everything and renders it simply impossible ever to corner a phrenologist. "It is not the size alone, but the *quality* of an organ, which determines it degree of power."

Oh! oh! I see.—The argument may be briefly stated thus by the Phrenologist: "Heads I win, tails you lose." Well, that's convenient.

It must be confessed that Phrenology has a certain resemblance to the Pseudo-sciences. I did not say it was a Pseudo-science.

I have often met persons who have been altogether struck up and amazed at the accuracy with which some wandering Professor of Phrenology had read their characters written upon their skulls. Of course, the Professor acquires his information solely though cranial inspections and manipulations,—what are you laughing at? (to the boarders,)—But let us just *suppose*, for a moment, that a tolerably cunning fellow, who did not know or care anything about Phrenology, should open a shop and undertake to read off people's characters at fifty cents or a dollar apiece. Let us see how well he could get along without the "organs."

I will suppose myself to set up such a shop. I would invest one hundred dollars, more or less, in casts of brains, skulls, charts, and other matters that would make the most show for the money. That would do to begin with. I would then advertise myself as the celebrated Professor Brainey, or whatever name I might choose, and wait for my first customer. My first customer is a middle-aged man. I look at him, – ask him a question or two, so as to hear him talk. When I have got the hang of him, I ask him to sit down, and proceed to fumble his skull, dictating as follows:

SCALE FROM 1 TO 10.

LIST OF FACULTIES FOR CUSTOMER.	PRIVATE NOTES FOR MY PUPIL. *Each to be accompanied by a wink*
Amativeness, 7.	Most men love the conflicting sex, and all men love to be told they do.
Alimentiveness, 8.	Don't you see that he has burst off his waistcoat-button with feeding, — hey
Acquisitiveness, 8.	Of course. A middle-aged Yankee.
Approbativeness, 7+	Hat well brushed. Hair ditto. Mark the effect of that plus sign.
Self-Esteem, 6.	His face shows that.
Benevolence, 9.	That'll please him.
Conscientiousness, 8 1/2	That fraction looks first-rate.
Mirthfulness, 7	Has laughed twice since he came in.
Ideality, 9	That sounds well.

Form, Size, Weight,
Color, Locality, Eventuality,
etc., etc., 4 to 6   Average everything that Color, Locality,
                                                cannot be guessed.

Eventuality, etc. etc.
And so the other faculties.

Of course, you know, that isn't the way the Phrenologists do. They go only by the bumps.—
What do you keep laughing so for? (to the boarders.) I only said that is the way I should practice
"Phrenology" for a living.

End of my Lecture. (Holmes 1892, III, 195–202)

With this piece in the widely circulating *Atlantic Monthly*, Holmes broke his silence
about phrenology in a public way, telling readers why he was assailing the phrenologists
and their bumpology. He pointed out how phrenologists from the start only accepted
favorable cases, ignoring or explaining away all contradictions or, when all else failed,
using safety nets such as, "It is not the size alone, but the quality of an organ, which
determines it degree of power." Thus, they were promoting a pseudoscience, because, as
his Professor explained, "Pseudo-science consists of nomenclature, with a self-adjusting
arrangement, by which all positive evidence, or such as favors its doctrines, is admitted,
and all negative evidence, or such as tells against it, is excluded."

Holmes was not the first physician to write about these methodological transgressions
and abuses. Peter Mark Roget (1779–1869), later of *Thesaurus* fame, did precisely this in
his 1818 piece on "Cranioscopy" for the *Encyclopedia Britannica*. With "such convenient
logic," Roget wrote, "it would be easy to prove anything,"—and for this reason "they will
be rejected as having proved nothing" (Roget [1824] 1818, 437; Kruger and Finger 2013).
Holmes did not, however, bring up Roget by name.

He also saw a need to give his readers a lesson in logic. He told them how it can be
"hard to prove a negative." Nonetheless, he continued, it "is not necessary to prove the
falsity of the phrenological statement … [but] only necessary to show that its truth is not
proved, and cannot be, by the common course of argument."

He further explained why the head-readers might seem remarkably skilled and accurate. First, he mentioned how a phrenologist would typically make statements that would
be applicable to just about anyone. And second, he brought up how these frauds would
attend to the clothes a person was wearing, whether the person seemed well-fed and
healthy, and how a person spoke and handled himself or herself for more personal details.
Small cranial bumps had nothing to do with the information they were spewing!

Thus, the Professor portrayed the head-readers as cheats or charlatans. Yet he did not
say that the brain-based doctrine lacked redeeming features. On the contrary, he told his
fellow boarders, "A Pseudo-science … may contain many truths, and even valuable ones."
Importantly, his Professor did not dismiss the idea of many independent practical faculties
of mind or Gall's conclusion that different parts of the cortex must have different
functions. In fact, Holmes's Professor endorsed the idea that the faculties that make us
most human are located in the front of the brain. But the notion that the cranium
faithfully reflects the growth of a myriad of underlying brain organs, each with a known
function, was another matter and almost certainly a delusion.

Holmes's 1861 Harvard lecture

Holmes now felt it necessary to share his thoughts about phrenology with his medical students. He did this as a part of his Harvard University lecture, "Border Lines of Knowledge in Some Provinces of Medical Science" (Holmes 1892, IX, 209–272). The year was 1861, the same year Paul Broca (1824–1880) began to present cases of brain damage supporting Gall's concept of cortical localization of function, now stripped of cranioscopy (Broca 1861, 1863; Finger 1994, 2000; Young 1970). The phrenology section of this lecture is devoid of the humor he had exhibited in his piece for the public, and it too is reproduced in full here.

> By the manner in which I spoke of the brain, you will see that I am obliged to leave phrenology *sub Jove*,—out in the cold,—as not one of the household of science. I am not one of its haters; on the contrary, I am grateful for the incidental good it has done. I love to amuse myself in its plaster Golgothas, and listen to the glib professor, as he discovers by his manipulations
>
> "All that disgraced my betters met in me."
>
> I loved to see square-headed, heavy-jawed Spurzheim make a brain flower out into a corolla of marrowy filaments, as Vieussens had done before him, and to hear the dry-fibred but human hearted George Combe teach good sense under the disguise of his equivocal system. But the pseudo-sciences, phrenology and the rest, seem to me only appeals to weak minds and the weak points of the strong ones. There is a *pica* or false appetite in many intelligences; they take to odd fancies in place of wholesome truth, as girls gnaw at chalk and charcoal. Phrenology juggles with nature. It is so adjusted as to soak up all the evidence that helps it, and shed all that harms it. It crawls forward in all weathers, like Richard Edgeworth's hygrometer. It does not stand at the boundary of our ignorance, it seems to me, but is one of the will-o'-the-wisps of its undisputed central domain of bog and quicksand. Yet I should not have devoted so many words to it, did I not recognize the light it has thrown on human actions by its study of congenital organic tendencies. Its maps of the surface of the head are, I feel sure, founded on a delusion, but its studies of individual character are interesting and instructive.
>
> The "snapping turtle" strikes after its natural fashion when it first comes out of the egg. Children betray their tendencies in their way of dealing with the breasts that nourish them; nay, I can venture to affirm, that long before they are born they teach their mothers something of their turbulent or quiet tempers.
>
> "Castor gaudet equis, ovo prognatus eodem Pugnis."
>
> Strike out the false pretensions of phrenology; call it *anthropology*; let it study man the individual in distinction from man the abstraction, the metaphysical or theological lay figure; and it becomes "the proper study of mankind," one of the noblest and most interesting of pursuits. (Holmes 1892, IX, 244–246)

Thus, Holmes now portrayed phrenology as a pseudoscience to his medical students, also informing them that phrenological maps were no more than a "delusion." He cited earlier anatomist Raymond Vieussens (1641–1715), as well as phrenologists Spurzheim and Combe, and again he could well have had Roget on his mind when he spoke of the doctrine being mired in a "bog and quicksand." Roget ([1824] 1818, 433) had portrayed phrenology as a poorly constructed edifice built of "flimsy materials" on a "sandy foundation."

Holmes did bring up two good things to come from the pseudoscience in this lecture. First, it drew more attention to inborn tendencies and propensities, a theme intrinsic to *Elsie Venner*, the novel he published in 1858, and one that would reappear in his two later "medicated novels," *The Guardian Angel* and *A Mortal Antipathy* (Holmes 1892, V–VII). Second, he felt that the doctrine was stimulating physicians and academicians to study character differences across the family of man, his "Anthropology," "the proper study of mankind." But even more than these things, it also included the mind and psychology, subjects that clearly fascinated him (Gibian 2001; Weinstein 2009).

In retrospect

Holmes died on October 7, 1894, and was buried in Boston's Mount Auburn Cemetery, where Spurzheim's headless body was interred in 1832. Sir William Osler (1908, 57), the doyen of medical humanities, would praise him "as the most successful combination which the world has ever seen of the physician and the man of letters."

Holmes was not, however, the only famous American to use humor to debunk phrenologists during the nineteenth century. Mark Twain (Samuel Langhorne Clemens; 1835–1910)—who grew up in tiny Hannibal, Missouri, and never attended college—also assailed the phrenologists, almost certainly influenced by Holmes, who quite literally beat him to the punch (Finger 2019). Twain turned on phrenology during the 1870s, after conducting a "little test." He had Lorenzo Fowler, Holmes's chosen phrenologist, read his head before and after identifying himself as the famous author (Finger 2020). Getting two very different readings, Twain proceeded to ridicule phrenology in novels (including *Tom Sawyer* and *Huckleberry Finn*) and in autobiographical pieces. Following what Holmes had done with *The Professor at the Breakfast-Table*, he even listed a few of his own supposed faculties, each followed by a hilarious comment, in his futuristic but never-completed novel *Eddypus* (Twain 1972).

Twain read the *Atlantic Monthly*, and he and Holmes met on several occasions. One was at a dinner party in 1877, where Twain related how he had unintentionally "stolen" one of Holmes's dedications for his first successful book (Twain 1880). This was 24 years before he began to write *Eddypus*, which, more than any other piece, revealed just how much what Holmes presented as a physician, a wit, and a critic of phrenology, seemed to have remained very much on his mind.

Both Holmes and Twain were committed to exposing and debunking worthless medical and scientific fads. Both were clearly dismayed by how the phrenologists were using trickery, and both used humor to educate the gullible public. Although neither supported "bumpology," both stopped short of condemning all of phrenology—seeing, for example, how it could lead to a better understanding of the human mind and brain. For Holmes in particular, the controversial doctrine drew needed attention to inborn tendencies, the vagaries of the mind, differences across people, and new ways of thinking about the brain. For Twain, it added to his colorful vocabulary while enriching his appreciation of individual differences, an important feature in his verbal portraits.

In closing, it is worth pondering what Holmes once stated about the human mind and those intent on profiting from its weaknesses:

> There is every reason to suppose that the existing folly will follow in the footsteps of the past, and after displaying a given amount of cunning and credulity in those deceiving and deceived, will drop from the public view like a fruit which has ripened into spontaneous rottenness, and be succeeded by the fresh bloom of some other delusion required by the same excitable portion of the community. (Holmes 1892, IX, 17)

Although he was referring to Elisha Perkin's (1741–1799) metallic "tractors" (see Miller 1935) in this lecture from 1842, we can only wonder whether he might already have had phrenology classified in the same way at this early date.

Notes

1. Holmes's published writings, lectures, and letters appear in his *Works*, published in 1892.
2. For a bound copy of the identical text (presumably the original document) from Fowler and Wells, but with a slightly earlier date, see Worth (1939).

Acknowledgments

Many people helped with this project and deserve thanks, including Jessica Murphy at Harvard, who sent me a list of the books Holmes donated to the Boston Medical Library; Paige Roberts of Andover Academy, who provided the script version of his head reading; and Ann-Marie Harris of the Berkshire Athenaeum/Library, who provided scans from the *Illustrated Instructor* that included Holmes's head reading. I would also like to thank Drs. Scott Podolsky and Paul Eling for their assistance and insights, and Dan Roe and Caleb Barr for their help perusing Holmes's notebooks at Harvard.

Disclosure Statement

No potential conflict of interest was reported by the author.

References

Ackerknecht E. H. 1956. P. M. A. Dumoutier et la collection phrénologique du Musée de l'Homme. *Bulletins et Mémoires de la Société d'Anthropologie de Paris* X° sér. 7:289–308. doi:10.3406/bmsap.1956.9731.
Ackerknecht E. H. 1967. *Medicine at the Paris Hospital.* Baltimore: Johns Hopkins Press.
Broca P. 1861. Remarques sur le siège de la faculté du langage articulé; suivies d'une observation d'aphémie (perte de la parole). *Bulletins de la Société Anatomique (Paris)* 6:330–357, 398–407.
Broca P. 1863. Localisation des fonctions cérébrales. Siège du langage articulé. *Bulletins de la Société d'Anthropologie* 4:200–04.
Broussais F. 1836. *Cours de Phrénologie.* Paris: JB Baillière.
Davies J. D. 1955. *Phrenology fad and science: A 19th-century American Crusade.* New Haven, CT: Yale University Press.
Dowling W. C. 2006. *Oliver Wendell Holmes in Paris: Medicine, theology, and the autocrat of the Breakfast Table.* Hanover, NH: University Press of New England.
Eling P., and S. Finger. 2019. Franz Joseph Gall on the cerebellum as the organ for the reproductive drive. *Frontiers in Neuroanatomy* 13:1–13. doi:10.3389/fnana.2019.00040.
Finger S. 1994. *Origins of neuroscience: A history of explorations into brain function.* New York: Oxford University Press.
Finger S. 2000. *Minds behind the brain: A history of the pioneers and their discoveries.* New York: Oxford University Press.
Finger S. 2019. Mark Twain's fascination with phrenology. *Journal of the History of the Behavioral Sciences* 55:99–121. doi:10.1002/jhbs.21960.
Finger S. 2020. Mark Twain's phrenological experiment: Three renditions of his "small test". *Journal of the History of the Neurosciences* 29:101–18. doi:10.1080/0964704X.2019.1690388.
Finger S., and P. Eling. 2019. *Franz Joseph Gall: Naturalist of the mind, visionary of the brain.* New York: Oxford University Press.

Fowler O. S., and L. N. Fowler. 1859. *Illustrated self-instructor in phrenology and physiology; with one hundred engravings, and a chart of the character.* New York: Fowler and Wells.

Gall F. J. 1835. *On the functions of the brain and each of its parts...* 6 vols, ed. N. Capen, trans. W. Lewis. Boston: Marsh, Capen and Lyon.

Gall F. J., and J. G. Spurzheim. 1810–19. *Anatomie et Physiologie du Système Nerveux en Général, et du Cerveau en Particulier.* 4 vols. and atlas. Paris: F. Schoell. [Gall was sole author on vols. III and IV].

Gibian P. 2001. *Oliver Wendell Holmes and the culture of conversation.* Cambridge: Cambridge University Press.

Gibian P. 2009. Doctor Holmes: The life in conversation. In *Oliver Wendell Holmes: Physician and man of letters*, ed. S. Podolsky and C. S. Bryan, 71–91. Sagamore Beach, MA: Science History Publications.

Holmes O. W. 1892. *The works of Oliver Wendell Holmes.* 15 vols. Boston: Houghton, Mifflin and Company.

Howe M. A. D. 1939. *Holmes of the breakfast-table.* London: Oxford University Press.

Hoyt E. P. 1979. *The improper Bostonian: Dr. Oliver Wendell Holmes.* New York: William Morrow & Company, Inc.

Jones R. M. 1973. American doctors and the Parisian medical world, 1830–1840. *Bulletin of the History of Medicine* 47:40–65, 177–204.

Jones R. M. 1978. *The Parisian education of an American Surgeon. Letters of John Mason Warren (1832–1835).* Philadelphia: American Philosophical Society.

Kruger L., and S. Finger. 2013. Peter Mark Roget: Physician, scientist, systematist; his *Thesaurus* and his impact on 19th-century neuroscience. In *Literature, neurology, and neuroscience: Historical and literary connections*, ed. S. Finger, A. Stiles, and F. Boller, 173–95. New York: Elsevier. Progress in Brain Research 203.

Laugée T. 2014. A morbid Pantheon: The birth of the Museum of the phrenological society of Paris. *Études Française* 49:47–61. doi:10.7202/1021202ar.

Lokensgard H. O. 1940. Oliver Wendell Holmes's "phrenological character.". *New England Quarterly* 13:711–18. doi:10.2307/360070.

Louis P. C. A. 1828. *Recherches sur les Effets de la Saignée dans Plusieurs Maladies Inflammatoires.* Paris: Migneret.

Magendie F. 1834. *Précis Élémentaire de Physiologie.* 4th ed. Brussels: H Dumont.

Miller W. S. 1935. Elisha Perkins and his metallic tractors. *Yale Journal of Biology and Medicine* 8:41–47.

Morse J. T. Jr. 1896. *Life and letters of Oliver Wendell Holmes.* 2 vols. Boston: HoughtonMifflin Co. [Also Vols. XIV and XV in Holmes's *Works*].

Osler W. 1908. *An Alabama student and other biographical essays.* New York: Oxford University Press.

Podolsky S., and C. S. Bryan, eds. 2009. *Oliver Wendell Holmes: Physician and Man of Letters.* Sagamore Beach, MA: Science History Publications.

Renneville M. 1998. Un musée oublié: Le cabinet phrénologique de Dumoutier. *Bulletins et Mémoires de la Société d'Anthropologie de Paris* Nouvelle sér. 10:477–84. doi:10.3406/bmsap.1998.2533.

Roget P. M. [1824] 1818. Cranioscopy. In *Supplement to the fourth, fifth, and sixth editions of the Encyclopaedia Britannica, Vol. 3*, 419–37. Edinburgh: Printed for A. Constable and Co. [bound in 1824].

Scudder H. 1901. *James Russell Lowell.* New York: AMS Press.

Shyrock R. H. 1947. *American medical research past and present.* New York: Commonwealth Fund.

Shyrock R. H. 1960. *Medicine and society in America, 1660–1860.* New York: New York University Press.

Stern M. B. 1971. *Heads and headlines: The phrenological fowlers.* Norman: University of Oklahoma Press.

Tilton E. M. 1947. *Amiable autocrat: A biography of Dr. Oliver Wendell Holmes.* New York: Henry Schumann.

Twain M. 1880. Remarks from at The Holmes Breakfast. *Atlantic Monthly* 45 (Suppl.):11–12.

Twain M. 1972. The secret history of Eddypus, the World Empire. In *Mark Twain's fables of man*, ed. J. S. Tuckey, 315–85. Berkeley: University of California Press.

Vimont J. 1832–35. *Traité de Phrénologie Humaine et Comparée*. Paris: J. B. Baillière.

Walsh A. A. 1976. Phrenology and the Boston medical community in the 1830s. *Bulletin of the History of Medicine* 50:261–73.

Warner J. H. 1998. *Against the spirit of system: The French impulse in nineteenth-century American Medicine*. Princeton, NJ: Princeton University Press.

Warren E. 1860. *The life of John Collins Warren, M.D., compiled chiefly from his autobiography and journals*. 2 vols. Boston: Ticknor and Fields.

Warren J. C. 1921. The collection of the Boston phrenological society—A retrospect. *Annals of Medical History* 3:1–11.

Weinstein M. A. 2009. Oliver Holmes's depth psychology. In *Oliver Wendell Holmes: Physician and man of letters*, ed. S. Podolsky and C. S. Bryan, 93–103. Sagamore Beach, MS: Science History Publications.

Worth W. 1939. The autocrat in profile. *Colophon* 1:49–56.

Young R. M. 1970. *Mind, brain and adaptation in the 19th century*. Oxford: Clarendon Press.

Index